Diacylglycerol Oil
Second Edition

Diacylglycerol Oil
Second Edition

Editors

Yoshihisa Katsuragi
Kao Corporation
Tokyo, Japan

Takuji Yasukawa
Kao Corporation
Tokyo, Japan

Noboru Matsuo
Kao Corporation
Tokyo, Japan

Brent D. Flickinger
Archer Daniels Midland Company
Decatur, Illinois

Ichiro Tokimitsu
Kao Corporation
Tokyo, Japan

Mark G. Matlock
Archer Daniels Midland Company
Decatur, Illinois

Urbana, Illinois

AOCS Mission Statement

To be a global forum to promote the exchange of ideas, information, and experience, to enhance personal excellence, and to provide high standards of quality among those with a professional interest in the science and technology of fats, oils, surfactants, and related materials.

AOCS Press, Urbana, IL 61802

ISBN 978-1-893997-68-4

Library of Congress Cataloging-in-Publication Data

Diacylglycerol oil / editors, Yoshihisa Katsuragi ... [et al.]. -- 2nd ed.
p. ; cm.
Includes bibliographical references and index.
ISBN 978-1-893997-68-4 (alk. paper)
1. Diglycerides--Physiological effect. 2. Diglycerides--Metabolism. I. Katsuragi, Yoshihisa. II. American Oil Chemists' Society.
[DNLM: 1. Diglycerides--physiology. 2. Diglycerides--metabolism. QU 85 D536 2008]

QP752.D54D536 2008
612'.01577--dc22

2008015810

Printed in the United States of America.
12 11 10 09 08 5 4 3 2

The paper used in this book is acid-free and falls within the guidelines established to ensure permanence and durability.

Preface to the Second Edition

It has been four years since the first edition of "Diacylglycerol Oil" appeared. During this period, research has progressed and publications on diacylglycerol oil have accumulated. The first edition was a marked success and there has been a demand for copies. Taking this opportunity, we have planned a second edition. Major revisions were made to the chapters on metabolism (Part II), health benefits (Part III) and regulatory status and safety (Part IV).

A remarkable achievement concerning the health benefits of diacylglycerol oil has been the research in pre-diabetics and diabetics. The study showed that the attenuation effect in the postprandial elevation of serum TG level after diacylglycerol vs. triacylglycerol oil consumption was more pronounced in subjects with insulin resistance than in subjects without insulin resistance. These results were included in Chapters 8 and 12. A new section was added to Chapter 8: "Postprandial Lipid Responses after DAG Oil Consumption in Pre-diabetics and Diabetics." The title of Chapter 12 was changed to more accurately reflect the contents: "Effects of Long-Term Ingestion of Diacylglycerol Oil on Various Clinical Conditions of Diabetes," and a new section was added: "DAG Oil Consumption Delays the Progression of Renal Failure in Type 2 Diabetic Patients with Nephropathy." The improvement of impaired glucose tolerance by diacylglycerol ingestion was also observed in animal studies (Chapter 9). A new chapter is titled "Anti-obesity Effect of Long-Term Consumption of Dietary Diacylglycerol in Experimental Animal Models," and a new section, "Improvement of Impaired Glucose Tolerance by DAG Ingestion," was added.

Since the first edition, the regulatory status of diacylglycerol oil also has progressed, and now the authorized countries to be placed on the market have expanded to the European Union, Canada, Australia/New Zealand, and Brazil. Chapter 16 has been revised to describe the current status of this topic. Regarding the safety of diacylglycerol oil, all the references published thus far, as well as all unpublished internal reports, were re-examined by the authors of Chapter 17. The new chapter includes all animal and human results, especially under clinical conditions. The intensive examination further confirmed the safety of diacylglycerol oil.

Diacylglycerol oil has received attention as a potential dietary tool for people with metabolic syndrome. Both intra-abdominal fat obesity, and impaired glucose tolerance—diabetes—are risk factors for coronary heart disease. The results of recent studies further show a usefulness of diacylglycerol oil for the prevention and improvement of metabolic syndrome caused by the accumulation of intra-abdominal fat.

Publication of several original research articles and review articles on diacylglycerol oil from independent researchers are among the notable indicators showing that diacylglycerol has received attention from researchers engaged in human nutrition and related fields. We hope that nutritionists, physicians, public health professionals, food scientists and technologists, and process engineers can use this book to update their expertise.

We are grateful to all of the contributors for thoroughly reviewing their chapters for the revision. Their unwavering involvement made it possible to revise many chapters in a short amount of time. We would like to express our deep appreciation to Mr. Magdi Mossoba, Chairperson of the AOCS Books and Special Publications Committee for accepting the production of the second edition. We also would like to express our sincere appreciation to Ms. Jodey Schonfeld, Senior Director, Product Development of AOCS, and to Mr. Brock Peoples, Production Editor of AOCS Press for their cooperation. We would like to express our gratitude to the staff at AOCS Press for their invaluable assistance, and for accepting many difficult demands.

Yoshihisa Katsuragi
Takuji Yasukawa
Noboru Matsuo
Brent D. Flickinger
Ichiro Tokimitsu
Mark G. Matlock

Preface to the 2004 Edition

Fat is an essential nutrient in the diet. In recent years, a growing number of scientists have been working intently to elucidate the nutritional properties of dietary fat, and our knowledge concerning the nutritional properties of dietary fats and oils has expanded. Health properties attributed to certain fatty acids, such as n-3 fatty acids and monounsaturated fatty acids, have become popular research topics. However, the nutritional properties of diacylglycerol, a natural component of various fats and oils and a well-known digestive product of triacylglycerol, have been little studied.

Since the late of 1980s, considerable new information concerning nutritional properties of diacylglycerol oil has accumulated. These properties can be divided roughly into two main parts. First, diacylglycerol is absorbed in the small intestine in much the same manner as triacylglycerol, but is less likely to be stored as body fat. Second, diacylglycerol suppresses postprandial triglyceride elevation compared with triacylglycerol. These properties are attributed primarily to the structure of the diacylglycerol molecule, rather than the fatty acid components.

After an overview in Part I, a review of the metabolism and health benefits of diacylglycerol in Parts II and III should be valuable information for the reader, especially nutritionists and public health professionals because obesity and the metabolic disorders caused by obesity are growing global problems. Safety and regulatory information, as shown in Part IV, should be helpful resources for all professionals. Edible oil, in addition to serving as a source of essential fatty acids, is important as a heating medium in cooking. Although diacylglycerol oil has unique health properties, cooking and other food application properties should also be considered. Part V discusses the physicochemical properties of diacylglycerol oil including cooking properties and unique application properties attributed to the presence of a hydroxyl group in the structure of diacylglycerol, something that is not present in triacylglycerol. Part VI provides a review of the manufacturing process of diacylglycerol oil. We hope that this book will be useful to all those interested in diacylglycerol oil, especially nutritionists, physicians, public health professionals, food scientists and technologists, and process engineers.

We are grateful to all of the contributors for their unwavering involvement and cooperation. Without their participation, this work would not have been possible. We would like to express our deep appreciation to the chairperson of the AOCS Books and Special Publications Committee for accepting this undertaking. Special thanks go to Dr. David Kritchevsky of the Wistar Institute for his kind reviews and valuable comments concerning metabolism and health benefits. We also would like to express our sincere appreciation to Ms. Mary Lane and Ms. Melissa Blankenship of AOCS Press for their cooperation. It has been a pleasure working with them. Finally, we would like to heartily express our gratitude to the publishing staff at AOCS Press and Ms. Yoko Mikawa of Kao Corporation for their invaluable assistance.

Yoshihisa Katsuragi
Takuji Yasukawa
Noboru Matsuo
Brent D. Flickinger
Ichiro Tokimitsu
Mark G. Matlock

Contents

Diacylglycerol Oil
Second Edition

Part I

Overview

Chapter 1

Diacylglycerols

Takuji Yasukawa and Yoshihisa Katsuragi

Global R&D Health Care, Kao Corporation, Sumida-ku, Tokyo 131–8501, Japan

Introduction

Fat Intake and Obesity

It is now well accepted that diets high in fat are risk factors for obesity and heart disease. Changes in the content of meals, and in dietary habits and lifestyles have contributed to the prevalence of these conditions. Public health agencies and professional health organizations have issued guidelines and recommended programs that encourage dietary modifications and exercise to improve overall health. Some of the common recommendations include a reduced consumption of high-energy diets containing high amounts of fat and sugar. Recently, the intake of good fat and a balancing of the types of fat in the diet were also recommended.

The World Health Organization (WHO) reported that more than one billion adults are overweight at present, and at least 300 million of these were clinically obese in 2000; worldwide, obesity is a growing problem, not only in industrialized countries, but also in urban areas of developing countries (WHO, 2003a). In the United States, obese and overweight adults [body mass index (BMI) ≥ 25 kg/m^2] comprised 66.3% of the total population for 2003-2004 (Ogden et al., 2006). In Japan, the National Nutrition Survey conducted in 2003 revealed that 27.8% of men ≥ 20 y old are obese (BMI ≥ 25 kg/m^2) (Research Association of Health and Nutrition Information, 2005). The prevalence of obesity has increased dramatically since 1990. However, the survey revealed that BMI in women in 2001 was decreased in all age groups, except those in their 70s, compared to 1999. Since the late 1980s, studies revealed that the risk of diseases such as diabetes and coronary artery disease as well as all cases of mortality increased in proportion to an above optimal increase in body adiposity (Manson et al., 1987; Lindsted et al., 1991; Pi-Sunyer, 1991; Seidell et al., 1996). In 1989, the "deadly quartet" theory was proposed, i.e., the risk of coronary heart disease (CHD) is increased in the presence of four abnormalities that include upper body obesity, glucose intolerance, hypertriglyceridemia, and hypertension (Kaplan, 1989). This was the first proposal in which obesity was incorporated as one of the multiple risk factors for CHD. In 1995, Matsuzawa reported that the complications of visceral fat accumulation, hypertension, glucose intolerance, and hypertriglyceridemia invite CHD risk. This is the so-called "visceral fat syndrome" in which visceral fat accumulation is the most important risk factor (Matsuzawa et al.,

1995). Visceral fat causes insulin resistance, leading to metabolic disorders. Thus, it has been demonstrated that the distribution of body fat is closely associated with homeostasis in the human body.

Recent large-scale clinical studies revealed that obesity and overweight lead to adverse metabolic effects on blood pressure, cholesterol, triglycerides, and insulin resistance (Ogden et al., 2006; Carey et al., 1997; Lamon-Fava et al., 1996; Rimm et al., 1995; Monkdad et al., 2001). This complication of multiple risk factors is commonly referred to as the "Metabolic syndrome" according to the WHO. WHO (Zimmet et al., 2001) and the National Cholesterol Education Program (ATP III) in the United States (National Cholesterol Education Program, 2001) announced their original or modified guidelines, respectively, in 2001, on the basis of various research results (Lakka et al., 2002). Obesity is currently considered to be one of the risk factors in their guidelines, and its prevention has become a primary goal.

Current Lipid Research

Changes in dietary habits and exercise are the most popular recommendations for preventing obesity and the metabolic syndrome (WHO, 2003b). In their diet choices, people should eliminate excess intake of fat and oil and should consume good fat, based on scientific evidence. In 1980s and 1990s, fat substitutes involving the use of starch, cellulose, protein, pectin, and dextrose were developed in the search for a substance that would provide a foodstuff with the same texture as fat (Wylie-Rosett, 2002). Salatrim (Wylie-Rosett, 2002; Smith et al., 1994) and Caprenin (Wylie-Rosett, 2002; Wardlaw et al., 1995) were developed as a low-energy fat that contains 5 vs. 9 kcal/g for conventional edible oil. These structural fats contain short- or medium-chain fatty acids, which partially replace the original long-chain fatty acids of triglycerides, introduced by the interesterification techniques. Olestra is another type of fat substitute, which consists of synthetic hexa-, hepta-, and octa- fatty acid esters of sucrose. It was developed by Procter & Gamble Company in the United States (Eldridge et al., 2002; Stubbs, 2001). The physicochemical properties of Olestra are similar to those of conventional fats, but it is resistant to hydrolysis by pancreatic lipase, and is barely absorbed in the digestive tract. As a result, it provides no energy. The use of Olestra was permitted in savory snacks in the United States. The specific physiologic property of medium-chain triglycerides (MCT), which generally consist of C8-10 fatty acids, was recently reported. MCT increase energy expenditure and may result in satiety, compared with long-chain triglycerides (St-Onge & Jones, 2002). The potential for weight control through MCT intake is currently being examined (Bach et al., 1996; Tsuji et al., 2001). It was reported recently that conjugated linoleic acid (CLA) consumption influences on lipid metabolism, reduces obesity, and inhibits cancer (Pariza et al., 2001; Brown & McIntosh, 2003).

The fat substitutes and structural fats described above are/were not always applicable for use in all fats and oil products because of issues of taste, smell, texture, physical chemistry, and safety. Therefore, further studies in these areas are expected in the near future.

Recently, health benefits of certain fatty acids were demonstrated in the area of lipid metabolism. Fish oils, rich in n-3 polyunsaturated fatty acids (n-3 PUFA) such as docosahexaenoic acid (DHA) and eicosapentaenoic acid (EPA), are thought to reduce platelet activation, lower plasma triglycerides and (vitamin K-dependent) coagulation factors, and/or decrease vascular tone, which may help to reduce arterial disease (Vanschoonbeek et al., 2003; de Lorgeril & Salen, 2003). Furthermore, it was reported that n-3-PUFA consumption has a blood pressure-lowering effect (Weisinger et al., 2001). The health benefits of α-linolenic acid (ALA), which is also a n-3 PUFA, were shown in the Lyon Diet Heart Study, which examined the rate of recurrence after a first myocardial infarction. The results of the study showed that a high consumption of oleic acid and ALA as a part of test diet significantly reduced total sudden cardiac deaths and acute myocardial infarctions (de Lorgeril & Salen, 2003; de Lorgeril et al., 1999). In particular, ALA consumption was correlated negatively with CHD risk. Thus, recent large-scale clinical studies are currently clarifying the nutritional significance of this fatty acid.

Studies on *trans* fatty acids have been ongoing for many years to investigate the relation between their consumption and human health. In 2002, at the request of the U.S. Food and Drug Administration (FDA), the Institute of Medicine/National Academy of Science (IOM/NAS) reported "Dietary reference intakes for *trans* fatty acids" (Institute of Medicine, 2002). The report concluded that the consumption of high levels of *trans* fatty acids leads to an increase in LDL cholesterol levels and a decrease in HDL cholesterol levels, both of which increase the risk of CHD. The report recommended that *trans* fatty acid consumption be kept as low as possible while consuming a nutritionally adequate diet. The U.S. FDA required that the product's Nutrition Facts label for fat and oil show the "*trans* fat amount per serving." As described above, almost all of the fat and oil studies were focused on the nutritional properties and the functionality of fatty acids in triacylglycerols.

In the late 1980s, diacylglycerol oil, with a focus on the nutritional properties of the glyceride structure rather than the fatty acid content, was developed. Diacylglycerol has two main health benefits. One is the suppression of postprandial serum triglyceride elevation (see Chapters 6-8); the other is the suppression of body fat accumulation (see Chapters 9-11). The metabolism, health benefits, safety, applications, and manufacturing process of diacylglycerol oil are discussed in this book.

Diacylglycerols

Structure of DAG

Diacylglycerol (DAG) occurs in two isoforms, i.e., 1, 2- (or 2, 3)-diacyl-*sn*-glycerol (1, 2-DAG) and 1, 3-diacyl-*sn*-glycerol (1, 3-DAG) as shown in Figure 1.1. The 1, 2-DAG isoform is recognized as a metabolic intermediate that is formed after the ingestion of triacylglycerol (TAG). The DAG contained in refined edible oils is largely converted to the 1, 3-isoform by migration of the acyl group during the high-temperature manufactur-

CH_2OCOR_1 / R_2OCO—C—H / CH_2OH — *sn*-1,2-DAG

CH_2OH / R_2OCO—C—H / CH_2OCOR_1 — *sn*-2,3-DAG

CH_2OCOR_1 / HO—C—H / CH_2OCOR_2 — *sn*-1,3-DAG

(*sn* : stereospecific numbering system)

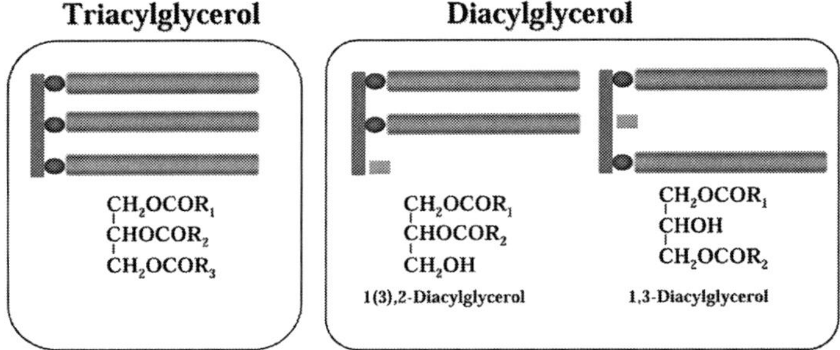

FIG. 1.1. Structure of diacylglycerols.

ing process (see Chapter 23). In most edible oils, ~70% (w/w) of the DAG is present as the 1, 3-isoform as shown in Table 1.1.

DAG in Nature

DAG is a natural component of glycerides in various fats and oils at levels up to 10% (w/w). Table 1.1 shows the content of DAG and other acylglycerols in edible oils of various origins (Perez-Camino et al., 1996; Kamal-Eldin & Appelqvist, 1994; Goh & Timms, 1985; Gopalkrishnan et al., 1987; Abdel-Nabey et al., 1992; D'lonzo et al., 1982; Barcel et al., 1985). In particular, olive oil has been used for a long time in the Mediterranean area. It was reported that some olive oil obtained in Majorca in Spain contained a high content of DAG (Barceló Mairata et al., 1985). DAG is also a main component of mono- and diglycerides, one of the popular emulsifiers used in foods. Therefore, DAG is normally present in plant oils and in fats and oil products that have been consumed by humans for a long period of time.

DAG Oil

DAG oil contains >80% DAG, <20% TAG, <3% monoacylglycerol (MAG), and small amounts of antioxidants and emulsifiers to maintain quality. DAG is generally made from fatty acids derived from vegetable oil and glycerin by using an immobilized 1, 3-regiospecific lipase. The ratio of the 1, 3-isoform to the 1, 2- or 2, 3-isoform is ~7:3

TABLE 1.1
Contents of Acylglycerols in Edible Oils of Various Origin[a]

	TAG	DAG			MAG	Others
		Total DAG	1,2-DAG	1,3-DAG		
	(g/100 g)					
Soybean[b]	97.9	1.0	ND	ND	0.0	1.1
Palm[b]	93.1	5.8	ND	ND	< DL	1.1
Cottonseed[b]	87.0	9.5	ND	ND	0.2	3.3
Corn[b]	95.8	2.8	1.5	2.9	< DL	1.4
Safflower[b]	96.0	2.1	1.2	2.7	< DL	1.9
Olive[b]	93.3	5.5	ND	ND	0.2	2.3
Olive (mol%)[c]	18–41	8–20				
Rapeseed[b]	96.8	0.8	ND	ND	0.1	2.3
Vegetable oil[d] (Soy + Canola)	98.3	1.7	0.6	1.1	< DL	ND
Canola[d]	97.1	2.9	1.0	1.9	< DL	ND
Sesame[d]	95.2	4.1	1.2	2.9	0.8	ND
Pure sesame[d]	95.5	3.9	1.2	2.7	0.6	ND
Rice[d]	92.4	7.6	2.4	5.2	< DL	ND
Rice germ[d]	91.2	8.8	2.7	6.1	< DL	ND
Corn germ[d]	95.5	4.5	1.5	2.9	< DL	ND
Pure grapeseed[d]	94.2	5.8	2.1	3.7	< DL	ND
DAG oil[e]	17.3	81.4	28.4	53.0	1.2	ND

[a]Abbreviations: ND, not determined; <DL, below detectable limit (<0.1 g/100g).
[b]Source: Abdel-Nabey et al., 1992; D'alonzo et al., 1982.
[c]Source: Barceló Mairata et al., 1985.
[d]Source: Nishide, T., unpublished data.
[e]Described in detail in Chapter 19.

due to acyl migration during the refining process. DAG oil as cooking oil is commercially available in Japan and in the United States, and its typical content is shown in Chapter 19.

The main nutritional functions of DAG oil vs. TAG oil are as follows: (i) a suppressive effect on postprandial serum triglyceride elevation, and (ii) a suppressive effect on body fat accumulation. DAG oil also has unique physicochemical properties caused by the presence of a free hydroxyl group in the structure compared with TAG oil, as shown in Chapter 18. The cooking properties of DAG oil are similar to those of conventional cooking oil (see Chapter 19).

Effect of DAG on Gastric Emptying Rate. DAG is an intermediate product in the digestion of TAG. When conventional oil is ingested, lipases in the saliva and gastric juice hydrolyze the fatty acid at position 1 or 3 of TAG. Hamosh and Scow reported on the lingual lipase specifically secreted in infancy (Hamosh & Scow, 1973), and assumed that the lipase has a biological function for efficiently digesting and absorbing fats and fatty oils in breast milk as an energy source. Fats in breast milk are modestly degraded by lingual lipase to DAG, transported to the digestive tract, and then readily digested and absorbed. These findings of Hamosh and Scow stimulated interest in the nutritional characterization of DAG.

We recruited subjects with weak gastrointestinal function to clarify the digestion of DAG, and performed sensory evaluations for the presence of heavy stomach feeling and heartburn. The ingestion of DAG was less likely to cause a heavy stomach feeling than that of TAG, although heartburn was not affected (unpublished data). To evaluate differences in postprandial stomach feeling between a DAG diet and a TAG diet, the influence on gastric emptying rate was investigated using radioisotopes; the gastric emptying rate of a diet containing DAG was significantly faster than that for a diet containing TAG, but was slower than that for a diet containing no oil (Yasunaga et al., 2000, see Chapter 3). This finding is consistent with our prediction, which led to subsequent DAG nutritional studies.

Digestion and Absorption. The energy from DAG, as measured by bomb calorimetry, is 38.9 kJ/g (9.29 kcal/g) and does not differ from that of common TAG 39.6 kJ/g (9.46 kcal/g) (Taguchi et al., 2001) (see Chapter 3). DAG as well as TAG is digested to monoglycerides and fatty acids by gastric lipase and pancreatic lipase in the stomach and in the small intestine, respectively. The main DAG digestive product is 1[3]-MAG because 1, 3-DAG is the main component, whereas the main digestive product of TAG is 2-MAG. The difference in the digestive products of DAG and TAG is associated with the ability of DAG to suppress body fat accumulation and postprandial triglyceride elevation. The digestive product reaches the small intestine and is absorbed by small intestinal epithelial cells. 1-MAG derived from 1, 3-DAG is not readily resynthesized to chylomicron triglycerides in the small intestinal epithelial cells compared with 2-MAG, derived from TAG (Yanagita et al., 2005; Kondo et al., 2003, see Chapters 3 and 6). Monoacylglycerol acyltransferase (MGAT) and diacylglycerol acyltransferase (DGAT) are key enzymes in forming chylomicron triglyceride from digestion products in intestinal mucosal cells. It was shown in some detail that 1, 3-DAG is not utilized extensively by DGAT as a substrate for chylomicron triglyceride synthesis because DGAT generally resynthesizes chylomicron triglycerides using 1 [or 3], 2-DAG as the substrate (Kondo et al., 2003). This may be a factor responsible for the suppression of postprandial triglyceride elevation (Taguchi et al., 2000; Tada et al., 2001; Tada & Yoshida, 2003; Tada, 2004; Tada et al., 2005; Takase et al., 2005; Ai et al., 2007; Watanabe et al., 1997a). Interestingly, Smith et al. (Smith et al., 2000) reported that a high-fat diet did not cause obesity in DGAT-deficient mice compared with normal mice. Because DGAT is less likely to use the digestive products produced by DAG ingestion, their observation might be related to the anti-obesity effect of DAG.

Effect of DAG on Energy Metabolism. Table 1.2 summarizes the effects of DAG on energy metabolism. It was reported that absorbed DAG increases the mRNA expression of β-oxidation-related enzymes and uncoupling proteins, which are related to thermogenesis in the small intestine of obese mice (Murase et al., 2002a). The same phenom-

TABLE 1.2
Summary of the Effects of DAG on Energy Metabolism

Animal study

- β-oxidation related enzyme activity increased in the liver and small intestine (Murase et al., 2002a; Murata et al., 1997; Murase et al., 2001; Kim et al., 2007; Meng et al., 2004).
- Messenger RNA expression of β-oxidation related enzymes increased in the liver and small intestine (Murase et al., 2002a; Murase et al., 2001).
- Messenger RNA expression of uncoupling proteins increased in the liver and small intestine (Murase et al., 2002; Murase et al., 2001).
- Energy expenditure enhanced and respiratory quotient decreased (Kimura al, 2006).
- Oxygen consumption increased (Watanabe et al., 1997).

Human

- Postprandial Respiratory quotient decreased (Saito et al., 2006)
- Respiratory quotient decreased and fat oxidation enhanced, however, no change in energy expenditure was observed (Kamphuis et al., 2003).

enon was also observed in the livers of rats (Murata et al., 1997) and obese mice fed a DAG diet (Murase et al., 2001). The activities of β-oxidation-related enzymes were also increased in rats. These findings indicate that the ingestion of DAG activates energy metabolism.

Obesity and overweight are the results of an imbalance between energy intake and energy consumption. To measure energy consumption levels under constant energy intake conditions is important for clarifying the mechanism of obesity. Animal studies showed that a DAG diet increases oxygen consumption (Kimura et al., 2006; Watanabe et al., 1997b). A recent report also revealed that the ingestion of DAG significantly decreased the respiratory quotient (RQ), which is calculated from consumed oxygen and expired carbon dioxide, and increased lipid oxidation compared with that of TAG in a human clinical study (Kamphuis et al., 2003; Saito et al., 2006) and in animals (Kimura et al., 2006). A decrease in RQ from the baseline indicates an increment of fat utilization as an energy source.

These findings from animal and human studies suggest that the activation of energy metabolism is a key issue in the mechanism by which DAG suppresses body fat accumulation and contributes to weight management. The effects of DAG oil on energy metabolism are examined in Chapters 4 and 5.

Efficacy. Table 1.3 summarizes the efficacy of DAG based on reports to date. In a double-blind, controlled human study performed in the United States for 6 mo, a significant reduction in body fat, especially visceral fat, as measured by DEXA (dual energy Xray absorptiometry), was observed in obese subjects (Maki et al., 2002). Another controlled study also showed a significant suppression in body fat accumulation for normal subjects in Japan who consumed 10 g DAG/d compared with TAG (Nagao et al., 2000). Both clinical studies used a restricted diet during the study. Long-term DAG ad libitum consumption studies for 1 or 2 y were conducted using subjects in a free-living environment,

TABLE 1.3
Summary of DAG Efficacy

Postprandial TG

Animals

Decreased plasma or serum TG levels (Watanabe et al., 1997a; Kristensen et al., 2006; Bauer et al., 2006)

Slow lymphatic transport of TG (Murata et al., 1994; Yanagita et al., 2004)

Humans

TG levels in Chyromicron or serum decreased in healthy subjects (Watanabe et al., 1997a; Taguchi et al., 2000; Tada et al., 2001)

TG levels in Chyromicron or serum decreased in subjects with insulin resistance or T2DM (Tada et al., 2005; Takase et al., 2005; Ai et al., 2007)

Remnant particle level decreased (Tada et al., 2001, Tada et al., 2005; Takase et al., 2005; Ai et al., 2007)

Body fat accumulation and weight

Aminals

Decreased body fat and weight (Watanabe et al., 1997a; Murase et al., 2001; Murase et al., 2002a; Murase et al., 2002b; Hase et al., 2001; Meng et al., 2004; Umeda et al., 2006)

Humans

Decreased body fat, especially visceral fat, and weight for overweight and normal subjects in Japan (Nagao et al., 2000; Kawashima et al., 2008) and for obesity in the United States (Maki et al., 2002) compared to a TAG diet.

Open labeled long-term *ad libitum* consumption study indicated a decrease in body weight (Yasukawa & Yasunaga, 2001; Katsuragi et al., 1999)

Waist circumferece and skin-fold thickness decreased in several long term studies (Yasukawa & Yasunaga, 2001; Katsuragi et al., 1999; Otsuki et al., 2004)

Decreased body fat such as total fat and visceral fat, however no change in weight was observed in obese children (Matsuyama et al., 2002; Matsuyama et al., 2007).

Fasting TG

Animals

Decreased (Hara et al., 1993; Meng et al., 2004; Ota et al., 2007)

Humans

Decreased in high initial TG subjects (Yasukawa & Yasunaga, 2001; Katsuragi et al., 1999) and diabetes (Yamamoto et al., 2001)

Cholesterol reduction with phytosterol

Animals

Lower serum cholesterol levels (Meguro et al., 2003; Ota et al., 2007; Fujii et al., 2007) and reduced atherosclerotic lesion (Ota et al., 2007; Fujii et al., 2001)

Humans

Decreased total cholesterol and LDL levels (Meguro et al., 2001; Takeshita et al., 2001; Takeshita et al., 2007; Goto et al., 1999)

Fasting blood sugar and HbA1c

Humans

Decreased HbA1c levels in diabetes (Yamamoto et al., 2001)

Delayed progression of renal failure in T2DM (Yamamoto et al., 2006b)

Anti-thrombotic effect

Animals

Attenuated arterial thrombosis (Ijiri et al., 2006a; Ijiri et al., 2006b) as described in Chapter 11. The objective of these studies was to investigate the weight-related efficacy of DAG under conditions of practical use. Hence, these studies used DAG cooking oil. In a placebo-controlled study, a significant reduction was observed in BMI, weight, and skinfold thickness for overweight subjects (Kawashima et al., 2008). An open-labeled, monadic study for 1 y indicated that ad libitum DAG consumption significantly decreased weight, BMI (Yasukawa & Yasunaga, 2001), waist circumference, and skinfold thickness (Yasukawa & Yasunaga, 2001; Katsuragi et al., 1999). Decreases in BMI and skinfold thickness were observed in another single-blind, monadic study over a 2-y period (Otsuki et al., 2004). The unique nutritional property of DAG consumption is its modest efficacy for weight reduction, particularly reduction in visceral fat. The efficacy of DAG tends to be observed more frequently in obese and overweight subjects, and weight maintenance was observed in low BMI subjects (Yasunaga et al., 2004).

Single-dose administration studies for normal subjects revealed that a DAG emulsion and meal containing DAG significantly suppressed postprandial chylomicron elevation in dose levels of 10, 20, and 44 g compared with TAG (Taguchi et al., 2000; Tada & Yoshida, 2003; Tada, 2004). A decrease in the amount of remnant lipoprotein was also observed in a single-dose administration study (Tada et al., 2001).

A significant decrease in fasting serum triglyceride levels (Yasukawa & Yasunaga, 2001; Katsuragi et al., 1999; Yamamoto et al., 2001; Yamamoto et al., 2006a; Yamamoto et al., 2006b), total cholesterol (Yasukawa & Yasunaga, 2001; Katsuragi et al., 1999), and hemoglobin A_{1c} levels (Otsuki et al., 2004; Yamamoto et al., 2001) (see Chapter 12) was observed in some but not all human clinical studies. Significant increases in HDL cholesterol were observed in long-term clinical studies over 1 y (Yasukawa & Yasunaga, 2001; Katsuragi et al., 1999; Otsuki et al., 2004) and in a clinical study of diabetes (Yamamoto et al., 2001) (see Chapter 12).

Health Benefits of Phytosterol-Rich DAG Oil. Phytosterols are a natural component of plant oils. Phytosterols as well as phytosterol fatty acid esters have a cholesterol-lowering effect in serum; therefore they are currently a subject of attention worldwide. The FDA recognized the cholesterol-lowering effect of plant sterols and their fatty acid ester forms, and approved the labeling of food products containing these compounds as "useful for lowering cholesterol." In Japan, the Ministry of Health, Labor and Welfare also approved three cooking oils containing phytosterols as "Food for Specified Health Use" as of the end of March 2004. Phytosterol-rich DAG oil is one of these products.

DAG is able to solubilize more plant sterols than TAG, (Meguro et al., 2001) as shown in Chapter 18. The efficacy of phytosterol-rich DAG oil is discussed in Chapters 13 and 14. The oil has a significant lowering effect on total and LDL cholesterol (Meguro et al., 2001; Meguro et al., 2003; Ota et al., 2007; Fujii et al., 2007; Takeshita et al., 2001; Goto et al., 1999). Phytosterol-rich DAG oil unites the nutritional properties of DAG and phytosterols. Concomitant use with statin (Takeshita et al., 2007) and soybean protein (Katsuyama et al., 2002), which also have cholesterol-lowering effects, is examined in Chapter 15.

Physicochemical Properties and Applications. Because DAG contains a hydroxyl group and has a lower molecular weight than TAG, critical points concerning its thermal properties such as melting and the smoke and flash points of DAG oil are different from those of TAG oil. The smoke and flash points of DAG oil are slightly lower than those of TAG (unpublished data). On the other hand, the melting point of DAG is slightly higher than that of TAG (unpublished data). Because of the removal of one fatty acid residue, resulting in a hydroxyl group, DAG is slightly more hydrophilic than TAG, and this may account for the physicochemical properties that differ from those of TAG, particularly the interfacial chemical properties. It is easy to make a W/O emulsion using DAG, even without any emulsifiers (Nakajima, 2004). This property is useful for making products such as low-energy spreads. The presence of a hydroxyl group also indicates the unique property of DAG that controls the crystal behavior of solid fat, when DAG is used as the solvent. This property is useful for improving the spreadability of margarine or shortening (unpublished data). In contrast, it is not easy to apply DAG to O/W food emulsions such as mayonnaise. For the production of DAG mayonnaise, the use of emulsifiers with a higher than normal hydrophile-lipophile balance may be required (Kawai et al., 2004). Considering the performance of DAG as a cooking oil, it appears that the taste and texture of the food products are comparable or better than those of food products containing TAG on the basis of sensory evaluations. Details regarding the basic properties and applications of DAG are discussed in Chapters 18-22.

Manufacturing Process. The performance requirements for the industrial manufacturing of DAG include high purity and low-cost production. DAG was prepared from fatty acids and glycerin by esterification using an immobilized 1, 3-regiospecific lipase (Watanabe et al., 2003). The manufacturing flow can be divided into two main processes, i.e., the production of DAG through enzymatic reactions and the refining process that is typically used in the production of edible oils. Enzymatic esterification is an efficient route for producing oil with a high DAG content. DAG obtained in the enzymatic reaction consists mainly of the 1, 3-isomer; some of the 1, 3-isomer is isomerized to the 1, 2-isomer during the refining process in which some heat is generated. Finally, DAG consists of ~70% 1, 3-isomer and 30% 1, 2-isomer.

Safety and Regulation. The safety of DAG has been established by various studies in humans, *in vitro*, and in animals as discussed in Chapter 17. The Ministry of Health and Welfare of Japan approved DAG oil as Food for Specified Health Use (FOSHU) in 1999. In 2000, the Japanese Society of Human Dry Dock, a professional association of Japanese physicians, recommended that DAG plays a role in improving human health. In the United States, DAG was submitted to the FDA as a Generally Recognized As Safe (GRAS) substance, and the FDA indicated that there was no question regarding safety (see Chapter 16).

Practical Use of DAG Oil. DAG oil is commercially available as a cooking oil (Econa Cooking Oil, Kao Corporation, Tokyo, Japan) since 1999. It is also sold in the United States under the name of Enova Oil since 2003.

Summary and Outlook for the Future

DAG, a natural oil component that has unique nutritional properties, "less likely elevates postprandial serum triglycerides and is less likely to become body fat compared to TAG." The physicochemical properties of DAG oil as a liquid oil are nearly the same as those of TAG oil. However, because DAG has a hydroxyl group in the molecule, some physicochemical properties differ from those of TAG. These specific properties, which are not associated with TAG, may require specific devices and technology to be able to apply DAG oil to food products; in some cases, however, this would be useful for the development of new products. DAG is an attractive new food material not only for its nutritional properties but also for its physicochemical properties.

Scientific and practical information regarding the nutritional properties, clinical efficacy, safety, manufacturing, and application technologies of DAG oil is summarized in this book and will be useful in increasing our understanding of the nature of DAG and the use of DAG oil to maintain a healthy body weight and prevent lifestyle-related diseases caused by obesity. Further analysis is expected to allow more scientific evidence and information to be accumulated concerning the efficacy, safety, mechanism of action, and application technologies that may provide new insights for managing a lifestyle-related chronic disease such as the metabolic syndrome.

References

Abdel-Nabey, A.A., Shehata, Y., Ragab, M.H., and Rossell, J.B. Glycerides of Cottonseed Oils from Egyptian and Other Varieties. *Riv. Ital. Sostanze Grasse* **1992,** *69,* 443-447.

Ai, M., Tanaka, A., Shoji, K., Ogita, K., Hase, T., Tokimitsu, I., and Shimokado, K. Suppressive effects of diacylglycerol oil on postprandial hyperlipidemia in insulin resistance and glucose intolerance. *Atherosclerosis,* **2007,** *195,* 225-236

Bach, A.C., Ingenbleek, Y., and Frey, A. The Usefulness of Dietary Medium-Chain Triglycerides in Body Weight Control: Fact or Fancy? *J. Lipid Res.* **1996,** *37,* 708-726.

Barceló Mairata, I., and Barceló Mairata, F. Analysis of the Lipid Composition of the Virgin Olive Oil from Majorca. *Fasc.* **1985,** *36,* 269-273.

Bauer, J.E., Nagaoka, D., Porterpan, B., Bigley, K., Umeda, T., and Otsuji, K., Postprandial lipolytic activities, lipids, and carbohydrate metabolism are altered in dogs fed diacylglycerol meals containing high- and low-glycemic-index starches. *J. Nutr,* **2006,** *136,* 1955S-1957S.

Brown, J.M., and McIntosh, M.K. Conjugated Linoleic Acid in Humans: Regulation of Adiposity and Insulin Sensitivity. *J. Nutr.* **2003,** *133,* 3041-3046.

Carey, V.J., Walters, E.E., Colditz, G.A., Solomon, C.G., Willett, W.C., Rosner, B.A., Speizer, F.E., and Manson, J.E. Body Fat Distribution and Risk of Non-Insulin-Dependent Diabetes Mellitus in Women. The Nurses' Health Study. *Am. J. Epidemiol.* **1997,** *145,* 614-619.

de Lorgeril, M., Salen, P., Martin, J.L., Monjaud, I., Delaye, J., and Mamelle, N. Mediterranean Diet, Traditional Risk Factors, and the Rate of Cardiovascular Complications After Myocardial Infarction: Final Report of the Lyon Diet Heart Study. *Circulation* **1999,** *99,* 779-785.

de Lorgeril, M., and Salen., P. Dietary Prevention of Coronary Heart Disease: Focus on Omega-6/ Omega-3 Essential Fatty Acid Balance. World Rev. *Nutr. Diet.* **2003,** *92,* 57-53.

D'lonzo, R.P., Kozarek, W.J., and Wade, R.L. Glyceride Composition of Processed Fats and Oils as Determined by Glass Capillary Gas Chromatography. *J. Am. Oil Chem. Soc.* **1982,** *59,* 2922-95.

Eldridge, A.L., Cooper, D.A., and Peters, J.C. A Role for Olestra in Body Weight Management. *Obes. Rev.* **2002,** *3,* 17-15.

Fujii, A., Allen, T.J., and Nestel, P.J. A 1,3-diacylglycerol-rich oil induces less atherosclerosis and lowers plasma cholesterol in diabetic apoE-deficient mice. *Atherosclerosis,* **2007,** *193,* 55-61.

Goh, E.M., and Timms, R.E. Determination of Mono- and Diglycerides in Palm Oil, Olein, and Stearin. *J. Am. Oil Chem. Soc.* **1985,** *62,* 730-734.

Gopalkrishnan, N., Narayanan, C.S., Mathew, A.G., and Arumughan, C. Lipid Composition of Coconut Cake Oil. *J. Am. Oil Chem. Soc.* **1987,** *64,* 539-541.

Goto, N., Mori, H., Katsuragi, Y., Toi, T., Yasukawa, T., and Shimasaki, H. Effects of Diacylglycerol Containing Phytosterol on Reducing Blood Cholesterol Concentration. *J. Jpn. Oil. Chem. Soc.* **1999,** *48,* 235-240.

Hamosh, M., and Scow, R.O. Lingual Lipase and Its Role in the Digestion of Dietary Lipid. *J. Clin. Invest.* **1973,** *52,* 88-95.

Hara, K., Onizawa, K, Honda, H., Otsuji, K., Ide T., and Murata, M. Dietary diacylglycerol-dependent reduction in serum triacylglycerol concentration in rats. *Ann. Nutr. Metab.* **1993,** *37,* 185-191.

Hase, T., Mizuno, T., Onizawa, K., Kawasaki, K., Nakagiri, H., Komine, Y., Murase, T., Meguro, S., Tokimitsu, I. Shimasaki, H., and Itakura, H. Effects of aLinolenic Acid Rich Diacylglycerol on Diet-Induced Obesity in Mice. *J. Oleo Sci.* **2001,** *50,* 701-710.

Ijiri, Y., Naemura, A., Yamashita, T., Ikarugi, H., Meguro, S., Tokimitsu, I., and Yamamoto, J., Mechanism of the antithrombotic effect of dietary diacylglycerol in atherogenic mice. *Pathophysiol. Haemost. Thromb.* **2006a,** *35* 380-387.

Ijiri, Y., Naemura, A., Yamashita, T., Meguro, S., Watanabe, H., Tokimitsu, I., and Yamamoto, J. Dietary diacylglycerol extenuates arterial thrombosis in apoE and LDLR deficient mice. *Thromb. Res.* **2006b,** *117,* 411-417.

Institute of Medicine/National Academy of Science, U.S.A. Letter Report on Dietary Reference Intakes for Trans Fatty Acids, 2002; (Internet address: http://www.iom.edu/CMS/5410.aspx)

Kamal-Eldin, A., and Appelqvist, L.A. Validation in Fatty Acid Composition of the Different Acyl Lipids in Seed Oil from Four Sesamum Species. *J. Am. Oil Chem. Soc.* **1994,** *71,* 135-139.

Kamphuis, M.M., Mela, D.J., and Westerterp-Plantenga, M.S. Diacylglycerols Affect Substrate Oxidation and Appetite in Humans. *Am. J. Clin. Nutr.* **2003,** *77,* 1133-1139.

Kaplan, N.M. The Deadly Quartet. Upper-Body Obesity, Glucose Intolerance, Hypertriglyceridemia, and Hypertension. *Arch. Intern. Med.* **1989,** *149,* 1514-1520.

Katsuragi, Y., Toi, T., and Yasukawa, T. Effects of Dietary Diacylglycerols on Obesity and Hyperlipidemia. *J. Jpn. Hum. Dry Dock* **1999,** *14,* 258-262.

Katsuyama, H., Honma, M., Mouri, K., Terada, N., Namioka, M., Mochizuki, Y., Tomita, M., Hamana, M., Funatsu K., Kondo S., Yamashita, T., Miyajima, E., Nakamura, H. Evaluation of efficacy and safety of Combination of FOSHU -Combination of Econa oil and soy protein. *Prog Med.* **2002,** *22,* 2782-5.

Kawai, S. Characterization of diacylglycerol oil mayonnaise emulsified using phospholipase A2-treated egg yolk. *J. Am. Oil Chem. Soc.* **2004,** *81,* 993-998

Kawashima, H., Takase. H., Yasunaga, K., Wakaki, Y., Katsuragi, Y., Mori, K., Yamaguchi, T., Hase, T., Matsuo, N., Yasukawa, T., Tokimitsu, I., and Koyama, W. One-year ad-libitum consumption of diacylglycerol as a part of regular diet results in modest weight loss in comparison with consumption of triacylglycerol control oil in overweight Japanese subjects. *J. Am. Diet. Assoc.* **2008,** *108,* 57-66.

Kim, H.J., Lee, K.T., Lee, M.K., Jeon, S,M,, Jung, U.J., Cho, Y.Y., and Choi, M.S. Hypolipidemic effect of dietary diacylglycerol oil in Sprague-Dawley rats fed a normal diet. *J Med Food.* **2007,** *10,* 60-66.

Kimura, S., Tsuchiya, H., Inage, H., Meguro, S., Matsuo, N., and Tokimitsu, I. Effects of dietary diacylglycerol on the energy metabolism. *Int. J. Vitam. Nutr. Res.* **2006,** *76,* 75-9.

Kondo, H., Hase, T., Murase, T., and Tokimitsu, I. Digestion and Assimilation Features of Dietary DAG in the Rat Small Intestine. *Lipids* **2003,** *38,* 25-30.

Kristensen, J.B., Jorgensen, H., and Mu, H., Diacylglycerol oil does not affect portal vein transport of nonesterified fatty acids but decreases the postprandial plasma lipid response in catheterized pigs. *J. Nutr.* **2006,** *136,* 1800-1805.

Lakka, H.M., Laaksonen, D.E., Lakka, T.A., Niskanen, L.K., Kumpusalo, E., Tuomilehto, J., and Salonen, J.T. The Metabolic Syndrome and Total and Cardiovascular Disease Mortality in Middle-Aged Men. *J. Am. Med. Assoc.* **2002,** *288,* 2709-2716.

Lamon-Fava, S., Wilson, P.W., and Schaefer, E.J. Impact of Body Mass Index on Coronary Heart Disease Risk Factors in Men and Women. The Framingham Offspring Study. *Circulation* **1996,** *16,* 1509-1515.

Lindsted, K., Tonstad, S., and Kuzma, J.W. Body Mass Index and Patterns of Mortality Among Seventh-Day Adventist Men. *Int. J. Obes.* **1991,** *15,* 397-406.

Maki, K.C., Davidson, M.H., Tsushima, R., Matsuo, N., Tokimitsu, I., Umporowicz, D.M., Dicklin, M.R., Foster, G.S., Ingram, K.A., Anderson, B.D., Frost, S.D., and Bell, M. Consumption of Diacylglycerol Oil as Part of a Reduced-Energy Diet Enhances Loss of Body Weight and Fat in Comparison with Consumption of a Triacylglycerol Control Oil. *Am. J. Clin. Nutr.* **2002,** *76,* 1230-1236.

Manson, J.E., Stampfer, M.J., Hennekens, C.H., and Willett, W.C. Body Weight and Longevity: A Reassessment. *J. Am. Med. Assoc.* **1987,** *257,* 353-358.

Matsuyama, T. Effect of Diacylglycerol in the Obese Children of Patients. *Pediatr. Jpn.* **2002,** *43,* 928-933.

Matsuyama, T., Shoji, K., Takase, H., Kamimaki, I., Tanaka, Y., Otsuka, A., Watanabe, H., Hase, T., and Tokimitsu, I. Effects of phytosterols in diacylglycerol as part of diet therapy on hyperlipidemia in children. *Asia Pac. J. Clin. Nutr.* **2007,** *16,* 40-8.

Matsuzawa, Y., Nakamura, T., Shimomura, I., and Kotani, K. Visceral Fat Accumulation and Cardiovascular Disease. *Obes. Res.* **1995,** *3,* 645S-647S.

Meguro, S., Higashi, K., Hase, T., Honda, Y., Otsuka, A., Tokimitsu, I., and Itakura, H. Solubilization of Phytosterols in Diacylglycerol Versus Triacylglycerol Improves the Serum Cholesterol-Lowering Effect. *Eur. J. Clin. Nutr.* **2001,** *55,* 513-517.

Meguro, S., Hase, T., Otsuka, A., Tokimitsu, I., and Itakura, H. Effect of Phytosterols in Dietary Diacylglycerol on Atherosclerosis in Cholesterol-Fed Rabbits. *Nutrition* **2003,** *19,* 670-675.

Meng, X., Zou, D., Shi, Z., Duan, Z., and Mao, Z., Dietary diacylglycerol prevents high-fat diet-induced lipid accumulation in rat liver and abdominal adipose tissue. *Lipids.* **2004,** *39,* 37-41.

Mokdad, A.H., Bowman, B.A., Ford, E.S., Vinicor, F., Marks, J.S., and Koplan, J.P. The

Continuing Epidemics of Obesity and Diabetes in the United States. *J. Am. Med. Assoc.* **2001,** *286,* 1195-1200.

Murase, T., Mizuno, T., Omachi, T., Onizawa, K., Komine, Y., Kondo, H., Hase, T., and Tokimitsu, I. Dietary Diacylglycerol Suppresses High Fat and High Sucrose Diet-Induced Body Fat Accumulation in C57BL/6J Mice. *J. Lipid Res.* **2001,** *42,* 372-378.

Murase, T., Aoki, M., Wakisaka, T., Hase, T., and Tokimitsu, I. Anti-Obesity Effect of Dietary Diacylglycerol in C57BL/6J Mice: Dietary Diacylglycerol Stimulates Intestinal Lipid Metabolism. *J. Lipid Res.* **2002a,** *43,* 1312-1319.

Murase, T., Nagasawa, A., Suzuki, J., Wakisaka, T., Hase, T., and Tokimitsu, I. Dietary α-Linolenic Acid-Rich Diacylglycerols Reduce Body Weight Gain Accompanying the Stimulation of Intestinal β-Oxidation and Related Gene Expressions in C57BL/KsJ-db/db mice. *J. Nutr.* **2002b,** *132,* 3018-3022.

Murata, M., Hara, K., and Ide, T. Alteration by Diacylglycerols of the Transport and Fatty Acid Composition of Lymph Chylomicrons in Rats. *Biosci. Biotechnol. Biochem.* **1994,** *58,* 1416-1419.

Murata, M., Ide, T., and Hara, K. Reciprocal Responses to Dietary Diacylglycerol of Hepatic Enzymes of Fatty Acid Synthesis and Oxidation in the Rat. *Br. J. Nutr.* **1997,** *77,* 107-121.

Nagao, T., Watanabe, H., Goto, N., Onizawa, K., Taguchi, H., Matsuo, N., Yasukawa, T., Tsushima, R., Shimasaki, H., and Itakura, H. Dietary Diacylglycerol Suppresses Accumulation of Body Fat Compared to Triacylglycerol in Men in a Double-Blind Controlled Trial. *J. Nutr.* **2000,** *130,* 792-797.

Nakajima, Y. Water-retaining ability of diacylglycerol. *J. Am. Oil. Chem. Soc.* **2004,** *81,* 907-912

National Cholesterol Education Program, Expert Panel on Detection, Evaluation, and Treatment of High Blood Cholesterol in Adults, Executive Summary of The Third Report of The National Cholesterol Education Program (NCEP) Expert Panel on Detection, Evaluation, and Treatment of High Blood Cholesterol in Adults (Adult Treatment Panel III). *J. Am. Med. Assoc.* **2001,** *285,* 2486-2497.

Ogden, C.L., Carroll, M.D., Curtin, L.R., McDowell, M.A., Tabak, C.J., and Flegal, K.M. Prevalence of Overweight and Obesity in the United States, 1999-2004. *J. Am. Med. Assoc.* **2006,** *295,* 1549-1555

Ota, N., Soga, S., Hase, T., Tokimitsu, I., and Murase, T., Dietary diacylglycerol induces the regression of atherosclerosis in rabbits. *J. Nutr.* **2007,** *137,* 1194-1199.

Otsuki, K., Mori, K., Takase, H., and Katsuragi, Y. Two Years, Long-term Effects of Dietary Diacylglycerols on the Risks of the Metabolic Syndrome. *J. Japan. Soc. Hum. Dry Dock* **2004,** *19,* 29-32.

Pariza, M.W., Park, Y., and Cook, M.E. The Biologically Active Isomers of Conjugated Linoleic Acid. *Prog. Lipids Res.* **2001,** *40,* 283-298.

Perez-Camino, M.C., Moreda, W., and Cert, A. Determination of Diacylglycerol Isomers in Vegetable Oils by Solid Phase Extraction Followed by Gas Chromatography on a Polar Phase. *J. Chromatogr.* **1996,** *A721,* 305-314.

Pi-Sunyer, F.X. Health Implications of Obesity. *Am. J. Clin. Nutr.* **1991,** *53,* 1595S-1603S.

Research Association of Health and Nutrition Information, ed. The National Health and Nutrition Survey in Japan, 2003, Ministry of Health, Labour and Welfare of Japan, **2005**; pp. 152, Daiichi-Shuppan, Tokyo.

Rimm, E.B., Stampfer, M.J., Giovannucci, E., Ascherio, A., Spiegelman, D., Colditz, G.A., and Willett, W.C. Body Size and Fat Distribution as Predictors of Coronary Heart Disease Among

Middle-Aged and Older US Men. *Am. J. Epidemiol,* **1995,** *141,* 1117-1127.

Saito, S., Tomonobu, K., Hase, T., and Tokimitsu, I. Effects of diacylglycerol on postprandial energy expenditure and respiratory quotient in healthy subjects. *Nutrition* **2006,** *22,* 30-5.

Seidell, J.C., Verschuren, W.M., van Leer, E.M., and Kromhout, D. Overweight, Underweight, and Mortality. A Prospective Study of 48, 287 Men and Women. *Arch. Intern. Med.* **1996,** *156,* 958-963.

Smith, R.E., Finley, J.W., and Leveille, G.A. Overview of Salatrim, a Family of Low-Calorie Fats. *J. Agric. Food Chem.* **1994,** *42,* 432-434.

Smith, S.J., Cases, S., Jensen, D.R., Chen, H.C., Sande, E., Tow, B., Sanan, D.A., Raber, J., Eckel, R.H., and Farese, R.V., Jr. Obesity Resistance and Multiple Mechanisms of Triglyceride Synthesis in Mice Lacking DGAT. *Nat. Genet.* **2000,** *25,* 87-90.

St-Onge, M.P., and Jones, P.J. Physiological Effects of Medium-Chain Triglycerides: Potential Agents in the Prevention of Obesity. *J. Nutr.* **2002,** *132,* 329-332.

Stubbs, R.J. The Effect of Ingesting Olestra-Based Foods on Feeding Behavior and Energy Balance in Humans. *Crit. Rev. Food Sci. Nutr.* **2001,** *41,* 363-386.

Tada, N., Watanabe, H., Matsuo, N., Tokimitsu, I., and Okazaki, M. Dynamics of Postprandial Remnant-Lipoprotein Particles in Serum After Loading of Diacylglycerols. *Clin. Chem. Acta* **2001,** *311,* 109-117.

Tada, N., and Yoshida, H. Diacylglycerol on Lipid Metabolism. *Curr. Opin. Lipidol.* **2003,** *14,* 29-33.

Tada, N. Physiological actions of diacylglycerol outcome. *Curr. Opin. Clin. Nutr. Metab. Care.* **2004,** *7,* 145-9.

Tada N., Shoji, K., Takeshita, M., Watanabe, H., Yoshida, H, Hase, T., Matsuo, N., and Tokimitsu, I. Effects of diacylglycerol ingestion on postprandial hyperlipidemia in diabetes. *Clinica Chimca Acta* **2005,** *353,* 87-94.

Taguchi, H., Watanabe, H., Onizawa, K., Nagao, T., Gotoh, N., Yasukawa, T., Tsushima, R., Shimasaki, H., and Itakura, H. Double-Blind Controlled Study on the Effects of Dietary Diacylglycerol on Postprandial Serum and Chylomicron Triacylglycerol Responses in Healthy Humans. *J. Am. Coll. Nutr.* **2000,** *19,* 789-796.

Taguchi, H., Nagao, T., Watanabe, H., Onizawa, K., Matsuo, N., Tokimitsu, I., and Itakura, H. Energy Value and Digestibility of Dietary Oil Containing Mainly 1,3-Diacylglycerol Are Similar to Those of Triacylglycerol. *Lipids* **2001,** *36,* 379-382.

Takase, H., Shoji, K., Hase, T., and Tokimistu, I., Effect of diacylglycerol on postprandial lipid metabolism in non-diabetic subjects with and without insulin resistance. *Atherosclerosis,* **2005,** *180,* 197-204.

Takeshita, M., Goto, N., Katsuragi, Y., and Yasukawa, T. Effects of Diacylglycerol Containing Plant Sterol on Risk Factors for Arteriosclerosis. *J. Jpn. Hum. Dry Dock* **2001,** *16,* 9-3.

Takeshita, M., Katsuragi, Y., Kusuhara, M., Higashi, K., Miyajima, E., Mizuno, K., Mori, K., Obata,T. Ohmori, R., Ohsuzu, F., Onodera, Y., Sano. Sawada, S., Tabata, S., Tokimitsu, I., Tomonobu, K., Yamashita, T., Yasukawa, T., Yonemura, A., and Nakamura H. Phytosterols dissolved in diacylglycerol oil reinforce the cholesterol-lowering effect of low-dose pravastatin treatment. *Nutr. Metab. Cardiovasc. Dis.* **2008,** (in press).

Tsuji, H., Kasai, M., Takeuchi, H., Nakamura, M., Okazaki, M., and Kondo, K. Dietary Medium-Chain Triacylglycerols Suppress Accumulation of Body Fat in a Double-Blind, Controlled Trial in Healthy Men and Women. *J. Nutr.* **2001,** *131,* 2853-2859.

Vanschoonbeek, K., de Maat, M.P., and Heemskerk, J.W. Fish Oil Consumption and Reduction of Arterial Disease. *J. Nutr.* **2003,** *133,* 657-660.

Wardlaw, G.M., Snook, J.T., Park, S., Patel, P.K., Pendley, F.C., Lee, M.S., and Jandacek, R.J. Relative Effects on Serum Lipids and Apolipoproteins of a Caprenin-Rich Diet Compared with Diets Rich in Palm Oil/Palm-Kernel Oil or Butter. *Am. J. Clin. Nutr.* **1995,** *61,* 535-642.

Watanabe, H., Onizawa, K., Taguchi, H., Kobori, M., Chiba, H., Naito, S., Matsuo, N., Yasukawa, T., Hattori, M., and Shimasaki, H. Nutritional Characterization of Diacylglycerols in Rats. *J. Jpn. Oil Chem. Sci.* **1997a,** *46,* 301-307.

Watanabe, H., Onizawa, K., Taguchi, H., Fujimori, N., Naito, S., Gotoh, N., Yasukawa, T., Hattori, M., and Shimasaki, H. Effect of Diacylglycerols on Lipid Metabolism in Humans. *J. Jpn. Oil Chem. Sci.* **1997b,** *46,* 309-317.

Watanabe, T., Shimizu, M., Sugiura, M., Sato, M., Kohori, J., Yamada, N., and Nakanishi, K. Optimization of Reaction Conditions for the Production of DAG Using an Immobilized 1,3-Regiospecific Lipozyme RMIM. J. Am. Oil Chem. Soc. 2003, 80, 1201-1207.

Weisinger, H.S., Armitage, J.A., Sinclair, A.J., Vingrys, A.J., Burns, P.L., and Weisinger, R.S. Perinatal Omega-3 Fatty Acid Deficiency Affects Blood Pressure Later in Life. *Nat. Med.* **2001,** *7,* 258-259.

World Health Organization. Controlling the Global Obesity Epidemic, 2003a; (Internet address: http://www.who.int/nutrition/topics/obesity/en/index.html)

World Health Organization. Diet, Nutrition and the Prevention of Chronic Diseases, 2003b; Series no. 916, pp. 54-134, WHO, Geneva.

Wylie-Rosett, J. Fat Substitutes and Health: An Advisory from the Nutrition Committee of the American Heart Association. *Circulation* **2002,** *105,* 2800-2804.

Yamamoto, K., Asakawa, H., Tokunaga, K., Watanabe, H., Matsuo, N., Tokimitsu, I., and Yagi, N. Long-Term Ingestion of Dietary Diacylglycerol Lowers Serum Triacylglycerol in Type II Diabetic Patients with Hypertriglyceridemia. *J. Nutr.* **2001,** *131,* 3204-3207.

Yamamoto, K., Takeshita, M., Tokimitsu, I., Watanabe, H., Mizuno, T., Asakawa, H., Tokunaga, K., Tatsumi, T., Okazaki, M., and Yagi, N., Diacylglycerol oil ingestion in type 2 diabetic patients with hypertriglyceridemia. *Nutrition,* **2006a,** *22,* 23-29.

Yamamoto, K., Tomonobu, K., Asakawa, H., Tokunaga, K., Hase, T., Tokimitsu, I., and Yagi N. Diet therapy with diacylglycerol oil delays the progression of renal failure in type 2 diabetic patients with nephropathy. *Diabetes Care* **2006b,** *29,* 417-419.

Yanagita T., Ikeda, I., Wang, Y.M., Nakagiri, H. Comparison of the lymphatic transport of radio-labeled 1,3-dioleoylglycerol and trioleoylglycerol in rats. *Lipids* **2004,** *39,* 827-32. Erratum in: *Lipids*. 2005, 40, 533.

Yasukawa, T., and Yasunaga, K. Nutritional Functions of Dietary Diacylglycerols. *J. Oleo Sci.* **2001**, 50, 427-432.

Yasunaga, K., Seo, Y., Katsuragi, Y., Oriuchi, N., Otake, H., Endo, K., and Yasukawa, T. Gastric Emptying Rate of Dietary Diacylglycerol, the 54th Annual Meeting of the Japanese Society of Nutrition and Food Science, Matsuyama, May 12-14, 2000.

Yasunaga, K., Glinsmann, W.H., Seo, Y., Katsuragi, Y., Kobayashi, S., Flickinger, B., Kennepohl, E., Yasukawa, T., and Borzelleca, J.F. Safety Aspects Regarding the Consumption of High-Dose Dietary Diacylglycerol Oil in men and Women in a Double-Blind Controlled Trial in Comparison with Consumption of a Triacylglycerol Control Oil. Food Chem. *Toxicol.* **2004,** *42,* 1419-1429.

Zimmet, P., Alberti, K.G., and Shaw, J. Global and Societal Implications of the Diabetes Epidemic. *Nature* **2001,** *414,* 782-787.

Part II

Metabolism

Section A

Digestion and Absorption

Chapter 2

Digestion and Absorption of Glycerides

Armand B. Christophe

Department of Internal Medicine, Division of Nutrition, Ghent University Hospital, Ghent, Belgium

Introduction

The major components of most food fats are triacylglycerols (TAG). TAG molecules can differ from each other in the fatty acids they contain and in the combination and/or stereochemical implantation of different fatty acids on the glycerol backbone. Other components with distinct nutritional effects that can be present in food fats are partial glycerides, phospholipids, sterols, fat-soluble vitamins and provitamins, antioxidants, and other fat-soluble components, some of which can affect gene expression.

By chemical and enzymatic processes, the structure of TAG can be modified and minor natural fat components such as monoacylglycerols (MAG), diacylglycerols (DAG), and phytosterol esters can be produced in large quantities. These have been shown to have special nutritional properties. This is also the case for TAG with changed stereospecific position or changed combination of fatty acids such as inter-esterified fats (Meijer & Weststrate, 1997), structured lipids (Nagata et al., 2003), or salatrims (Jang et al., 1998). DAG and MAG also have nutritional effects different from TAG with the same fatty acid composition.

The structural effects of food fats on several of their physiologic, nutritional, and pathophysiologic properties can be understood on the basis of a combination of their effect of gastric emptying, specificity of enzymes that are involved in lipid digestion, rate of uptake of their digestion products, and further metabolism of the latter. In this chapter, integrated aspects of digestion, absorption, and interactions of the absorbed products in the intestinal cell will be discussed.

Metabolism of Dietary Fat

Gastric Phase of Fat Digestion and Absorption

Except for experimental settings, fats are always consumed as part of a meal, which also contains nonfat constituents such as carbohydrates, proteins, or fiber. These components may affect phenomena related to fat digestion and absorption. Thus, results of experiments using fat loading or infusion of fat emulsions in the stomach or in the duodenum may not be fully applicable to real-life situations.

Physicochemical Reactions in the Stomach. An average Western diet contains ~100-150 g food fat, with ~97% of this as TAG. The remainder is made up of phospholipids (48 g/d, mainly phosphatidylcholine), cholesterol, other sterols, and minor lipid components. When they are the only lipids present in an aqueous environment, TAG form unstable emulsion droplets, whereas phosphatidylcholines form relatively stable lamellar structures. When TAG and phospholipids are present together in proportions as found in the diet, the latter act as an emulsifier by coating the TAG droplets, thus forming more stable emulsions. Proteins and their peptic digestion products stabilize TAG emulsions in the stomach and in the intestinal lumen (Meyer et al., 1976). Gastric lipolysis helps to increase emulsification in the stomach (Armand et al., 1994). Stabilized emulsion droplets are substrates for gastric lipase. A lower initial droplet size facilitates gastric and intraduodenal digestion (Armand et al., 1999). When viscous soluble fiber is present in the diet, the size of the emulsion particles is increased, and their subsequent rate of lipolysis is reduced (Pasquier et al., 1996).

DAG and MAG with thermodynamic equilibrium composition (*eq*-DAG and *eq*-MAG) are easily prepared industrially in large quantities. Such DAG are on the market as edible oils in Japan and in the United States. *eq*-DAG and *eq*-MAG are also used in smaller quantities as emulsifiers in some foods. *eq*-MAG, alone or in combination with medium-chain triglycerides (MCT), were used successfully to correct digestive malabsorptions and consequences thereof (Verdonk et al., 1978). Long-term effects of *eq*-MAG (Christophe & Verdonk, 1977; Christophe & Verdonk, 1980) and *eq*-DAG feeding (described in several chapters of this book) and postprandial effects of loading with these special fats were documented (see below). When DAG are fed as part of a meal, they probably also form small emulsion droplets in the stomach. MAG form in an excess of water lamellar liquid crystalline phases at body temperature (Chung & Caffrey, 1995). Their swelling can be affected by the presence of other substances (Engblom, 1996). When MAG are present with TAG in excess, they will likely form co-emulsions.

Gastric Digestion. In humans, the hydrolysis of TAG starts with the action of gastric lipase. Gastric lipase preferentially splits primary ester bonds, although there may be some hydrolysis from fatty acids esterified at the *sn*-2 position (Tiruppathi & Balasubramanian, 1982). It does not act on phospholipids or on cholesteryl esters nor is its activity inhibited by phospholipids (Jensen et al., 1994). With TAG as substrate, gastric lipase catalyzes the hydrolysis of the fatty acid at the *sn*-3 position preferentially to that at the *sn*-1 position (Rogalska et al., 1990) and prefers TAG with short and medium chain lengths. Thus, TAG with such fatty acids in the *sn*-3 position as are present in bovine milk and its products are preferential substrates. As a result of product inhibition by free fatty acids (FFA), the main digestion products of gastric lipase are FFA and *sn*-1, 2 DAG when TAG are the substrate (Pafumi et al., 2002). It was demonstrated that *sn*-1, 3 DAG also comprise substrates for gastric lipase but that with these substrates, the enantioselectivity for the *sn*-3 position no longer exists (Carriere et al., 1997). In a monolayer system, *sn*-1, 3- and *sn*-2,3 dicaprin are equally good substrates for human gastric lipase, but there is discrimination against 1, 2 dicaprin (Rogalska et al., 1993). It is likely that in co-emulsions of DAG with TAG, the former would be enriched in the outer layers of the fat droplets. Whether this would result in preferential hydrolysis is not known. It is not clear whether *sn*-1(3) MAG, which would be formed by gastric lipolysis of *sn*-1, 3 DAG or fed as such, would be substrates for gastric lipase.

It was demonstrated that medium-chain fatty acids, which are set free in the stomach, can be absorbed directly into the portal blood and thus affect hepatic metabolism (Levy et al., 1984). The specific activity of gastric lipase in vivo is comparable to that of pancreatic lipase for liquid fats and higher for solid fats (Carriere et al., 2000). Of course, pancreatic lipase output is much higher than gastric lipase output. High-fat diets increase gastric lipase output in humans (Armand et al., 1995). About 10% of the dietary fatty acids are liberated by gastric lipase from a test meal that contains TAG with long-chain fatty acids, but considerably higher amounts from a test meal that contains TAG with short- or medium-chain fatty acids (Carriere et al., 1993).

Even for long-chain fatty acid-containing TAG, this corresponds to the breakdown of about one third of the TAG into DAG and FFA in the stomach. Furthermore, there is evidence that gastric lipase is also active in the upper small intestine. The importance of gastric digestion is that its hydrolysis products affect the excretion of some hormones involved in fat digestion, thus helping to stabilize the surface of the TAG emulsion. This promotes the binding of pancreatic colipase later in the small intestine, thus making it a better substrate for pancreatic lipase (Dahim & Brockman, 1998). There is a compensatory increase in gastric lipase when pancreatic lipase is low. Nonetheless, under these conditions, the capacity for fat absorption remains limited when usual food fats are fed. It can be expected that under these conditions, the lipids of choice to enable feeding more fat and provide for energy and essential fatty acids would be semisynthetic triglycerides with a short-chain fatty acid at the *sn*-3 position and long-chain fatty acids that include essential fatty acids at the other positions.

Gastric Emptying. If gastric digestion is allowed to proceed long enough, the fat mixture that will be emptied into the upper small intestine will have an equilibrium composition governed by product inhibition (Pafumi et al., 2002). If not, its composition will depend on the rate of gastric digestion and on the time the dietary fat was retained in the stomach. The physicochemical and chemical forms in which dietary fats are consumed affect the rate of gastric emptying. Fats that are ingested as emulsions are partitioned into an aqueous phase and an oil phase, whereas fats consumed as nonemulsions form an oil phase. Fats in the aqueous phase are readily emptied from the stomach, whereas fats in an oil phase are emptied much more slowly (Vian et al., 1982). The chemical composition of the oil phase also influences the rate of emptying. The delaying effect depends on the saturation (Maes et al., 1998) and the chain length (Hunt & Knox, 1968) of the fatty acids. No data are available concerning the effect of the chemical structure of the fat if any. In healthy individuals, the rate of gastric emptying can be rate limiting for fat absorption (Maes et al., 1996). Fat digestion has an important role in the regulation of postprandial gastric acid secretion and fat emptying in humans. Lipase inhibition accelerates gastric emptying of both the solid and fat phases of a mixed meal (Borovicka et al., 2000).

No experimental data are available on the rate of gastric emptying of DAG and MAG relative to that of TAG. This may well depend on the physicochemical form in which they are present in the stomach (Borel et al., 1994).

Small Intestinal Phase

Lipids in the Lumen of the Small Intestine. The lipids that are emptied from the stomach into the small intestine are diluted by lipids of endogenous origin (~10-25 g/d when fasting). Endogenous lipids originate primarily from sloughed intestinal cells and biliary secretions. The amount of phospholipids of biliary origin (7-22 g/d, mainly phosphatidylcholine) is considerably greater than that of dietary origin (3-8 g/d) (Northfield & Hofman, 1975). The same is also often true for cholesterol (~1 g/d of endogenous origin). In bile, bile salt-phospholipid mixed micelles exist in a cylindrical arrangement (Hjelm et al., 1995). In the intestinal lumen, bile salts and phospholipid molecules are likely to partly adsorb to the lipid emulsion droplets.

Overview of Reactions in the Small Intestine. The main reactions occurring simultaneously or quasi-simultaneously in the small intestine are digestion of the glycerides in the emulsion phase, isomerization of partial glycerides in the emulsion phase, transfer of the digestion products from the emulsion phase into micellar phases, digestion and isomerization in the micellar phase, and uptake of digestion products mainly from micellar phases. Thus, the relative amounts of the different digestion products that will be taken up by the intestinal cells will depend on the relative rates of these different reactions.

Lipolysis. The major enzymes involved in fat digestion in the intestinal lumen are pancreatic lipase, pancreatic phospholipase A_2, and pancreatic cholesterol esterase, the last-mentioned also named bile-salt-dependent lipase. These enzymes work in concert with each other to yield the final reaction products.

Human pancreatic lipase catalyzes the equilibrium reaction between ester formation and hydrolysis. It is specific for primary alcohols and esters of such alcohols. Thus, under certain conditions, pancreatic and other lipases can be used for the synthesis of TAG with certain stereospecific properties (Xu et al., 2000). Lipases act on the lipid/water interface. In the beginning of and during fat digestion, the lipid/water interface is rich in water and the digestion products are continuously removed from the interface favoring lipolysis. Pancreatic lipase splits off fatty acids from the *sn*-1 and *sn*-3 position of TAG. It is more active against *sn*-1, 2- and *sn*-2, 3 DAG thus formed than against TAG (Lykidis et al., 1995). Thus, it can break down the TAG into *sn*-1,2- and *sn*-2, 3 DAG, which can be further hydrolyzed into *sn*-2 MAG. These are no longer substrates for pancreatic lipase. Pancreatic lipase is less active against polyunsaturated fatty acids with a double bond close to the ester bond, in particular when these are located in the *sn*-3 position (Bottino et al., 1967) and clearly prefers the *sn*-1 position of both TAG and DAG (Carriere et al., 1997). Thus, if there would be neither isomerization nor absorption of DAG and MAG, the ultimate digestion products of TAG would be a 2:1 mixture of FFA and *sn*-2 MAG. However, both reactions do occur (see below).

Phospholipase A_2 hydrolyzes the fatty acid esterified at the *sn*-2 position in phosphatidylcholine to yield lysophosphatidylcholine and FFA (Nalbone et al., 1980). This reaction is of importance for the digestion and uptake of neutral glycerides because the presence of phospholipids in the intestinal lumen retards TAG digestion (Wickham et al., 2002), affects the partitioning of lipid digestion products between the oil and micellar phase (Saunders & Sillery, 1976), and slows down the uptake of lipid digestion products from the micellar phase (Surpuriya & Higuchi, 1974). If sphingolipids were present in the diet, they are broken down by sphingomyelinase and the resulting ceramide by ceramidase. Carboxyl ester lipase hydrolyzes sterol esters. It prefers substrates present in the micellar state but has low activity against MAG (Lindstrom et al., 1988). Cholesterol is absorbed as free sterol, and the effects of phytostanol esters and phytosterol esters are mediate by the free phytostanol/sterol.

Isomerization Reactions. "Linear" partial glycerides are thermodynamically favored over "Branched" partial glycerides. MAG with equilibrium isomeric composition consist of ~45% *sn*-1-, 45% *sn*-3- and 10% *sn*-2 isomers (Chirstophe & Verdonk, 1977). For DAG, the isomer ratio is ~ 70:30 (Murase et al., 2001) for *sn*-1, 3- and *sn*-1, 2- plus *sn*-2, 3 DAG respectively. Isomerization of partial glycerides yields products with primary ester bonds, which are substrates for pancreatic lipase, thus resulting in more complete digestion with the liberation of glycerol. Because MAG are readily absorbed, the rate of isomerization relative to that of uptake is important for further digestion. The isomerization rate increases with unsaturation and decreases with chain length (Bowswinkel et al., 1996). Nonetheless, spontaneous isomerization of long-chain fatty acids containing *sn*-2 MAG in model systems is low compared with their intestinal uptake (Lyubachesskaya et al., 2000); ~15-20% of *sn*-2 MAG would isomerize in the intestinal lumen (Skipski et al., 1959).

Uptake of Glycerides by the Intestinal Cells. For a lipid to be a candidate for absorption, it must be able to reach the absorptive surface of the small intestine. This is the surface of the microvilli. Geometrically, they are so close to each other that emulsion droplets are too large to come between them. Moreover, an "unstirred" water layer lines the microvilli. Passage through this layer is rate limiting for fat absorption. MAG and FFA leave the emulsion/water interface where they were formed by the action of pancreatic lipase and form mixed micelles with conjugated bile acids. Only in this form can they easily access the absorptive surface of the small intestine and be absorbed. To my knowledge, no data are available on the effect of isomer structure or fatty acid composition of the MAG on the rate at which they are transferred from the emulsion phase to the micellar phase. Partitioning of fatty acids between the emulsion interface and the micellar phase depends on pH. When pH is low, such as in the absence of pancreatic juice, protonated fatty acids remain at the interface of the emulsion and inhibit further digestion, even when pancreatic lipase is administered. The interior of the MAG/FFA/conjugated bile salt mixed micelle is a lipophilic environment in which lipophilic substances such as cholesterol, carotenes, fat-soluble vitamins, and so on can dissolve. It is in this way that they can reach the absorptive surface and are candidates for uptake (Borel, 2003). Lysophosphatidylcholines formed by the digestion of phosphatidylcholines favor micellar solubilization and uptake of fat digestion products (Rampone & Long, 1977) and other dietary lipophilic substances (Sugawara et al., 2001). Phytosterols can compete with cholesterol for incorporation into the micelles and thus interfere with cholesterol absorption.

Lysophosphatidylcholine itself is taken up by the intestinal cell, affects intestinal metabolism (O'Doherty et al., 1974), and can be converted into phosphatidylcholine again. There is evidence that phosphatidylcholine can be taken up intact and excreted as part of intestinal lipoproteins in the lymph (Zierenberg et al., 1979).

When *eq*-DAG are fed in a diet, TAG are also inevitably present. It is expected on theoretical grounds that they will form emulsion droplets in the stomach together with the TAG. The *rac*-1, 2 isomers are normal intermediates of TAG digestion. The *sn*-1, 3 isomers, which are the major isomers in *eq*-DAG, would be formed from TAG only in very small quantities by isomerization of their intermediary *sn*-1, 2 (2, 3) DAG digestion products. If pancreatic lipase preferentially attacks the *sn*-1 position of 1, 3 DAG as it does in TAG, the major first digestion product would be *sn*-3 MAG, but *sn*-1 MAG would be formed as well. The type of isomer formed would be of no consequence if both MAG isomers, which are substrates for pancreatic lipase, would be broken down completely to glycerol and fatty acids before absorption. However, if there would be substantial absorption of the intact MAG, their isomeric composition could be of importance because it has been shown that they are handled differently in the intestinal cells (see below).

The composition of the mixed micelles that exist in the intestinal lumen and several of their physicochemical properties were described (Mansbach et al., 1975). It cannot be excluded that after *eq*-DAG or *eq*-MAG feeding, the geometrical structure of the mixed micelles could be different than after TAG feeding and could affect the extent of micellar solubilization and uptake of other lipophilic substances; however, experimental data are lacking.

Reaching the absorptive surface of the small intestinal cells is not enough for uptake. Bile salts, for instance, are not absorbed at this site, and absorption of phytosterols is very limited. After TAG feeding, long-chain fatty acids that were present in the *sn*-1, 3 position are taken up mainly as FFA, whereas those at the *sn*-2 position are taken up mainly as *sn*-2 MAG. In contrast to long-chain fatty acids, medium-chain fatty acids are mainly but not exclusively taken up as FFA whatever their position in the medium-chain or mixed-chain TAG ("structured lipids") that were present in the diet.

There is some evidence to suggest that small amounts of TAG and DAG can be taken up without prior digestion. After consumption of MCT, small amounts of these triglycerides were found in ascites fluid from a patient with an obstruction of the thoracic duct (Christophe et al., 1982), suggesting that small amounts of TAG (at least MCT) can be taken up without prior intestinal digestion. Postprandial and long-term effects after oral administration of MAG suggest that part of them may be taken up intact (see below). Some of the postprandial effects after feeding of DAG suggest that part of them may be taken up as MAG.

Intestinal Metabolism of Fat Digestion Products

Intestinal Handling of FFA. In the cells of the small intestine, several enzymes can act on free but not on esterified fatty acids. In contrast to MAG, FFA are diluted with fatty acids originating from the plasma free fatty acid pool (Nilsson & Becker, 1995). After activation, they can be oxidized, elongated, chain desaturated, and converted into complex lipids (Nilsson & Becker, 1995; Huang et al.,1995). The relative rates of the different metabolic routes depend on the nature of the fatty acids and on the presence of other components in the intestinal cells (Chen & Nillson, 1994). Conversion of saturated fatty acids into monounsaturates when they are absorbed in free form, i.e., when they were present in the outer position of the dietary TAG, could contribute to their lower hypercholesterolemic effect than when originally present at the inner position. FFA also affect intestinal gene expression (Murase et al., 2001) and the production of apolipoproteins; these effects are fatty acid dependent (Sanderson & Naik, 2000). When *sn*-2 MAG are simultaneously present in the intestinal cells, as after TAG feeding, FFA are converted mainly by a multienzyme complex in TAG that are excreted as chylomicrons. The excess of FFA reacts with *sn*-3 glycerophosphate to form lysophosphatidic acids, which are further converted into phosphatidic acids. Dephosphorylation results in *sn*-1, 2 DAG, which are precursors of both TAG and phospholipids. This results in the formation of intestinal lipoproteins of smaller size (Christophe & Verdonk, 1978).

Intestinal Handling of α-Monoglycerides. What will happen after influx of 1(3) *sn*-MAG in the intestinal cell is less clear. A series of competing reactions must be considered such as complete digestion, acylation, transacylation, phosphorylation, or intact excretion out of the intestinal cells. Lipases that hydrolyze MAG are present in the intestinal cells but their action seems to be reduced by other intestinal cell constituents (De Jong & Hulsmann, 1978). Some of them show obvious fatty acid specificity (Hanel & Gelb, 1995). Both *sn*-1- and *sn*-3 MAG can be acylated to form *sn*-1,3 DAG. This transformation is fatty acid and species dependent (Bierbach, 1983). Nonetheless, *sn*-1,3 DAG are poor substrates for intestinal TAG synthesis (Bierbach et al., 1983; Lehner et al., 1993). *Sn*-1- but not *sn*-3 MAG can be phosphorylated into lysophosphatidic acid (Paris & Clement, 1969) and possibly be metabolized further as described above. If MAG would escape intestinal metabolization and could leave the cell intact, they could be transported to the liver by portal blood bound to albumin (Duff et al., 2001). Portal absorption of MAG with medium-chain fatty acids was documented (Tso et al., 1995). It was demonstrated that cultured liver cells convert *sn*-1 glycerol monoethers (*sn*-1 MAG analogs) into phospholipids and triglycerides (Mackenzie et al., 1976).

There is some indirect evidence that after MAG feeding, intact *sn*-1 MAG or their phosphorylated derivatives reach the liver with retention of the *sn*-1 ester bond. After long-term feeding of *eq*-monolinoleate, carbon number analysis (separation and quantization according to the sum of the chain lengths of the three fatty acids present) of fasting serum TAG revealed an increased ratio of TAG species with carbon number 54 relative to 52. The former TAG combine fatty acids with chain length 18/18/18, whereas the latter have 16/18/18 as their chain length combination (Christophe & Verdonk, 1980). This finding was interpreted as follows. In the liver, TAG synthesis starts with *sn*-3 glycerophosphate as the initial fatty acid acceptor. *Sn*-3 glycerophosphate acyltransferase shows fatty acid chain length selectivity, incorporating ~60% fatty acids with 16 carbon atoms and 40% fatty acids with 18 carbon atoms at the *sn*-1 position. The other enzymes involved in hepatic synthesis of TAG that will be secreted in plasma (Yang et al., 1995) incorporate mainly unsaturated fatty acids with 18 carbon atoms (~ 80%). As a result, the main chain length combination of fatty acids in TAG in fasting plasma will be 16/18/18 as was found experimentally.

By feeding *sn*-1 MAG, the fatty acid selectivity of *sn*-3 glycerophosphate acyltransferase could be bypassed on the condition that these were phosphorylated and thus converted into lysophosphatidic acids. Phosphorylation of *sn*-1 monoglycerides by intestinal and liver cell homogenates was demonstrated (Paris & Clement, 1969. Mackenzie et al.,1976). If these *sn*-1 MAG-derived lysophospholipids would be further metabolized in the same way as the *sn*-3 glycerophosphate-derived lysophospholipids,

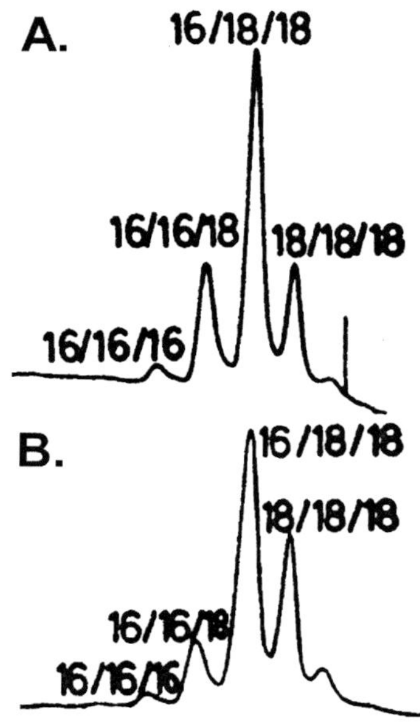

FIG. 2.1. Carbon number composition of serum triglycerides after long-term feeding of a diet with usual food fats (mainly triglycerides) (panel A) or a diet in which the visible food fat was replaced by *eq*-monolinoleate (panel B). Conditions: 3% JXR, 80 cm; He 40 mL/min; Ts = 150°C; r = 4°/min; TE = Tdetector = 330°C; on column injection.

TAG would be formed with the main fatty acid combination X/18/18, with X as the chain length of the fatty acid in the *sn*-1 position of the MAG. If this reasoning is correct, there should be a preferential increase in the fraction of TAG with the chain length combination 18/18/18 if *eq*-monolinoleate was fed. This was indeed the case. The carbon number composition of fasting serum triglycerides after long-term natural fat or *eq*-MAG feeding is given in Figure 2.1.

If the concept that the fatty acid specificity of *sn*-3 acyltransferase can be circumvented by feeding *sn*-1 MAG that contain at the *sn*-1 position a fatty acid that normally is not incorporated into lysophosphatidic acid is correct, TAG with unusual structures may be formed. Such "plasma TAG with unusual structure," which may also occur after DAG feeding in case these are partly taken up as MAG, may be turned over at different rates and consequently affect their fasting concentration.

Postprandial Effects

After TAG feeding, long-chain fatty acids leave the intestinal cell mainly as TAG incorporated into intestinal lipoproteins with the size of chylomicrons. These are transported *via* the lymphatic route. When they enter the blood stream, there is an increase in chylomicronemia and triglyceridemia, and the fatty acid composition of plasma TAG changes toward that of the fat fed (Sakr et al., 1997). Chylomicron-TAG are hydrolyzed by lipoprotein lipase, resulting in the formation of remnants and the liberation of FFA. As a consequence, the composition of the nonesterified fatty acids in plasma also changes toward that of the fat fed. Chylomicron remnants are taken up by the liver. Medium-chain fatty acids provided in the diet as MCT or as structured lipids are transported mainly as FFA by the portal vein (Mu & Hoy, 2000).

After isomeric *eq*-MAG loading, the well-known postprandial chylomicronemia, triglyceridemia, and shift of the fatty acid composition of serum triglycerides and nonesterified fatty acids toward that of the fat fed are much less pronounced than after natural fat feeding (Christophe & Verdonk, 1972). However, when MAG are fed together with FFA, effects similar to these of TAG feeding occur. This finding agrees with the finding that the availability of long-chain FFA in the intestine is critical for chylomicron synthesis (Minich et al., 1997). MAG feeding in normal volunteers does not result in increased fecal fat, indicating that these fats are absorbed as completely as TAG. In patients with pancreatogenic digestive malabsorption, they are better absorbed (Verdonk et al.,1978). MAG loading does not result in the appearance of MAG or DAG in blood in the postprandial state.

The reduced hypertriglyceridemia and chylomicronemia after MAG loading (Christophe & Verdonk, 1978; Christophe & Verdonk, 1972) in combination with the absence of increased stool fats or the appearance of MAG or DAG in blood in the postprandial state suggest that at least part of the fatty acids fed as MAG is not transported by the lymphatic system but by the portal vein. Influx of these components into the liver could initiate a cascade of reactions, which may explain the appearance of "unusual" triglycerides in the circulation (see above), lower adiposity, and lower insulinemia and leptinemia (Murase et al., 2001). The concept of portal transport of MAG based on find-

TABLE 2.1
Postprandial Lipemic and Lipoproteinemic Effects of Feeding TAG[a], *eq*-DAG[b], *eq*-MAG[a], and a 2:1 Mixture of FFA + *eq*-MAG[a]

	Fat fed				
Effects	TAG[a]	DAG[b]	MAG[a]	FFA	MAG[a] + FFA
Chylomicronemia	+++	++	±	+	++
VLDL increase	+	ND	−	+++	+
Triglyceridemia	+++	++	±	+++	+++
On fatty acids[c]	+++	ND	+	+++	+++

[a]With the fatty acid composition of sunflower seed oil (Christophe & Verdonk,1978; Christophe & Verdonk, 1972).
[b]With the fatty acid composition of olive oil (Banno et al., 2002).
[c]Change of the fatty acid composition of fasting plasma triglycerides toward that of the fat fed
ND, no data available.

ings in humans contrasts with findings in rats. When docosahexaenoic acid was administered intragastrically in emulsion form in lymph-cannulated rats as MAG, lymphatic output (mainly as TAG) was higher than when administered as TAG (Banno et al., 2002). It is not known whether this is due to the differences in experimental set-up or to species differences.

Loading MAG together with FFA results in TAG effects in the postprandial state (Table 2.1). Loading with FFA results in effects that are different from these obtained after loading with MAG or TAG (Table 2.1). *eq*-DAG feeding has postprandial effects distinct from feeding TAG with the same fatty acid composition. For instance, in rat liver, decreased activities of enzymes involved in fatty acid synthesis and increased activity in those involved in fatty acid oxidation were found after DAG compared with TAG feeding. It is conceivable that *sn*-1, 3 DAG will be partially digested and taken up as 1(3)-MAG and FFA. Indeed some of their postprandial effects mimic those of isomeric *eq*-MAG feeding, e.g., a lower increase in chylomicron TAG (Taguchi et al., 2000). The postprandial lipemic/lipoproteinemic effects of feeding TAG, DAG and MAG are summarized in Table 2.1.

References

Armand, M., Borel, P., Dubois, C., Senft, M., Peyrot, J., Salducci, F., Lafont, H., and Lairon, D. Characterisation of Emulsions and Lipolysis of Dietary Lipids in the Human Stomach. *Am. J. Physiol.* **1994**, *266*, G372-G381.

Armand, M., Hamosh, M., DiPalma, J.S., Gallagher, J., Benjamin, S.P., Philpott, J.R., Lairon, D., and Hamosh, P. Dietary Fat Modulates Gastric Lipase Activity in Healthy Humans. *Am. J. Clin. Nut.* **1995**, *62*, 74-80.

Armand, M., Pasquier, B., Andre, M., Borel, P., Senft, M., Peyrot, J., Salducci, F., Portugal, H., Jaussan, V., and Lairon, D.Digestion and Absorption of 2 Fat Emulsions with Different Droplet Sizes in the Human Digestive Tract. *Am. J. Clin. Nutr.* **1999**, *70*, 1096-1106.

Banno, F., Doisaki, S., Shimizu, N., and Fujimoto, K. Lymphatic Absorption of Docosahexaenoic Acid Given as Monoglyceride, Diglyceride, Triglyceride and Ethyl Ester in Rats. *J. Nutr. Sci. Vitaminol.* **2002**, *48*, 30-35.

Bierbach, H. Triacylglycerol Biosynthesis in Human Small Intestinal Mucosal Acyl-CoA:Monoglyceride Acyltransferase. *Digestion* **1983**, *28*,138-147.

Borel, P. Factors Affecting Intestinal Absorption of Highly Lipophilic Food Microconstituents (Fat Soluble Vitamins, Carotenoids and Phytosterols). *Clin. Chem.Lab. Med.* **2003**, *41*, 979-994.

Borel, P., Armand, M., Pasquier, B., Senft, M., Dutot, G., Melin, C., Lafont, H., and Lairon, D. Digestion and Absorption of Tube-Feeding Emulsions with Different Droplet Sizes and Compositions in the Rat. *J. Parenter. Enteral Nutr.* **1994**, *18*, 534-543.

Borovicka, J., Schwizer, W., Guttman, G., Hartmann, D., Kosinski, M. Wast, C., Bischof Delaloye, A., and Fried, M. Role of Lipase in the Regulation of Postprandial Gastric Acid Secretion and Emptying of Fat in Humans: A Study with Orlistat, a Highly Specific Lipase Inhibitor. *Gut* **2000**, *46*, 774-781.

Boswinkel, G., Derksen, J.P.T, vant Riet, K., and Cuperus, F. P. Kinetics of Acyl Migration in Monoglycerides and Dependence on Acyl Chain Length. *J. Am. Oil Chem. Soc.* **1996**, *73*, 707-711.

Bottino, N.R., Vandenburg, G.A., and Reriser, R. Resistance of Certain Long-Chain Polyunsaturated Fatty Acids of Marine Oils to Pancreatic Lipase Hydrolysis. *Lipids* **1967**, *2*, 489-493.

Carriere, F., Barrowman, J.A., Verger, R., and Laugier, R. Secretion and Contribution of Lipolysis by Gastric and Pancreatic Lipases During a Test Meal in Humans. *Gastroenterology* **1993**, *105*, 876-888.

Carriere, F., Renou, C., Lopez, V., De Caro, J., Ferrato, F., Lengsfeld, H., De Caro, A., Laugier, R., and Verger, R. The Specific Activities of Human Digestive Lipases Measured from the In Vivo and In Vitro Lipolysis of Test Meals. *Gastroenterology* **2000**, *119*, 946-960.

Carriere, F., Rogalska, E., Cudry, C., Ferrato, F., Laugier, R., and Verger, R. In Vivo and in Vitro Studies on the Stereoselective Hydrolysis of Triglycerides and Diglycerides by Gastric and Pancreatic Lipases. *Bioorg. Med. Chem.* **1997**, *5*, 429-435.

Chen, Q., and Nillson, A. Interconversion of Alpha-Linolenic Acid in Rat intestinal Mucosa: Studies In Vivo and in Isolated Villus and Crypt Cells. *J. Lipid Res.* **1994**, *35*, 601-609.

Christophe, A., and Verdonk, G. Postprandial Effects of Feeding Triglycerides or Their Digestion Products on the Quantity and Fatty Acid Pattern of the Major Serum Lipid Classes. *Arch. Intern. Physiol. Biochim.* **1972**, *80*, 954-955.

Christophe, A., and Verdonk, G. Effect of Substituting Monoglycerides for Natural Fats in the Diet on Fasting Serum Lipids in Man. *Biochem. Soc. Trans.* **1977**, *5*, 1041-1043.

Christophe, A., and Verdonk, G. Postprandial Changes in Light Scattering Intensity of Serum Lipoproteins of Different Size After Single Fat Feedings in Man, in *Protides of the Biological Fluids, Proceedings of the 25th Colloquium* (Peeters, H., ed.) 1978; pp. 367-369, Pergamon Press, New York.

Christophe, A., and Verdonk, G. Feeding Monoglycerides as Food Fat Results in an Unusual Serum Triglyceride Carbon Number Composition Compatible with the Idea of Portal Monoglyceride Absorption with Structure Retention. *Arch. Intern. Physiol. Biochim.* **1980**, *88*, B221-B222.

Christophe, A., Verdonk, G., Mashali, M., and Sandra, P. Fatty Acid Chain Length Combinations in Ascitic Fluid Triglycerides Containing Lymphatic Absorbed Medium Chain Fatty Acids. *Lipids* **1982**, *17*, 759-761.

Chung, H., and Caffrey, M. Polymorphism, Mesomorphism and Metastability of Monoelaidin in Excess Water. *Biophys. J.* **1995**, *69*, 1951-1963.

Dahim, M., and Brockman, H. How Colipase-Fatty Acid Interactions Mediate Absorption of Pancreatic Lipase to Interfaces. *Biochemistry* **1998**, *37*, 8396-8377.

De Jong, B.J., and Hulsmann, B.C. Monoacylglycerol Hydrolase Activity of Isolated Rat Small Intestinal Cells. *Biochim. Biophys. Acta* **1978**, *27*, 36-46.

Duff, S.M., Kalambur, S., and Boyle-Roden, E. Serum Albumin Binds β- and α-Monoolein In Vivo. *J. Nutr.* **2001**, *131*, 774-778.

Engblom, J. Bicontinuous Cubic Phase: A Model for Investigating the Effect on Lipid Bilayer Due to a Foreign Substance. *Chem. Phys. Lipids,* **1996**, *84*, 155-164.

Hanel, A.M., and Gelb, M.H. Multiple Enzymatic Activities of the Human Cytosolic 85-kDa Phospholipase A_2: Hydrolytic Reactions and Acyl Transfer to Glycerol. Biochemistry **1995**, *20*, 7807-7818.

Hjelm, R.P., Schteingart, C.D., Hofman, A.F., and Sivia, D.S. Probing Self Assembly in Biological Mixed Colloids by Sans, Deuteration, and Molecular Manipulation. *J. Phys. Chem.* **1995**, *99*, 16395-16400.

Huang, Y.S., Lin, J. W., Koba, K., and Anderson, S.N. n-3 and n-6 Fatty Acid Metabolism in Undifferentiated and Differentiated Human Intestine Cell Line (Caco-2). *Mol. Cell. Biochem.* **1995**, *18*, 121-130.

Hunt, J.N., and Knox, M.T. A Relation Between the Chain Length of Fatty Acids and the Slowing of Gastric Emptying. *J. Physiol.* **1968**, *194*, 327-336.

Jang, Y., Han, J.J., Kwon, S.J., and Rhee, J.S. Low-Calorie Structured Lipid Synthesis by Lipase-Catalyzed Transesterification. *Ann. N.Y. Acad. Sci.* **1998,** *864*, 313-317.

Jensen, R.G., deJong, F.A., Lambert-Davis, L.G., and Hamosh, M. Fatty Acid and Positional Selectivities of Gastric Lipase from Premature Human Infants: In Vitro Studies. *Lipids* **1994**, *29*, 433-435.

Lehner, R., Kuksis, A., and Itabashi, Y. Stereospecificity of Monoacylglycerol and Diacylglycerol Acyltransferases from Rat Intestine as Determined by Chiral Phase High Performance Liquid Chromatography. *Lipids* **1993**, *28*, 29-34.

Levy, E., Goldstein, R., Stankievicz, H., Hager, E., and Freier, S. Gastric Handling of Medium-Chain Triglycerides and Subsequent Metabolism in the Suckling Rat. *J. Pediatr. Gastroenterol. Nutr.* **1984**, *5*, 784-789.

Lindstrom, M.B., Sternby, B., and Borgstrom, B. Concerted Action of Carboxyl Ester Lipase and Pancreatic Lipase During Lipid Digestion In Vitro: Importance of the Physicochemical State of the Substrate. *Biochim. Biophys. Acta* **1988**, *959*, 178-184.

Lykidis, A., Mougios, V., and Arzoglou, P. Kinetics of the Two-Step Hydrolysis of Triacylglycerol by Pancreatic Lipases. *Eur. J. Biochem.* **1995**, *230*, 892-898.

Lyubachesskaya, G., and Boyle-Rode, E. Kinetics of 2 Monoacylglycerol Acyl Migration in Model Chylomicra. *Lipids* **2000**, *35*, 1353-1358.

Mackenzie, C.G., Reiss, O.K., Moritz, E., and Mackenzie, J.B. Metabolism of Glycerol Monoethers in Cultured Liver Cells and Implications for Monoglyceride Pathways. *J. Lipid Res.* **1976**, *17*, 629-636.

Maes, B.D., Ghoos, Y.F., Geypens, B.J., Hiele, M.I., and Rutgeerts, P.J. Relation Between Gastric Emptying Rate and the Rate of Intraluminal Lipolysis. *Gut* **1996**, *38*, 23-27.

Maes, B.D., Hiele, M.I, Geypens, B.J., Ghoos, Y.F., and Rutgeerts, P.J. Gastric Emptying of Liquid, Solid and Oil Phase of a Meal in Normal Volunteers and Patients with Bilroth II Gastrojejustomy. *Eur. J. Clin. Investig.* **1998**, *28*, 197-204.

Mansbach, C.M., Cohen, R.S., and Leff, P.B. Isolation and Properties of the Mixed Lipid Micelles Present in Intestinal Content During Fat Digestion in Man. *J. Clin Investig.* **1975**, *56*, 78-191.

Meijer, G.W., and Weststrate, W.A. Interesterification of Fats in Margarine: Effect on Blood Lipids, Blood Enzymes and Hemostasis Parameters. *Eur. J. Clin. Nutr.* **1997**, *51*, 527-534.

Meyer, J.H., Stevenson, E.A., and Watts, H.D. The Potential Role of Protein in the Absorption of Fat. *Gastroenterology* **1976**, *70*, 232-239.

Minich, D.M., Vonk, R.J., and Verkade, H.J. Intestinal Absorption of Essential Fatty Acids Under Physiological and Essential Fatty Acid-Deficient Conditions. *J. Lipid Res.* **1997**, *38*, 1709-1721.

Mu, H., and Hoy, C.E. Effect of Different Medium-Chain Fatty Acids on Intestinal Absorption of Structured Triacylglycerols. *Lipids* **2000**, *35*, 83-89.

Murase, T., Kondo, H., Hase, T., Tokimitsu, I., and Saito, M. Abundant Expression of Uncoupling Protein-2 in the Small Intestine: Up-Regulation by Dietary Fish Oil and Fibrates. *Biochim Biophys. Acta* **2001**, *1530*, 15-22.

Murase, T., Mizuno, T., Omachi, T., Onizawa, K., Komine, Y., Kondo, H., Hase, T., and Tokimitsu, I. Dietary Diacylglycerol Suppresses High Fat and High-Sucrose Diet Induced Body Fat Accumulation in C57BL/6J Mice. *J. Lipid Res.* **2001**, *42*, 372-378.

Nagata, J., Kasai, M., Watanabe, S., Ikeda, I., and Saito, M. Effects of Highly Purified Structured Lipids Containing Medium-Chain Fatty Acids and Linoleic Acid on Lipid Profiles in Rats. *Biosci. Biotechnol. Biochem.* **2003**, *67*, 1937-1943.

Nalbone, G., Lairon, D., Charbonnier-Augeire, M., Vigne, J.L., Leonardi, J., Chabert, C., Hauton, J.C., and Verger, R. Pancreatic Phospholipase A_2 Hydrolysis of Phosphatidylcholines in Various Physicochemical States. *Biochim. Biophys. Acta* **1980**, *620*, 612-625.

Nilsson, A., and Becker, B. Uptake and Interconversion of Plasma Unesterified n-3 Polyunsaturated Fatty Acids by the GI Tract of Rats. *Am. J. Physiol.* **1995**, *268*, G732-G738.

Northfield, T.C., and Hofman, A.M. Biliary Lipid Output During Three Meals and an Overnight Fast. *Gut* **1975**, *16*, 1-17.

O'Doherty, P.J.A., Yousef, I.M., and Kuksis, A. Effect of Phosphatidylcholine on Triacylglycerol Synthesis in Rat Intestinal Mucosa. *Can. J. Biochem.* **1974**, *52*, 726-733.

Pafumi, Y., Lairon, D., de la Porte, P.L., Juhel, C., Storch, J., Hamosh, M., and Armand, M. Mechanisms of Inhibition of Triacylglycerol Hydrolysis by Human Gastric Lipase. *J. Biol. Chem.* **2002**, *277*, 28070-28079.

Paris, R., and Clement, G. Biosynthesis Oflysophosphatidic Acid from 1-Monoolein and ATP by Subcellular Particles from Intestinal Mucosa. *Proc. Soc. Exp. Biol. Med.* **1969**, *131*, 363-365.

Pasquier, B., Armand, M., Castelain, C., Guillon, F., Borel, P., Lafont, H., and Lairon, D. Emulsification and Lipolysis of Triacylglycerols Are Altered by Viscous Soluble Dietary Fibres in Acidic Gastric Medium In Vitro. *Biochem. J.* **1996**, *314*, 269-275.

Rampone, A.J., and Long, L.R. The Effect of Phosphatidylcholine and Lysophosphatidylcholine on the Absorption and Mucosal Metabolism of Oleic Acid and Cholesterol In Vitro. *Biochim. Biophys. Acta* **1977**, *486*, 500-510.

Rogalska, E., Ransac, S., and Verger, R. Stereoselectivity of Lipases. II. Stereoselective Hydrolysis of Triglycerides by Gastric and Pancreatic Lipases. *J. Biol. Chem.* **1990**, *265*, 20271-20276.

Rogalska, E., Ransac, S., and Verger, R. Controlling Lipase Stereoselectivity Via the Surface Pressure. *J. Biol. Chem.* **1993**, *268*, 792-794.

Sakr, S.W., Attia, N., Haourigoui, M., Paul, J.L., Soni, T., Vacher, D., and Girard-Globa, A. Fatty Acid Composition of an Oral Load Affects Chylomicron Size in Human Subjects. *Br. J. Nutr.* **1997**, *77*, 19-31.

Sanderson, I.R., and Naik, S. Dietary Regulation of Intestinal Gene Expression. *Annu. Rev. Nutr.* **2000**, *20*, 311-338.

Saunders, D.R., and Sillery, J. Lecithin Inhibits Fatty Acid and Bile Salt Absorption from Rat Small Intestine In Vivo. *Lipids* **1976**, *11*, 830-832.

Skipski, V. P., Morehouse, M.G., and Duel, H.J. The Absorption in the Rat of a Dioleoyl-2-Deuteriostearoyl Glyceride C^{14}. *Arch. Biochem. Biophys.* **1959**, *81*, 93-104.

Sugawara, T., Tushiro, M., Zhang, H., Nara, E., Ono, H., and Nagao, A. Lysophosphatidylcholine-Enhanced Carotenoid Uptake from Mixed Micelles by Caco-2 Human Intestinal Cells. *J. Nutr.* **2001**, *131*, 2921-2927.

Surpuriya, V., and Higuchi, W.I. Enhancing Effect of Calcium Ions on Transport of Cholesterol from Aqueous Sodium Taurocholate-Lecithin Micellar Phase to Oil Phase. *J. Pharm. Sci.* **1974**, *63*, 1325-1327.

Taguchi, H., Wanatabe, H., Onizawa, K., Nagao, T. Gotoh, N. Yasukawa, Tsushima, R., Shimazaki, H., and Itakura, H. Double-Blind Controlled Study on the Effects of Dietary Diacylglycerol on Postprandial Serum and Chylomicron Triacylglycerol Responses in Healthy Humans. *J. Am. Coll. Nutr.* **2000**, *19*, 789-796.

Tiruppathi, C., and Balasubramanian, K.A. Purification and Properties of an Acid Lipase from Human Gastric Juice. *Biochim. Biophys. Acta* **1982**, *712*, 692-697.

Tso, P., Karlstad, M. D., Bistrian, B.R., and DeMichelle, S.J. Intestinal Digestion, Absorption, and Transport of Structured Triglycerides or Cholesterol in Rats. *Am. J. Physiol.* **1995**, *268*, G568-G577.

Verdonk, G., Christophe, A., and Geers, R. Possibilities of Semi-Synthetic Fats for Human Nutrition and Dietetics; New Concepts in the Physicopathology of Lipid Assimilation, in *Symposium on the Pharmacological Effects of Lipids;* American Oil Chemists' Society, Champaign, IL. (Kabara, J., ed.) 1978; pp. 191-202.

Vian, R., Vigneron, N., Najean, Y., and Bernie, J.J. Gastric Emptying and Intragastric Distribution of Lipids in Man. A New Scintigraphic Method of Study. *Dig. Dis. Sci.* **1982**, *27*, 705-711.

Wickham, M., Wilde, P., and Fillery-Travis, A. A Physicochemical Investigation of Two Phosphatidylcholine/Bile Salt Interphases: Implications for Lipase Activation. *Biochim. Biophys. Acta* **2002**, *1580*, 110-122.

Xu, X., Fomuso, L.B., and Akoh, C.C. Synthesis of Structured Triacylglycerols by Lipase-Catalysed Acidolysis in Packed Bed Bioreactor, *J. Agric. Food Chem.* **2000**, *48*, 3-10.

Yang, L.Y., Kuksis, A., Meyer, J.J., and Steiner, G. Biosynthesis of Chylomicron Triacylglycerols by Rats Fed Glyceryl or Alkyl Esters of Menhaden Oil Fatty Acids. *J. Lipid Res.* **1995**, *36*, 125-136.

Zierenberg, O., Odenthal, J., and Betzing, H. Incorporation of PPC into Serum Lipoproteins After Oral or I.V. Administration. *Atherosclerosis* **1979**, *34*, 259-276.

Chapter 3

Digestion and Absorption of Diacylglycerol

Hiroyuki Watanabe[a] **and Ichiro Tokimitsu**[b]

[a]Department of Health Science, Kochi Women's University, 5-12 Eikokuji-cho, Kochi 780-8515, Japan and [b]Health Care Food Research Laboratories, Kao Corporation, 2-1-3 Bunka Sumida-ku, Tokyo 131-8501, Japan

Introduction

Most edible oils contain diacylglycerol (DAG) as a minor constituent. Although the isoform of DAG that occurs in the process of triacylglycerol (TAG) digestion by lingual or pancreatic lipase is 1, 2- or 2, 3-DAG, a substantial fraction of DAG in edible oils is converted to 1,3-DAG by acyl migration (Crossley et al., 1959). We reported previously that the serum triglyceride concentration of rats fed DAG oil that is composed mainly of 1, 3-DAG was significantly lower than that of rats fed common TAG oil of a similar fatty acid composition (Hara et al., 1993). We also showed that the rate of lymphatic transport of triglyceride as chylomicrons was significantly retarded in rats that had been intragastrically infused with a DAG oil emulsion compared with a TAG oil emulsion (Murata et al., 1994). We reported recently that the long-term ingestion of dietary DAG oil, in contrast to TAG oil, reduces the accumulation of body fat in humans (Nagao et al., 2000; Maki et al., 2002). In the digestive tract, 1, 3-DAG is hydrolyzed either to 1 (or 3)-monoacylglycerol (MAG) and fatty acids or to glycerol and fatty acids via the intermediate 1-MAG (Watanabe et al., 1997), whereas TAG is hydrolyzed to 2-MAG and fatty acids. On the basis of these results, we hypothesized that the limited availability of 2-MAG for reesterification retards chylomicron-triglyceride transport in rats that have been infused with the DAG oil emulsion.

Before progressing to further studies on the possible mechanisms for these phenomena, it is essential to investigate precisely the energy value and bioavailability of DAG compared with those for TAG with the same fatty acid composition. When the overall absorption rate of DAG and TAG oil is similar, it is still possible that the processes of digestion and assimilation of DAG differ from those of TAG. A detailed analysis of the digestion products and metabolites originating from DAG or TAG might provide further mechanistic insights into the functions and role of DAG.

The altered digestion and absorption of DAG, on the other hand, may have some influence on the absorption of fat-soluble vitamins. Because the physicochemical properties of DAG are similar to those of TAG, it has the potential to be used in cooking, frying, shortening, and for dressings. Edible oils are carriers of fat-soluble vitamins, which are essential for growth and normal functions of the human body. These vitamins are generally absorbed from the gastrointestinal tract only when bile is present. Fat-soluble vitamins and dietary fat are emulsified by bile acids. During the digestion of dietary fats, micelles containing lipophilic compounds are formed. The absorption rate of fat-soluble vitamins and lipophilic compounds depends on the degree to which they are partitioned between oil and water (Jandacek, 1982). Therefore, it is important to determine the effect of DAG oil consumption on the fat-soluble vitamin status although the contribution of the altered digestion products of 1, 3-DAG in these micellar properties might be small.

This chapter summarizes the results of studies on energy values and the absorption coefficients of DAG and TAG oils. The effects of dietary DAG consumption on the absorption of fat-soluble vitamins and on gastric emptying time are also discussed because they are intimately related to the digestion process of dietary fats.

Energy Values of DAG and TAG Oils with a Similar Fatty Acid Composition

Fats and oils provide ~9 kcal/g metabolizable energy compared with 4 kcal/g for proteins and carbohydrates (Atwater, 1903. Maynard, 1944). These values have been approved and are widely used to calculate energy intake from various foodstuffs. These values are generalized, and the precise amount of energy contained in an individual food may vary depending on its composition and structure. In situations such as specific experiments, therapeutics, and special diet programs, precise determination of food energy values is required. These include experiments directed at the development of foods that involve less fat accumulation in the body.

It is thus necessary to determine the energy value of the fat and its digestibility when the effect of a dietary fat on energy metabolism is studied. Taguchi et al. (Taguchi et al., 2001) compared the energy values of DAG and TAG oils by calculation and by means of a bomb calorimeter, and also compared the absorption coefficient or digestibility of DAG and TAG oils to assess the contribution of energy and digestibility differences to the functional differences observed in these dietary oils, as described in other chapters of this book.

The fatty acid composition and acylglycerol composition of each of the oils used in the studies for the first two sections in this chapter are shown in Table 3.1. The fatty acid composition of the DAG oil was very similar to that of the blended TAG oil. The DAG concentration of the DAG oil was 87.0 g/100 g, and the ratio of 1(3), 2- to 1, 3-DAG was 32:68. Using the approach outlined in the literature (Livesey, 1984. Peters, 1991), Taguchi et al. (Taguchi et al., 2001) calculated the heat energy values of both DAG oil and TAG oil.

The energy values for the fatty acids were estimated using the equation $-\Delta Hc = 0.653n - 0.166d - 0.421$, where $-\Delta Hc$ is the heat of combustion in MJ/mol, n is the number of carbon atoms/molecule fatty acid, d is the number of double bonds/fatty acid, as described by Livesey (Livesy, 1984). The energy value used for glycerol was 18.0 kJ/g (Peters et al., 1991). The heat of esterification was neglected because it amounts to only 3.8 kJ/mol of ester bond as determined with methyl stearate (Adriaanse et al.,1965). In the case of tristearin, this value corresponds to 13 J/g.

TABLE 3.1
Data Used to Calculate the Energy Values for Triacylglycerol and Diacylglycerol Oils

	Weight fraction	
Component	DAG oil	TAG oil
Fatty acid	(g/100 g)	
16:0	2.4	6.0
18:0	0.7	2.2
18:1	28.0	29.1
18:2	60.3	57.8
18:3	5.6	2.5
20:0	<0.05	0.4
20:1	0.2	0.6
22:0	<0.05	0.3
22:1	<0.05	0.2
24:1	<0.05	0.2
Other	2.8	0.7
Acylglycerol		
TAG	10.7	97.2
DAG	87.0	1.1
1(3), 2-DAG	27.8	ND[a]
1, 3-DAG	59.2	ND
Monoacylglycerol	0.82	<0.05
Free fatty acid	ND	ND

[a]ND, not determined.

The theoretical energy values calculated from the amount of fatty acids and glycerol released from 1 g of test oil and from the energy value for each fatty acid are shown in Table 3.2. The energy values determined using a bomb calorimeter for the DAG oil and the TAG oil are also shown in Table 3.2. The combustion energies measured by bomb calorimetry were in good agreement with the calculated values. The energy value of the DAG oil was ~98% of the TAG oil. Because the fat energy content in a practical diet is only a part of the total energy, this energy difference (2%) between the oils would be expected to result in only a negligible difference in the total energy value of the diet. Because the animal diet used in the experiment of the next section contained 38.6 energy% of fat, this difference in energy value (2%) between the oils will produce an 0.8% difference in the energy value of the diet in the animal experiment. The 0.8% difference in the energy value between the DAG diet and the TAG diet corresponds to 3.6 kcal/100 g. This difference is far less than the energy required to detect any differences in the growth of rats at the maximum sensitivity. In the human trial by Nagao et al. for example (Nagao et al., 2000), the contribution of the energy difference to the results was much less (<0.1%) because the amount of the test oil substituted was only 10 of 42 g of the daily total fat consumed. Our results in conjunction with the previous study suggest that the energy difference between DAG and TAG is negligible.

TABLE 3.2
Theoretical and Experimental Energy Values of Diacylglycerol and Triacylglycerol Oils

		Diacylglycerol oil		Triacylglycerol oil	
Fatty acid	Gross energy (kJ/g)	Weight fraction (g/g oil)	Contribution to energy value (kJ/g oil)	Weight fraction (g/g oil)	Contribution to energy value (kJ/g oil)
16:0	39.1	0.023	0.9	0.058	2.3
18:0	39.9	0.007	0.3	0.021	0.8
18:1	39.5	0.263	10.4	0.280	11.1
18:2	39.3	0.567	22.2	0.557	21.8
18:3	38.9	0.053	2.0	0.024	0.9
20:0	40.5	0.000	0.0	0.004	0.2
20:1	40.2	0.002	0.1	0.006	0.2
22:0	41.0	0.000	0.0	0.003	0.1
22:1	40.7	0.000	0.0	0.002	0.1
24:1	40.9	0.000	0.0	0.002	0.1
Fatty acids		0.913	35.9	0.956	37.6
Glycerol	18.0	0.145	2.6	0.105	1.9
Total		1.000	38.5	1.000	39.5
Experimental			38.9		39.6

Absorption Coefficient of DAG Oil (Animal Study)

The rat is the most frequently used animal model to predict the digestibility of various food components. Bach Knudsen et al. (Bach Knudsen et al., 1994) demonstrated that the digestibility of non starch polysaccharides and most of their monosaccharide residues was higher in humans than in rats, whereas the digestibility values for protein, energy, and fat in general were very similar for the two species. Wisker et al. (Wisker et al., 1996) also reported that the digestibilities of protein, fat, and nonstarch polysaccharides in a low-fiber diet and diets containing coarse or fine whole-meal rye were comparable in rats and humans. It should, therefore, be reasonable to compare the digestibility of DAG oil and TAG oil using a rat model.

Taguchi et al. (Taguchi et al., 2001) calculated the absorption coefficient of DAG and TAG oils from a lipid analysis of the diet and 3-d fecal excretion. Male Sprague-Dawley rats (n = 16; 5 wk old) were divided into 2 groups of 8 so that the body weight of each group was approximately equal; they were then transferred to individual metabolic cages. Rats in one group were fed a DAG oil diet [20 g/100 g DAG] and the others were fed a TAG oil diet [20 g/100 g TAG] (DAG group and TAG group), respectively. Feces were collected and pooled during the last 3 study days (day 13-15). The digestibility coefficient is defined as the percentage of ingested fat that is not excreted in the feces (Apgar et al., 1987. Mitchell et al., 1989. Monsma et al., 1996).

The overall movement of fats in the rat during the last 3 d of the experiment are shown in Table 3.3. No significant differences between the diet groups were found for the following variables: food intake, fat intake, dried feces mass, fat excretion, fat content

TABLE 3.3
Fat Intake and Excretion over 3 d in Rats Fed Either the Diacylglycerol or Triacylglycerol Diet[a]

Variable	Diacylglycerol group	Triacylglycerol group
	[g/(3 d·rat)]	
Diet intake	49.99 ± 1.23	50.12 ± 1.74
Fat intake	9.67 ± 0.24	9.67 ± 0.33
Dried feces mass	3.15 ± 0.17	3.35 ± 0.14
Fat excretion	0.354 ± 0.043	0.352 ± 0.022
	(g/100 g)	
Fat content of dry feces	10.49 ± 0.38	11.18 ± 1.03
	(%)	
Fat absorption coefficient	96.3 ± 0.42	96.3 ± 0.26

[a]Values are means ± SE (n = 8).

of dry feces, and fat absorption coefficients. To determine the effect of dietary DAG on the lipid composition of feces, they analyzed the fecal lipid profile. The groups did not differ with respect to the pattern of the fecal lipids, as evidenced by GC.

Similar fat digestibility (95.44%, 95.62% for DAG and TAG, respectively) has been reported by Meng et al. (Meng et al., 2004) using Sprague-Dawley rats. In addition, Murase et al. determined the digestibility of dietary DAG in mice that accumulated less body fat during a period of 5 mo compared with the TAG diet (Murase et al., 2001). They measured the amount of lipid in the feces of mice fed a high-fat (30 g/100 g) diet. Although the amount of fecal lipid in mice fed a high-fat diet increased in both the DAG and TAG groups, compared with that of the low-fat diet-fed mice, the amount of fecal lipid in the high-DAG-fed mice was nearly equivalent to that in the high-TAG-fed mice. These results indicate that the rate of absorption of dietary DAG oil does not differ significantly from that of TAG oil.

These results suggest that the physiologic differences between the DAG oil and TAG oil observed in animals and humans described in other chapters are caused by differences in metabolic fate after absorption by gastrointestinal epithelial cells. This encourages further studies on the mechanism of the function of dietary DAG as it relates to the reduction of body fat accumulation in animals and humans.

Digestion of DAG and the Fate of the Digestion Products (Animal Study)

TAG is digested into 2-MAG and free fatty acids (FFA) in the intestinal lumen and then resynthesized to triglyceride in the intestinal mucosal cells after absorption. It was shown previously that triglyceride is resynthesized in the mucosa, mainly via the 2-MAG pathway through two steps of enzyme action—the acylation of 2-MAG and the subsequent acylation of 1,2(2,3)-DAG (Clark & Huebscher, 1960; Clark & Huebscher, 1961). However, the metabolic pathway of 1,3-DAG, a major constituent of dietary DAG oil in the small intestine, has not been clarified.

Metabolites of dietary TAG and DAG found in the stomach and small intestine during the digestive process have been studied in rats (Osaki et al., 2005). Osaki et al. investigated the metabolic fate of dietary TAG and DAG by analyzing their digestion products in the stomach and small intestine. Male Wistar rats (7 wk old, 189 ± 7.8 g) were divided into two groups, so that the body weight of each group was approximately equal, and transferred to individual metabolic cages. A diet containing 10% TAG or DAG oil was fed ad libitum for 9 d. After 18 h of fasting, each diet was re-fed ad libitum for 1h. At defined time points, 0, 1, and 4 h after re-feeding the test diets, the rats were anesthetized with diethyl ether and killed by withdrawing blood from the abdominal aorta. The stomach and small intestine (60 cm from the pylorus) were removed. The weights of the contents of the stomach and small intestine were measured, and the acylglycerol and FFA levels were analyzed by GC 0, 1, and 4h after the 1h re-feeding. The amounts of re-fed diet ingested and the gastric and small intestinal content did not differ between the two diet groups.

In the TAG diet group, TAG and DAG, especially 1(3), 2-DAG, were the main components. In addition, 1, 3-DAG and 1(3)-MAG were present in the stomach, and the 1, 3-DAG levels increased over time after the re-feeding period. In the DAG diet group, the main products in the stomach were DAG, MAG, FFA, and TAG. There were significantly greater amounts of 1, 3-DAG, 1(3)-MAG, and FFA in the stomach in the DAG diet group compared with the TAG diet group. The amount of FFA in the stomach relative to the amount of ingested TAG plus DAG in the DAG diet group was higher than that in the TAG diet group. Acylglycerol and FFA levels were considerably lower in the small intestine than in the stomach. These results indicate that in the stomach, where acyl migration might occur, the digestive products already differ when ingestion of TAG oil is compared to ingestion of DAG oil, and that of the two, DAG might be more readily digested by lingual lipase. Furthermore, almost all of the dietary lipid was absorbed, irrespective of the structure of the acylglycerol present in the small intestine.

Kondo et al. (Kondo et al., 2003) investigated the metabolic features of DAG in the small intestine with regard to the digestion pathway in the lumen and the triglyceride synthesis pathway in the mucosa using a rat model to clarify the mechanisms underlying the beneficial effects of the DAG oil described in other chapters of this book.

An intraduodenal infusion was performed according to the method of Mansbach and Nevin, using male Wistar rats at 6–10 wk of age (Mansbach & Nevin, 1998). When the rats were infused duodenally with [carboxyl-^{14}C] trioleoylglycerol (TO), 76.0% (mean, n = 4) of the labeled TO was digested in 10 min after infusion, and 1, 2(2, 3)-dioleoylglycerol (DO) (9.9%), 2-monooleoylglycerol (MO) (28.2%), and oleic acid (OA) (36.8%) were generated in the intestinal lumen as the major products of digestion (Fig. 3.l). In contrast, when 1,3-[carboxyl-^{14}C]DO was infused, the degree of DO digestion (mean: 79.8%, n = 4) did not differ from that of TO, and 1(3)-MO (25.6%) and oleic acid (50.6%) were generated mainly as the digestion products in the lumen (Fig. 3.1). The high oleic acid/MO ratio in the lumen infused with DAG suggests that substantial amounts of l(3)-MAG were further converted to FFA and glycerol in the lumen after the infusion of DAG.

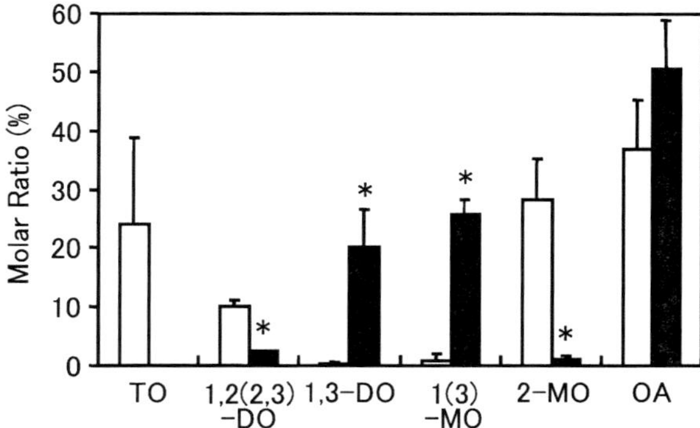

FIG. 3.1. Analysis of digestion products of [carboxyl-^{14}C] trioleoylglycerol (TO) and 1,3-[carboxyl-^{14}C] dioleoylglycerol (DO) in the rat intestinal lumen. Rats were intraduodenally injected with a TAG emulsion containing [carboxyl-^{14}C] TO (□) or a DAG emulsion containing 1,3-[carboxyl-^{14}C] DO (■). MO, monooleoylglycerol; OA, oleic acid. Values are means ± SD; n = 4. *Different from TAG infusion, $P < 0.001$.

Furthermore, lipid metabolites in rat intestinal mucosal cells infused with DAG or TAG were determined. To achieve a steady state of digestion, emulsions of TAG oil or DAG oil were infused into the duodenum for 5 h. After 5 h, an emulsion supplemented with ^{14}C-labeled linoleic acid was infused and the incorporation of the ^{14}C label into the mucosal lipids was determined. Ten minutes after the ^{14}C-infusion, 59.8 ± 9.1% (n = 6) of the ^{14}C was incorporated into the mucosal lipids of the TAG-infused rats, whereas a smaller amount of ^{14}C label was incorporated in the case of DAG-infused rats [42.9 ± 7.8% (n = 6), $P < 0.001$].

The incorporation of the ^{14}C label into the mucosal triglyceride was decreased in the DAG-infusion group compared with the TAG-infusion group. In the TAG-infusion group, 91.5 ± 1.2% of the incorporated radioactivity was detected in triglyceride, whereas 86.2 ±1.4% was detected in triglyceride in the DAG-infusion group (Fig. 3.2). This finding suggests that DAG digestion products were less efficiently synthesized into triglyceride in the intestinal mucosa compared with TAG digestion products.

In contrast to the lower incorporation of the ^{14}C label into the mucosal triglyceride in the DAG-infusion group, a higher proportion of radioactivity was detected in 1, 3-DAG, MAG, and FFA in the DAG-infusion group compared with the TAG-infusion group. The radioactivity of 1, 3-DAG accounted for 1.1% of the total amount of incorporated radioactivity in the TAG-infusion group, whereas that in the DAG-infusion group accounted for 4.5% (Fig. 3.2), resulting in a 4.4-fold increase in the ratio ($P < 0.001$).

The incorporation of label into MAG was also significantly higher in the DAG-infusion group (DAG infusion: 0.4%; TAG infusion: 0.1%) (Fig. 3.2). The separation of MAG isomers by TLC using boric acid–impregnated plates revealed that 57.3 ± 2.1% of MAG in the mucosal lipids was 1(3)-MAG in the DAG-infusion group. However, 1(3)-MAG was not detected in the mucosal lipids of the TAG-infusion group.

Moreover, the amounts of labeled glycerides in the mucosal cells were strikingly different between the treatment groups. Compared with an infusion with TAG, a 4.5-fold higher amount of 1, 3-DAG accumulated in the intestinal mucosa after a DAG infusion (Fig. 3.3), whereas the amount of 1, 2(2, 3)-DAG did not differ between the TAG and DAG infusion groups.

To identify the pathway for the synthesis of 1, 3-DAG, everted sacs prepared from a rat duodenum were cultured with labeled FFA in the presence or absence of MAG. When the everted intestinal sacs were cultured with [1-^{14}C]linoleic acid in the presence of 1-MO, 1, 3-DAG synthesis in the mucosa of the everted sacs accounted for 39.6% of the total DAG synthesis (Fig. 3.4). In contrast, when the everted sacs were incubated in the presence of 2-MO and labeled linoleic acid, mucosal 1, 3-DAG synthesis accounted for only 11.0% of the total DAG synthesis. In the absence of MO (a culture with glycerol, oleic acid, and labeled linoleic acid), on the other hand, the synthesis of 1, 3-DAG was 2.8% of the total DAG synthesis. This indicates that after the intraduodenal infusion of DAG, some of the 1(3)-MAG is absorbed from the lumen into the mucosal cells without further hydrolysis and then esterified to 1, 3-DAG. A substantial amount of 1, 3-DAG was also produced in mucosal cells after TAG infusion (Figs. 3.2, 3.3). The 1, 3-DAG might be synthesized via the acylation of 1(3)-MAG generated by the isomerization of 2-MAG.

A measurement of diacylglycerol acyltransferase (DGAT) activity in the microsome fraction prepared from intestinal mucosa revealed that the isomeric 1, 3-DO was not utilized substantially as a substrate for triglyceride synthesis compared with 1, 2-DO (Fig. 3.5). The amount of triglyceride produced in the presence of 1, 3-DO was ~10% or less of that in the presence of 1, 2-DO.

DGAT is a microsomal enzyme that catalyzes the last step in the synthesis of triglyceride (Bell & Coleman, 1980. Lehner & Kuksis, 1996). The reaction involves the acylation of DAG, which is supplied mainly by the esterification of 2-MAG in the intestine after TAG ingestion. Thus far, two species of DGAT, DGAT-1 and DGAT-2, have been identified, and the small intestine expresses both of these (Cases et al., 1998.Cases et al., 2001). The low substrate specificity of mucosal DGAT for 1, 3-DAG may be a factor that is responsible for the accumulation of 1, 3-DAG and the decrease in triglyceride synthesis in intestinal mucosal cells after a DAG infusion. These findings suggest that the nature by which 1, 3-DAG is digested and assimilated in the intestine may be responsible for the reduction in postprandial serum triglyceride levels by dietary DAG.

Effect of DAG Oil Consumption on Absorption of Fat-Soluble Vitamins (Human Study)

Fats are digested in three different and coordinated processes: emulsification, hydrolysis of the substrate, and micellar solubilization of the hydrolysis products in the aqueous medium of the intestinal contents (Friedman & Nylund, 1980). Therefore, the absorption of fat-soluble vitamins and other lipophilic compounds by enterocytes may be influenced by various dietary factors. A decrease in serum 25-hydroxyvitamin D, for example, was reported for humans who ingest large amounts of dietary fiber (Batchelor & Compston,

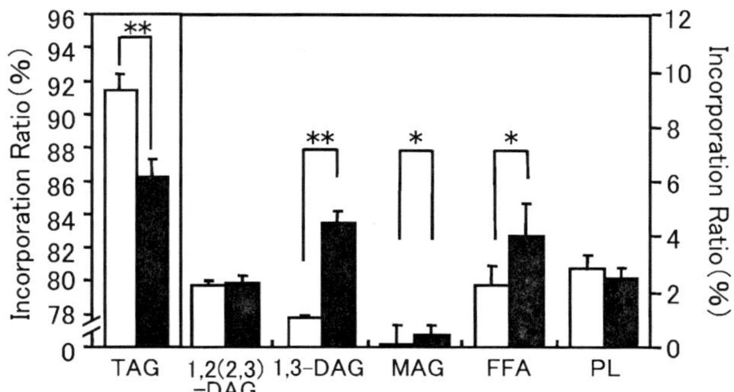

FIG. 3.2. Incorporation of [1-^{14}C] linoleic acid into lipids in rat intestinal epithelial cells after the infusion of DAG or TAG. A lipid emulsion containing 30 mmol/L TAG (□) or 45 mmol/L DAG (■) was infused at a rate of 4.5 mL/h into the duodenum in rats. After 5 h, [1-^{14}C] linoleic acid (2.2 × 10^{6} dpm) was added to the infusion emulsion. At 10 min after the initiation of [1-^{14}C] linoleic acid infusion, the intestinal epithelial cells were collected. The percentages of radioactivity in TAG (triglyceride), DAG, MAG, FFA, and phospholipids (PL) were determined. Values are means ± SD; n = 4, $*P < 0.01$, $**P < 0.001$.

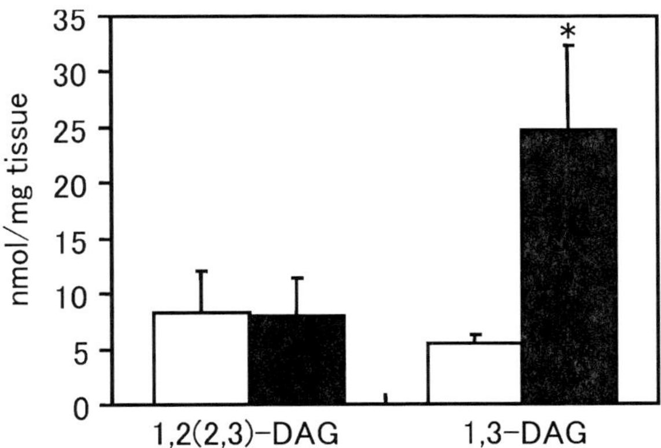

FIG. 3.3. Analysis of the amount of 1,2-DAG or 1,3-DAG in rat intestinal epithelial cells after intraduodenal infusion. Rats were intraduodenally infused with TAG (open bar) or DAG (solid bar) as described in the legend for Figure 3.2. Values are means ± SD; n = 4; *different from TAG infusion, $P < 0.05$.

1983). The type of dietary fat, consisting of hydrolyzable TAG, can also affect the extent of the absorption of cholesterol and, presumably, other lipophilic compounds as well (Sylven & Borgstrom, 1969). Nonabsorbable lipophilic substances such as sucrose polyesters affect the absorption of fat-soluble molecules (Jandacek, 1982).

Thus, Watanabe et al. (Watanabe et al., 2001) examined the issue of whether the DAG oil, when administered over a long period to humans, has an altered effect on the

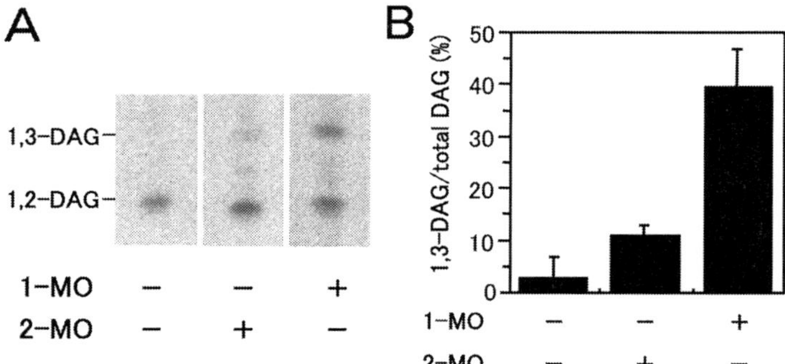

FIG. 3.4. Incorporation of [^{14}C] linoleic acid into 1,3-DAG in epithelial cells of rat intestinal everted sacs. Everted sacs from a rat intestine were incubated with [1-^{14}C] linoleic acid in the presence or the absence of *sn*-1-MO (1-MO) and *sn*-2MO (2-MO). The extracted lipids were fractionated by TLC using chloroform/acetone (95:4) as the developing solvent. A representative TLC pattern is shown in panel A. The percentage of radioactive 1,3-DAG to the total radioactivity of DAG is shown in panel B. Values are means ± SD; n = 3. For abbreviations, see Figure 3.1.

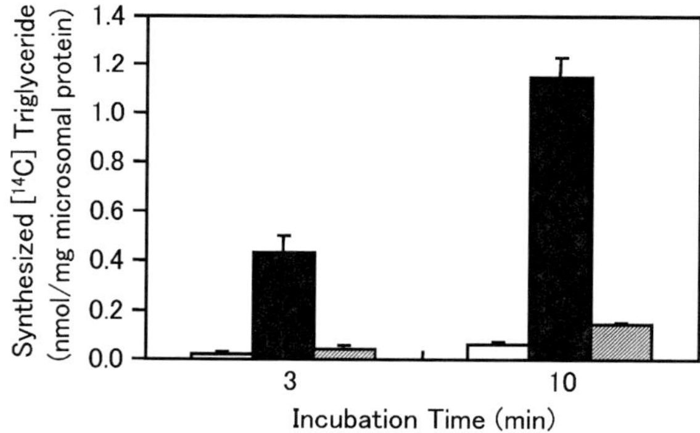

FIG. 3.5. Utilization of diacylglycerol isomers for triglyceride synthesis. The microsomal fraction (0.5 mg/mL) obtained from the rat intestinal mucosa was incubated with *sn*-1,2-DO (solid bar) or 1,3-DO (hatched bar). The open bar represents the control incubated without DO. Values are means ± SD; n = 3. For abbreviations, see Figure 3.1.

bioavailability of fat-soluble vitamins, compared with the TAG oil. Serum concentrations of fat-soluble vitamins were used as indicators. This study was a parallel, placebo-controlled, double-blind design. Healthy men aged 27-47 y participated in this study.

Subjects were randomly assigned to two groups; one group was given DAG oil (n = 15) and the other group was given TAG oil (n = 12). The dose of the test oil was 20 g/d in mayonnaise or an emulsion drink. The subjects were instructed not to change their daily energy intake or physical activity, to avoid excessive intakes of alcohol and food, and not to take any medication or vitamin supplements during the study. The physical characteristics of the subjects at the beginning of the study are shown in Table 3.4. According to the diet record analysis, no significant differences in dietary intake were observed between the baseline and test periods in either group. All laboratory values remained within the normal ranges during the study. Body weight, waist circumference, hip circumference, and serum parameter changes indicative of test food–related effects were not observed during the study.

At 4, 8, and 12 wk, blood samples were drawn from fasting subjects and serum concentrations of vitamins A, E, and D were determined. Vitamin A was measured as retinol by the method of Thompson et al. (Thompson et al., 1973). Vitamin E levels were calculated as the sum of α-, ß-, and γ-tocopherol by the method of Abe et al. (Abe et al., 1975). Three species of vitamin D were measured by the modified method of Shepard et al. (Shepard et al., 1979).

Changes in serum concentrations of fat-soluble vitamins are shown in Tables 3.5–3.8. Although there were some fluctuations from the initial values during the study period in the serum concentrations of vitamin A, vitamin E, 25-hydroxyvitamin D, and 1-α, 25-dihydroxyvitamin D, differences between the two groups at each point were not significant. From these results, they concluded that DAG has no effect on the absorption of fat-soluble vitamins.

The absorption of fat-soluble vitamins during the digestion of dietary lipids is inhibited by the ingestion of the gastrointestinal lipase inhibitor, orlistat (Melia et al., 1996), and a noncaloric fat substitute, olestra (Koonsvitsky et al., 1997). These observations indicate that the role of edible oils in the absorption of fat-soluble vitamins is very important as a nutritional element, and DAG, in contrast to these lipid metabolism modifying agents, was shown not to affect the absorption of fat-soluble vitamins.

Faster Gastric Emptying of Food Containing DAG Oil (Human Study)

Before the health benefits of DAG were discovered in the early stages of the study of dietary DAG, information was available indicating that DAG had a light taste, was less

TABLE 3.4
Baseline Characteristics of Study Subjects and Their Responses After the Experiment[a]

	Baseline		12 wk	
	DAG group	TAG group	DAG group	TAG group
n	15	12	15	12
Age (y)	33.8 ± 4.5	33.3 ± 5.7		
Weight (kg)	65.6 ± 2.2	66.0 ± 2.6	65.7 ± 2.2	66.5 ± 2.6
Height (cm)	169.4 ± 1.4	171.9 ± 1.3		
Serum triglyceride (mg/dL)	90.8 ± 10.6	79.3 ± 9.2	85.4 ± 11.2	75.0 ± 10.3
Serum total cholesterol (mg/dL)	188.4 ± 6.7	184.6 ± 8.4	184.6 ± 8.4	184.8 ± 6.7

[a]Values are means ± SE.

TABLE 3.5
Mean Serum Concentration of Vitamin A During the Experiment[a]

	0 wk	4 wk	8 wk	12 wk	P^b		
					Lipid	Time	Interaction
DAG	665 ± 105	629 ± 77	640 ± 88	630 ± 92			
Group	(100)	(95.6 ± 12.2)	(97.5 ± 15.1)	(95.7 ± 14.2)			
TAG	641 ± 117	606 ± 161	616 ± 130	609 ± 180	0.679	0.332	1.000
Group	(100)	(94.2 ± 16.8)	(96.9 ± 16.8)	(94.3 ± 19.4)			

[a]Values are means ± SE (ng/mL serum); n = 15 for DAG and 12 for TAG. The percentage of the initial value is given in parentheses.
[b]*P*-values are for main effects and interactions by a two-way ANOVA between two groups.

TABLE 3.6
Mean Serum Concentration of Vitamin E During the Experiment[a]

	0 wk	4 wk	8 wk	12 wk	P^b		
					Lipid	Time	Interaction
DAG	1.27 ± 0.26	1.10 ± 0.15**	1.13 ± 0.20**	1.26 ± 0.18			
Group	(100)	(87.9 ± 9.3)	(89.8 ± 12.7)	(100.4 ± 11.7)			
TAG	1.21 ± 0.24	1.15 ± 0.30*	1.10 ± 0.29**	1.17 ± 0.22	0.155	0.619	0.759
Group	(100)	(94.5 ± 11.0)	(90.5 ± 12.4)	(97.2 ± 7.5)			

[a]Values are means ± SE (mg/100 mL serum); n = 15 for DAG and 12 for TAG. The percentage of the initial value is given in parentheses. Asterisks indicate different from wk 0: *$P < 0.05$, **$P < 0.01$.
[b]*P*-values are for main effects and interactions by a two-way ANOVA between two groups.

TABLE 3.7
Mean Serum Concentration of 25-Hydroxyvitamin D During the Experiment[a]

	0 wk	4 wk	8 wk	12 wk	P^b		
					Lipid	Time	Interaction
DAG	27.4 ± 8.7	29.4 ± 6.6	30.8 ± 6.9	26.8 ± 7.9			
Group	(100)	(115.5 ± 43.0)	(123.4 ± 55.7)	(106.2 ± 50.6)			
TAG	24.4 ± 7.1	26.5 ± 4.8	29.0 ± 6.2*	24.8 ± 4.5	0.090	0.072	0.986
Group	(100)	(113.0 ± 24.7)	(125.2 ± 42.9)	(105.6 ± 24.2)			

[a]Values are means ± SE (ng/mL serum); n = 15 for DAG and 12 for TAG. The percentage of the initial value is given in parentheses. *Different from wk 0, $P < 0.05$.
[b]*P*-values are for main effects and interactions by a two-way ANOVA between two groups.

TABLE 3.8
Mean Serum Concentration of 1-α, 25-Dihydroxyvitamin D During the Experiment[a]

	0 wk	4 wk	8 wk	12 wk	P^b		
					Lipid	Time	Interaction
DAG	30.9 ± 7.7	29.8 ± 7.6	41.0 ± 6.6**	34.9 ± 7.3			
Group	(100)	(100.7 ± 34.0)	(141.3 ± 45.7)	(120.9 ± 48.0)			
TAG	30.9 ± 8.4	30.6 ± 9.6	44.4 ± 9.1**	40.9 ± 9.1	0.100	0.109	0.523
Group	(100)	(102.9 ± 34.4)	(150.7 ± 46.5)	(139.5 ± 44.6)			

[a]Values are means ± SE (pg/mL serum); n = 15 for DAG and 12 for TAG. The percentage of the initial value is given in parentheses. **Different from wk 0, $P < 0.01$.
[b]*P*-values are for main effects and interactions by a two-way ANOVA.

heavy on the stomach, and had a less oily texture. Through these sensory evaluations,we hypothesized that the gastric emptying time of DAG oil might be shorter than that of TAG oil. To test this hypothesis, Yasunaga et al. (Yasunaga et al., 2000) measured the halftime for gastric emptying of the ingested test food cooked with DAG or TAG oil.

This study was a double-blind and crossover trial with six subjects (5 men and 1 woman, 28–34 y old). All subjects were healthy with no history of gastrointestinal tract disease and were not taking any medications. They fasted for 4 h before starting the experiments and ingested the test food within 5 min. The test food was a scrambled egg containing 50 g of whole egg and test oil and a loaf of bread. The dose of test oil was 30 g/60 kg body weight. The DAG oil contained ~88% DAG by weight and the remainder was TAG with a trace of monoglycerides and sterols. The TAG oil was prepared by mixing rapeseed and safflower oil so that the fatty acid composition was similar to that of DAG oil. The control test food was prepared without adding the test oil. To measure gastric emptying time, technetium 99 human macroaggregated albumin (Tc-99 MAA) was incorporated into the scrambled egg so as to locate by a scintigram the position of ingested food in the gastrointestinal tract. Immediately after ingestion, the subjects reclined on their backs in a bed with a scintillation detector for 180 min. The scintigram was taken from both forward and backward of the gastric region at 10-min intervals.

The radioactivity in the stomach compensated for the disintegration of technetium, whose half-life is 6.01 h. The time course for gastric emptying after the ingestion of the test food is shown in Figure 3.6. All of the ingested radioactivity was detected in the

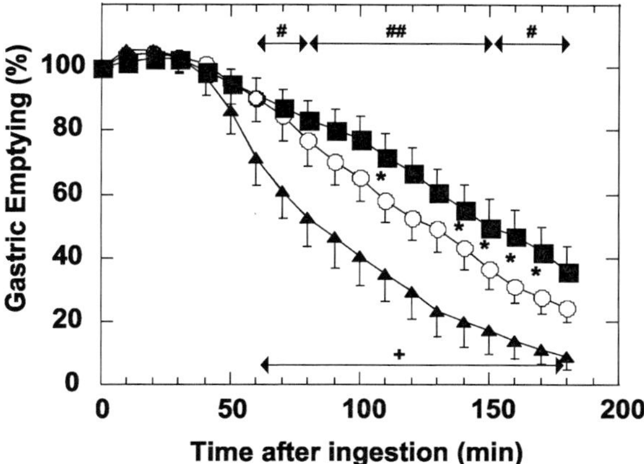

FIG. 3.6. Time course for gastric emptying after the ingestion of test foods with DAG oil (○), TAG oil (■) and control food without oil (▲). Gastric emptying was measured by a scintigram using 99Tc human macroaggregated albumin incorporated into the test food. Each value was expressed as percent of initial value and compensated for disintegration of Tc. *Significant difference between DAG oil and TAG oil ($P < 0.05$); #significant difference between TAG oil and Control (#$P < 0.05$; ##$P < 0.01$); +significant difference between DAG oil and Control (+$P < 0.05$).

stomach during the initial 30 min after ingestion. Approximately 40 min after ingestion, the test food passed into the duodenum, and the amount of food in the stomach decreased linearly for a period of up to 3 h. The control food (no oil) passed most rapidly from the stomach. The amount of food with DAG oil remaining in the stomach was significantly less than that of TAG oil at several time points, suggesting that the food with DAG oil was emptied from the stomach faster than that of TAG oil. The average halftime ($t_{1/2}$) of gastric emptying of the ingested food was calculated. The halftime values for DAG oil, TAG oil, and the control were 125, 155, and 98 min, respectively. The halftime of the DAG-containing food was significantly less than that of TAG-containing food ($P < 0.05$).

The complaint of heaviness on the stomach was well correlated with the slow gastric emptying time, indicating that the subjects with shorter gastric emptying time had fewer gastric complaints after ingestion of the test food. This suggests that the gastric emptying time is related to the sensation of satiety and supports, at least in part, the hypothesis that DAG oil is light in the stomach because of its faster gastric emptying.

Summary

In this chapter, we discussed the characteristics of DAG in terms of energy value, absorption coefficient, digestion products, effect on the absorption of fat-soluble vitamins, and stomach-emptying time compared with those of TAG. The energy value of DAG oil containing 87 wt% DAG and TAG oil with a similar fatty acid composition was 38.9 kJ/g (9.29 kcal/g) and 39.6 kJ/g (9.46 kcal/g), respectively. The difference was <2%. Therefore, the difference in the energy value in the practical diet was considered to be negligible, and the nutritional differences between DAG oil and TAG oil appear to reside in the metabolic fate of DAG after absorption by enterocytes. DAG oil and TAG oil did not differ in other features including apparent absorption coefficients and the effect on fat-soluble vitamin status. However, the digestion products of DAG oil, containing mainly 1, 3-DAG, were distinct from those of TAG oil with regard to 1-MAG production as a major intermediate during the process of DAG digestion. These different metabolic processes after the ingestion of 1, 3-DAG oil in enterocytes may be one of the mechanisms responsible for the suppressed postprandial increase in triglyceride-rich lipoproteins and for the suppressed body fat accumulation as a long-term effect compared with TAG ingestion. In addition, faster gastric emptying time was another characteristic feature of DAG oil. Although this effect may be related to the sensory evaluation that DAG is light in the stomach, the mechanism for this is not known at present.

References

Abe, K., Yuguchi, Y., and Katsui, G. Quantitative Determination of Tocopherols by High-Speed Liquid Chromatography. *J. Nutr. Sci. Vitaminol.* **1975**, *21*, 183-188.

Adriaanse, N., Dekker, H., and Coops, J. Heats of Combustion of Normal Saturated Fatty Acids and Their Methyl Esters. *Recueil* **1965**, *84*, 393-407.

Apgar, J.L., Shively, C.A., and Tarka S.M., Jr. Digestibility of Cocoa Butter and Corn Oil and

Their Influence on Fatty Acid Distribution in Rats. *J. Nutr.* **1987**, *117*, 660-665.

Atwater, W.O. The Demands of the Body for Nourishment and Dietary Standards. *Annu. Rep. Storrs Agric. Exp. Stn.*, **1903**, No. 15, 123-146.

Bach Knudsen, K.E., Wisker, E., Daniel, M., Feldheim, W., and Eggum, B.O. Digestibility of Energy, Protein, Fat and Non-Starch Polysaccharides in Mixed Diets: Comparative Studies Between Man and the Rat. *Br. J. Nutr.* **1994**, *71*, 471-487.

Batchelor, A.J., and Compston, J.E. Reduced Plasma Half-Life of Radio-Labelled 25-Hydroxyvitamin D3 in Subjects Receiving a High-Fibre Diet. *Br. J. Nutr.* **1983**, *49*, 213-216.

Bell, R.M., and Coleman, R.A. Enzymes of Glycerolipid Synthesis in Eukaryotes. *Annu. Rev. Biochem.* **1980**, *49*, 459-487.

Cases, S., Smith, S.J., Zheng, Y.W., Myers, H.M., Lear, S.R., Sande, E., Novak, S., Collins, C., Welch, C.B., Lusis, A.J., Erickson, S.K., and Farese, R.V. Jr. Identification of a Gene Encoding an Acyl CoA:Diacylglycerol Acyltransferase, a Key Enzyme in Triacylglycerol Synthesis. *Proc. Natl. Acad. Sci. USA* **1998,** *95*, 13018-13023.

Cases, S., Stone, S.J., Zhou, P., Yen, E., Tow, B., Lardizabal, K.D., Voelker, T., and Farese, R.V., Jr. Cloning of DGAT2, a Second Mammalian Diacylglycerol Acyltransferase, and Related Family Members. *J. Biol. Chem.* **2001,** *276*, 38870-38876.

Clark, B., and Huebscher, G. Biosynthesis of Glycerides in the Mucosa of the Small Intestine. *Nature* **1960,** *185*, 35-37.

Clark, B., and Huebscher, G. Biosynthesis of Glycerides in Subcellular Fractions of Intestinal Mucosa. *Biochim. Biophys. Acta* **1961,** *46*, 479-494.

Crossley, A., Freeman, I.P., Hudson, J.F., and Pierce, J.H. Acyl Migration in Diglycerides. *J. Chem. Soc.,* **1959,** 760-764.

Friedman, H. I., and Nylund, B. Intestinal Fat Digestion, Absorption and Transport. A Review. *Am. J. Clin. Nutr.* **1980,** *33*, 1108-1139.

Hara, K., Onizawa, K., Honda, H., Otsuji, K., Ide, T., and Murata, M. Dietary Diacylglycerol-Dependent Reduction in Serum Triacylglycerol Concentration in Rats. *Ann. Nutr. Metab.* **1993,** *37*, 185-191.

Jandacek, R.J. The Effect of Nonabsorbable Lipids on the Intestinal Absorption of Lipophils. *Drug Metab.Rev.* **1982,** *13*, 695-714.

Kondo, H., Hase, T., Murase, T., and Tokimitsu, I. Digestion and Assimilation Features of Dietary DAG in the Rat Small Intestine. *Lipids* **2003,** *38*, 25-30.

Koonsvitsky, B.P., Berry, D.A., Jones, M.B., Lin, P.Y., Cooper, D.A., Jones, D.Y., and Jackson, J.E. Olestra Affects Serum Concentrations of α-Tocopherol and Carotenoids but Not Vitamin D or Vitamin K Status in Free-Living Subjects. *J. Nutr.* **1997,** *127*, 1636S-1645S.

Lehner, R., and Kuksis. A. Biosynthesis of Triacylglycerols. *Prog. Lipid Res.* **1996,** *35*, 169-201.

Livesey, G. The Energy Equivalents of ATP and the Energy Values of Food Proteins and Fats. *Br. J. Nutr.* **1984,** *51*, 15-28.

Maki, K.C., Davidson, M.H., Tsushima, R., Matsuo, N., Tokimitsu, I., Umporowicz, D.M., Dicklin, M.R., Foster, G..S., Ingram, K.A., Anderson, B.D., Frost, S.D., and Bell, M. Consumption of Diacylglycerol Oil as Part of a Reduced-Energy Diet Enhances Loss of Body Weight and Fat in Comparison with Consumption of a Triacylglycerol Control Oil. *Am. J. Clin. Nutr.* **2002,** *76*, 1230-1236.

Mansbach, C.M., II, and Nevin, P. Intracellular Movement of Triacylglycerols in the Intestine. *J.*

Lipid Res. **1998,** *39*, 963-968.

Maynard, L.A. The Atwater System of Calculating the Caloric Value of Diets. *J. Nutr.* **1944,** *28*, 443-452.

Melia, A.T., Koss-Twardy, S.G., and Zhi, J. The Effect of Orlistat, an Inhibitor of Dietary Fat Absorption, on the Absorption of Vitamins A and E in Healthy Volunteers. *J. Clin. Pharmacol.* **1996,** *36*, 647-653.

Meng, X., Zou, D., Shi, Z., Duan, Z., Mao, Z. Dietary diacylglycerol prevents high-fat diet-induced lipid accumulation in rat liver and abdominal adipose tissue. *Lipids* **2004,** *39*, 37-41.

Mitchell, D.C., McMahon, K.E., Shively, C.A., Apgar, J.L., and Kris-Etherton, P.M. Digestibility of Cocoa Butter and Corn Oil in Human Subjects: A Preliminary Study. *Am. J. Clin. Nutr.* **1989,** *50*, 983-986.

Monsma, C.C., Gallaher, D.D., and Ney, D.M. Reduced Digestibility of Beef Tallow and Cocoa Butter Affects Bile Acid Excretion and Reduces Hepatic Esterified Cholesterol in Rats. *J. Nutr.* **1996,** *126*, 2028-2035.

Murase, T., Mizuno, T., Omachi, T., Onizawa, K., Komine, Y., Kondo, H., Hase, T., and Tokimitsu, I. Dietary Diacylglycerol Suppresses High Fat and High Sucrose Diet-Induced Body Fat Accumulation in C57BL/6J Mice. *J. Lipid Res.* **2001,** *42*, 372-378.

Murata, M., Hara, K., and Ide, T. Alteration by Diacylglycerols of the Transport and Fatty Acid Composition of Lymph Chylomicrons in Rats. *Biosci. Biotech. Biochem.* **1994,** *58*, 1416-1419.

Nagao, T., Watanabe, H., Goto, K., Onizawa, K., Taguchi, H., Matsuo, N., Yasukawa, T., Tsushima, R., Shimasaki, H., and Itakura, H. Dietary Diacylglycerol Suppresses Accumulation of Body Fat Compared to Triacylglycerol in Men in a Double-Blind Controlled Trial. *J. Nutr.* **2000,** *130*, 792-797.

Osaki, N., Meguro, S., Yajima, N., Matsuo, N., Tokimitsu, I., Shimasaki, H. Metabolites of dietary triacylglycerol and diacylglycerol during the digestion process in rats. *Lipids* **2005,** *40*, 281-286.

Peters, J.C., Holcombe, B.N., Hiller, L.K., and Webb, D.R. Caprenin 3. Absorption and Caloric Value in Adult Humans. *J. Am. Coll. Toxicol.* **1991,** *10*, 357-367.

Shepard, R.M., Horst, R.L., Hamstra, A.J., and Deluca, H.F. Determination of Vitamin D and Its Metabolites in Plasma for Normal and Anephric Men. *Biochem. J.* **1979,** *182*, 55-69.

Sylven, C., and Borgstrom, B. Intestinal Absorption and Lymphatic Transport of Cholesterol in the Rat: Influence of the Fatty Acid Chain Length of the Carrier Triglycerides. *J. Lipid Res.* **1969,** *10*, 351-355.

Taguchi, H., Nagao, T., Watanabe, H., Onizawa, K., Matsuo, N., Tokimitsu, I., and Itakura, H. Energy Value and Digestibility of Dietary Oil Containing Mainly 1,3-Diacylglycerol Are Similar to Those of Triacylglycerol. *Lipids* **2001,** *36*, 379-382.

Thompson, J.N., Erdody, P., and Maxwell, W.B. Simultaneous Fluorometric Determination of Vitamins A and E in Human Serum and Plasma. *Biochem. Med.* **1973,** *8*, 403-414.

Watanabe, H., Onizawa, K., Naito, S., Taguchi, H., Goto, N., Nagao, T., Matsuo, N., Tokimitsu, I., Yasukawa, T., Tsushima, R., Shimasaki, H., and Itkura, H. Fat-Soluble Vitamin Status Is Not Affected by Diacylglycerol Consumption. *Ann. Nutr. Metab.* **2001,** *45*, 259-264.

Watanabe, H., Onizawa, K., Taguchi, H., Kobori, M., Chiba, H., Naito, S., Matsuo, N., Yasukawa, T., Hattori, M., and Shimasaki, H. Nutritional Characterization of Diacylglycerols in Rats.

J. Jpn. Oil. Chem. Soc. **1997,** *46*, 301-307.

Wisker, E., Bach Knudsen, K.E., Daniel, M., Feldheim, W., and Eggum, B.O. Digestibilities of Energy, Protein, Fat and Nonstarch Polysaccharides in a Low Fiber Diet and Diets Containing Coarse or Fine Whole Meal Rye Are Comparable in Rats and Humans. *J. Nutr.* **1996,** *126*, 481-488.

Yasunaga, K., Seo, Y., Katsuragi, Y., Oriuchi, N., Otake, H., Endo, K., and Yasukawa, T. Gastric Emptying of Dietary Diacylglycerol from Stomach, 54th Annual Meeting of the Japanese Society of Nutrition and Food Science, Matsuyama, 2000, May 12-14, p. 205.

Section B

Influence on the Utilization of Lipids for Energy Metabolism

Chapter 4

Activation of Lipid Metabolism and Energy Expenditure by Dietary Diacylglycerol

Takatoshi Murase[a] and Shuichi Kimura[b]

[a]Biological Science Laboratories, Kao Corporation, Ichikai-machi, Haga-gun, Tochigi 321-3497, Japan and [b]Graduate School of Human Life Science, Showa Women's University, Setagaya-ku, Tokyo 154-8533, Japan

Introduction

Compared with triacylglycerol (TAG), which has a similar fatty acid composition, 1, 3-diacylglycerol (DAG) reduces postprandial hyperlipidemia (Murata et al., 1994; Hara et al., 1993; Taguchi et al., 2000). Further, long-term intake of DAG suppresses the accumulation of body fat in humans and animals (Murase et al., 2001; Murase et al., 2002a; Nagao et al., 2000; Maki et al., 2002). These changes in lipid metabolism after DAG ingestion may be a direct or indirect result of the differences in small intestinal metabolism caused by the structural characteristics of DAG (Murase et al., 2002a; Kondo et al., 2003). This chapter reviews insights into DAG-induced activation of small intestinal and hepatic lipid metabolism, and enhancement of whole-body energy expenditure.

Effect of Diacylglycerol on Small Intestinal Lipid Metabolism

The majority of ingested TAG is hydrolyzed into 2-monoacylglycerol (2-MAG) and fatty acids in the gastrointestinal tract. These substances are then absorbed into the intestinal epithelium and immediately resynthesized into TAG through the 2-MAG pathway. The hydrolysis of 1, 3-DAG, on the other hand, is presumed to form 1(or 3)-MAG and fatty acids, which are likely absorbed into the epithelium or further hydrolyzed into glycerol and fatty acids. Thus, the metabolic pathway of 1, 3-DAG probably differs from that of TAG, and this difference affects intestinal function, particularly the lipid-metabolizing system. The small intestine is affected initially and directly by dietary components. An analysis of the effect of dietary DAG on intestinal lipid metabolism indicated that DAG activates lipid catabolism in the small intestine (Murase et al., 2002a).

In C57BL/6J mice, a model of diet-dependent obesity, the accumulation of body fat is reduced after long-term intake of DAG compared with TAG (Chapter 9). To elucidate the mechanisms underlying the effects of dietary DAG, we analyzed the biochemical changes caused by the ingestion of DAG and TAG, particularly in the early stage of obesity (Murase et al., 2002a). Mice were fed a low-TAG diet (5% TAG), high-TAG diet (30% TAG), or high-DAG diet (15% TAG +15% DAG), and the intestinal and hepatic ß-oxidation activities of fatty acids were measured after 10 d. There was no significant change in the liver, but small intestinal ß-oxidation activity was markedly increased (Fig. 4.1). Northern blot analysis of the expression of molecules related to lipid metabolism in the small intestine indicated that mRNA expression of acyl-CoA oxidase (ACO) and medium-chain acyl-CoA dehydrogenase (MCAD), which are involved in the ß-oxidation of fatty acids, was significantly increased in the high-DAG diet group (Fig. 4.2).

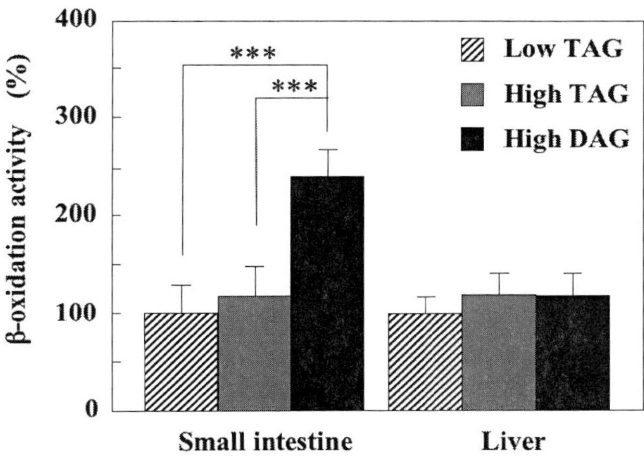

FIG. 4.1. Effects of dietary diacylglycerol on ß-oxidation in the small intestine (Murase et al., 2002a). ***$P < 0.001$.

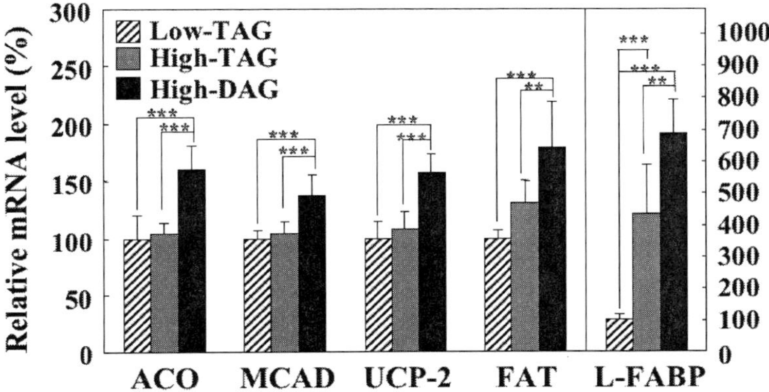

FIG. 4.2. Effects of dietary diacylglycerol on the expression of genes involved in lipid metabolism in the small intestine (Murase et al., 2002a). Abbreviations: ACO, acyl-CoA oxidase; MCAD, medium-chain acyl-CoA dehydrogenase; UCP-2, uncoupling protein-2; FAT, fatty acid translocase; L-FABP, L-fatty acid binding protein. **$P < 0.01$; ***$P < 0.001$.

In addition, mRNA expression of uncoupling protein-2 (UCP-2), fatty acid translocase (FAT), and L-fatty acid binding protein (L-FABP) was also significantly increased, suggesting that increased small intestinal ß-oxidation activity is caused by increased gene expression of lipid metabolism-related molecules. Such activation of small intestinal lipid metabolism by DAG was also confirmed in C57BL/KsJ-*db/db* mice fed diets with α-linolenic acid (ALA) as the major fatty acid component (Murase et al., 2002b; Murase et al., 2005), i.e., ALA-DAG and ALA-TAG. When mice were fed ALA-DAG for 1 mo, both small intestinal ß-oxidation activity and mRNA levels of related genes (ACO, MCAD, UCP-2) were significantly higher than those in the ALA-TAG-fed mice.

These findings indicate that DAG ingestion activates the small intestinal lipid metabolism at a relatively early stage. The contributions of muscle and liver to energy consumption are considered to be relatively large. On the other hand, the small intestine is one of the largest organs in the body and is active in ß-oxidation of fatty acids (Nemali et al., 1988; Kelly et al., 1989). We recently reported that intestinal lipid metabolism is associated with susceptibility to obesity (Kondo et al., 2006). Thus, the small intestine may contribute to energy expenditure, particularly postprandial energy expenditure, and activation of small intestinal lipid metabolism by DAG may therefore affect energy balance and the fate of ingested lipids.

We investigated the activation of small intestinal ß-oxidation by TAG and DAG with various fatty acid compositions. Although the degree of activation varied with the molecular species of composite fatty acids, DAG activated small intestinal ß-oxidation (data not shown), suggesting that its structural characteristics are key to DAG activity. Based on the finding that on day 10 of ingestion, there were no differences between the high-TAG and high-DAG groups in the concentrations of blood components such as insulin and leptin, or in the expression of genes related to lipid metabolism in tissues other than the small intestine, DAG-induced activation of small intestinal lipid metabolism was likely a direct influence of DAG or its metabolites, and not due to humoral factors.

To clarify the mechanism of activation of small intestinal lipid metabolism by DAG, we compared the metabolic pathway of lipids in the small intestine using ^{14}C-TAG and ^{14}C-DAG, as shown in Figure 4.3 (Murase et al., 2002a). The details are described in Chapter 2. In brief, TAG was hydrolyzed to 2-MAG via 1, 2(2, 3)-DAG and absorbed by the small intestinal epithelium. In contrast, DAG was likely hydrolyzed mainly to fatty acids via 1(3)-MAG and absorbed. Accordingly, the ratio of fatty acids present in the small intestinal lumen was significantly higher after DAG administration than after TAG administration (Fig. 4.3A). The pattern of ^{14}C-labeled lipids in the small intestinal mucosa was also analyzed, and the ratios of 1(3)-MAG, 1, 3-DAG, and free fatty acids (FFA) were significantly higher in the DAG treatment group (Fig. 4.3B).

The FFA increase induced by DAG ingestion in the small intestinal mucosa may be involved in the DAG-induced activation of small intestinal lipid metabolism. The gene expression of many lipid metabolism-related molecules such as ACO, MCAD, and L-FABP is regulated by peroxisome proliferator-activated receptors (PPAR) (Schoonjans et al., 1996; Li & Glass, 2004).

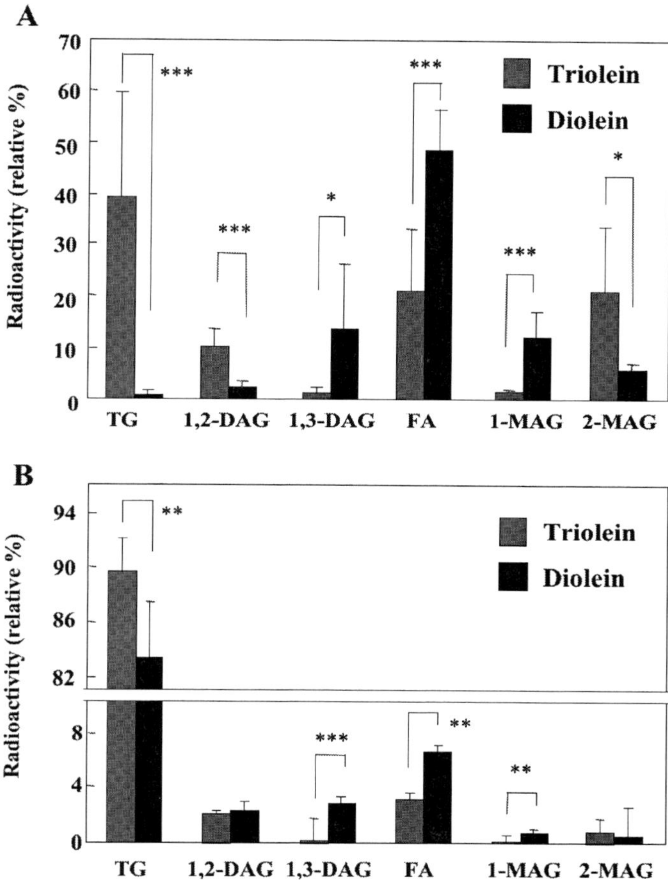

FIG. 4.3. (A) Analysis of digestion products of 1,3-[carboxyl-^{14}C] diolein or [carboxyl-^{14}C] triolein in the intestinal lumen (Murase et al., 2002a). (B) Analysis of ^{14}C-labeled lipids in the intestinal mucosa (Murase et al., 2002a). $^{*}P < 0.05$, $^{**}P < 0.01$, $^{***}P < 0.001$.

PPARα is highly expressed in the small intestine; fatty acids act as ligands and activate PPARα, inducing the expression of lipid metabolism-related genes. The fatty acids produced by DAG ingestion may act as signal molecules and activate the lipid metabolism in a PPAR-mediated manner. Further investigation is required, however, to elucidate the detailed mechanism of activation of small intestinal lipid metabolism, the involvement of PPAR in the mechanism, and the contribution of the small intestine to systemic energy metabolism.

Although the increase in FFA in the small intestinal mucosa may reflect the amount of fatty acids produced in the small intestinal lumen, the increase might also be explained by the DAG metabolic pathway in the intestinal epithelium. Both the FFA and 1, 3-DAG ratios are increased in the small intestinal mucosa. The substrate specificity of diacylglycerol acyltransferase (DGAT) for 1, 3-DAG is low, and 1, 3-DAG is not readily resynthesized to TAG in vitro (Lehner & Kuksis; 1993). Further, the 1, 3-DAG precursor, 1(3)-MAG, is also less likely to be used as a substrate for monoacylglycerol acyltransferase (MGAT) (Cao et al., 2003; Cheng et al., 2003). We therefore hypothesized that these characteristic metabolites of DAG produced in the intestinal mucosa, together with differences in their levels and substrate specificity for the enzymes related to their metabolism, transiently increase the fatty acid concentration in the small intestinal mucosa. Figure 4.4 shows the possible metabolic fate of DAG in the small intestine.

Because 2-MAG is not produced directly after ingestion of DAG, and DAG metabolites are less likely to be used as the substrates of MGAT and DGAT responsible for the 2-MAG pathway, the contribution of the α-glycerophosphate (GP) pathway may be higher after DAG ingestion than after TAG ingestion. TAG synthesized by the 2-MAG pathway is rapidly used for chylomicron synthesis. In contrast, TAG synthesized via the α-GP pathway is stored in the cytoplasm, which is a pool with slow turnover (Yang & Kukusis, 1991). TAG is subsequently hydrolyzed and reesterified to produce TAG in microsomes, and is then incorporated into chylomicrons. Accordingly, fatty acids at positions 1 and 3 are likely to be hydrolyzed by lipase after DAG ingestion, thus increasing fatty acid production. Simultaneously, DAG metabolites do not readily enter the 2-MAG pathway because of the absence of 2-MAG, thereby increasing the relative ratio of the α-GP pathway; this may result in slower and possibly lower secretion of chylomicron-TAG into lymph and in the activation of the small intestinal lipid-metabolizing system.

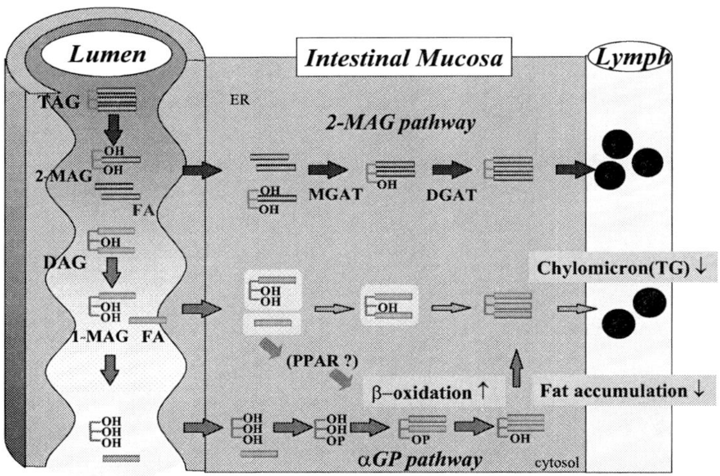

FIG. 4.4. Possible metabolic fate of dietary diacylglycerol in the small intestine.

Effect of Diacylglycerol on Hepatic Lipid Metabolism

Hepatic lipid metabolism is also activated by DAG. In Zucker fatty rats, the accumulation of TAG in the liver was reduced after 1 mo ingestion of ALA-DAG compared with ALA-TAG or the control group (Murase et al., 2005). Supplementation with ALA-DAG significantly upregulated β-oxidation activity, and ACO and MCAD mRNA levels in the liver. These findings indicate that ALA-DAG might be useful for preventing fatty liver formation and this effect could be closely related to the stimulation of lipid catabolism in the liver.

Murata et al. (Murata et al., 1997) extensively analyzed fat accumulation in the liver, the activity of hepatic lipid-metabolizing enzymes, and the fatty acid composition of lipids in the liver of rats fed diets containing 0% to 9.4% DAG for 2 to 3 wk. Ingestion of DAG decreased the serum TAG concentration and significantly decreased the liver TAG content. At the same time, the activities of enzymes involved in fatty acid synthesis, such as glucose-6-phosphate dehydrogenase (G6PD), malic enzyme (ME), and fatty acid synthase (FAS), were significantly decreased (Fig. 4.5). In contrast, in the DAG ingestion group, the oxidation activity of palmitoyl-CoA in the mitochondria and peroxisomes was significantly increased, by 25% to 40% of the control value (Fig. 4.6). The activities of enzymes related to ß-oxidation of the fatty acids acyl-CoA dehydrogenase, ACO, enoyl-CoA hydratase, and 3-hydroxyacyl-CoA dehydrogenase also increased dose-dependently with DAG (Fig. 4.7).

The activities of the enzymes related to the metabolism of unsaturated fatty acids containing *cis*-double bonds, i.e., 2, 4-dienoyl-CoA reductase and Δ^3, Δ^2-enoyl-CoA isomerase, also increased with increased doses of DAG. There were no significant dif-

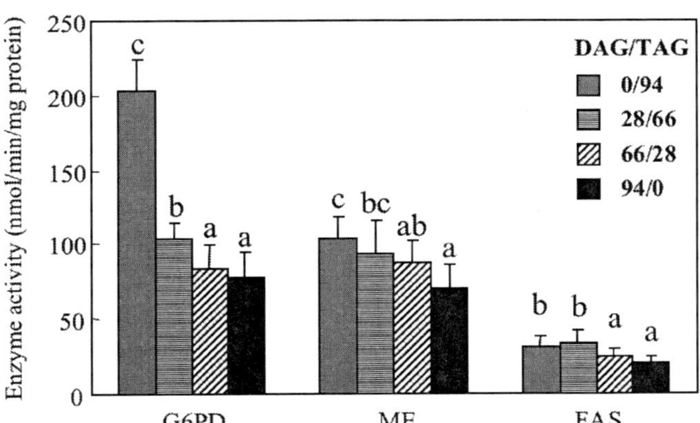

FIG. 4.5. Activities of enzymes involved in fatty acid synthesis in the liver of rats fed diets containing different proportions of DAG and TAG. DAG/TAG shows the amount of dietary fatty acids present as DAG and TAG (g/kg diet) (Murata et al., 1997). Abbreviations: G6PD, glucose-6-phosphate dehydrogenase; ME, malic enzyme; FAS, fatty acid synthase.

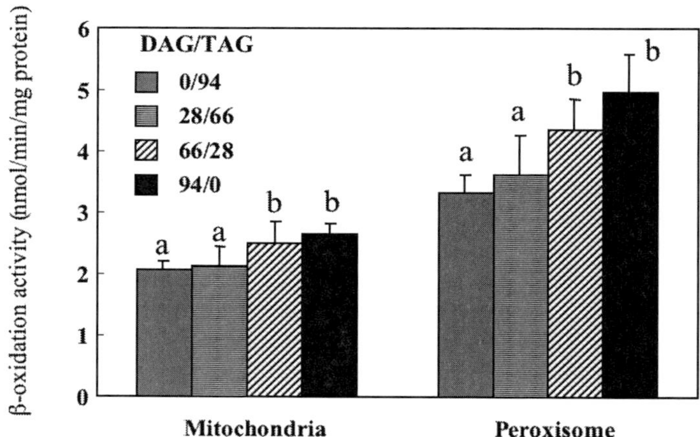

FIG. 4.6. Rate of oxidation of palmitoyl-CoA substrate in rat liver homogenate from rats fed diets containing different proportions of DAG and TAG. DAG/TAG shows the amount of dietary fatty acids as DAG and TAG (g/kg diet) (Murata et al., 1997).

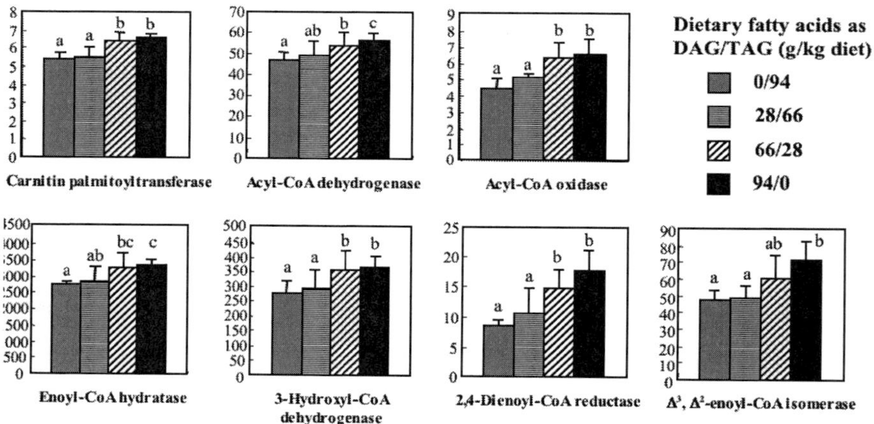

FIG. 4.7. Activities of enzymes of fatty acid oxidation [nmol/(min·mg protein)] in the liver of rats fed diets containing different proportions of DAG and TAG (Murata et al., 1997).

ferences in the activities of the mitochondrial marker enzyme, succinate dehydrogenase; the peroxisome marker enzyme, catalase; microsomal NADPH-cytochrome C reductase; or the cytosol marker enzyme, lactate dehydrogenase, between the TAG and DAG groups. The absence of changes in marker enzyme activities suggested that DAG induces ß-oxidation-related enzymes without proliferation of cell organelles.

Murata et al. (Murata et al., 1997) analyzed the fatty acid composition of lipids in the liver in rats fed TAG and DAG diets with a similar fatty acid composition and found differences between the two groups. Substitution of TAG with DAG decreased the ratio of oleic acid in liver TAG and increased the ratio of dietary linoleic acid. The ratios of docosapentaenoic acid and docosahexaenoic acid to liver phospholipids were slightly but significantly increased by DAG ingestion. In contrast, the ratios of saturated fatty acids and oleic acid to liver phospholipids were decreased by DAG ingestion. These phenomena may be closely related to lipid metabolism in the liver. A DAG-induced decrease in hepatic fatty acid synthesis may increase the relative use of dietary linoleic acid for the substrate of lipid synthesis, thus increasing linoleic acid in liver TAG. The increase in the relative use of α-linolenic acid may also have increased the activities of 2,4-dienoyl-CoA reductase and Δ^3, Δ^2-enoyl-CoA isomerase, which are related to the metabolism of α-linolenic acid (Fig. 4.7).

Meng and colleagues reported that ingestion of a 20% DAG diet for 8 wk significantly reduced body weight gain and liver TAG accumulation compared with rats fed 20% TAG (Meng X et al., 2004). At 8 wk, acyl-CoA carnitine acyltransferase (ACAT) activity in the liver of rats fed on the DAG diet was increased by 40%, and DGAT activity was decreased by 17%. It was also reported that high-fat diet-induced mRNA expression and enzyme activity of microsomal triacylglycerol transfer protein (MTP) in the liver were suppressed by the ingestion of DAG (Taguchi et al., 2002).

These results indicate that hepatic lipid metabolism is modulated by the dietary consumption of DAG. The precise mechanism of the regulation of hepatic lipid metabolism is not yet clear. Watanabe et al. speculated that the increased release of FFA into the portal circulation by DAG ingestion enhances ß-oxidation in the liver (Watanabe et al., 2001). Hormonal regulation or a central nervous system-mediated pathway in the regulation of lipid metabolism by DAG remains to be elucidated.

Activation of the hepatic β-oxidation system and decreases in the activities of fatty acid synthesizing enzymes induced by DAG ingestion would contribute to a reduction in body fat accumulation and of hepatic lipid.

Effect of Diacylglycerol on Whole Body Energy Metabolism

Based on findings that energy value and bioavailability do not differ between DAG and TAG (see Chapter 3) and the amount of food consumption during the experiments did not differ significantly between treatment groups, the differences in fat accumulation observed in both animal and human experiments is likely due to different amounts of energy expenditure after the intake of these respective fats. The increase in mRNA for energy metabolism-related molecules and the stimulation of enzyme activities responsible for ß-oxidation are obviously consistent with the stimulation of energy expenditure, which results in the suppression of diet-induced obesity.

As mentioned above, DAG rich in α-linolenic acid as the constituent fatty acid has a potent stimulatory effect on lipid metabolism (Murase et al., 2002b). Watanabe et al. (Watanabe et al., 2001) investigated the effects of ALA-DAG ingestion on the oxidation

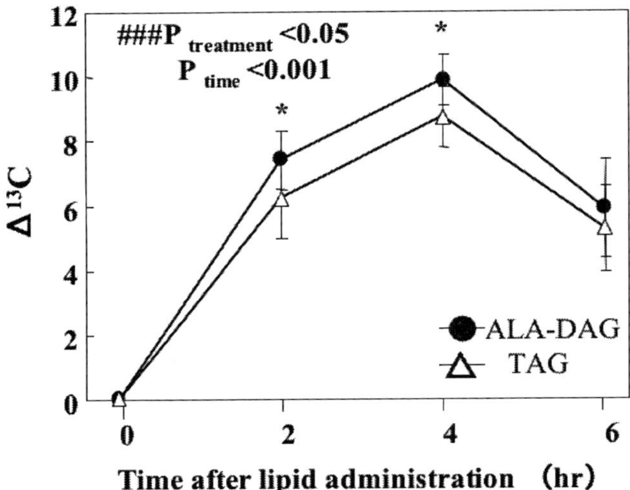

FIG. 4.8. Changes in ^{13}C in carbon dioxide exhaled by rats after a single dose of lipid. Data are means ± SD (n = 8). *Different from TAG group (Student's t-test; $P < 0.05$). ###Different from DAG group (repeated-measures ANOVA; P(treatment) <0.05, P(time)< 0.001) (Watanabe et al., 2001).

of dietary fat. Male Sprague-Dawley rats (4-wk old) were fed an ALA-DAG diet for 3 wk. They were then food-deprived for 6 h and administered lipids containing [1-^{13}C] tripalmitin. All of the gas exhaled by the rats was collected at 2-h intervals up to 6 h postadministration of [1-^{13}C] tripalmitin. As a control study, they conducted the same experiment using rats fed TAG for 3 wk. The ^{13}C content in the carbon dioxide expired from the ALA-DAG-fed rats was significantly greater ($P < 0.05$) than that from the TAG-fed rats up to 4 h after [1-^{13}C] tripalmitin administration (Fig. 4.8). This result indicated that the long-term ingestion of DAG, compared with that of TAG, enhances dietary fat oxidation in rats.

Some reports indicate that DAG not only stimulates energy expenditure after long-term ingestion, but is also readily metabolized after a single ingestion. Watanabe et al. (Watanabe et al., 1997) fed 6-wk-old rats food containing 10% TAG for 7 d; they were then food deprived for 18 h while they were housed in expiratory air metabolism cages. The rats were then orally administered lipid emulsion containing either diolein or triolein, and the oxygen level in the expired air was measured. Oxygen consumption increased after oral administration of the diolein or triolein emulsion. The oxygen consumption increased for up to 20 min after administration of either emulsion. Thereafter, the oxygen consumption in the diolein group increased more than that in the triolein group. Although it is not known whether fat or carbohydrate is responsible for these increases in oxygen consumption, these effects may explain, at least in part, the suppression of diet-induced obesity by DAG.

Kimura et al. (Kimura et al., 2006) investigated the difference in energy metabolism

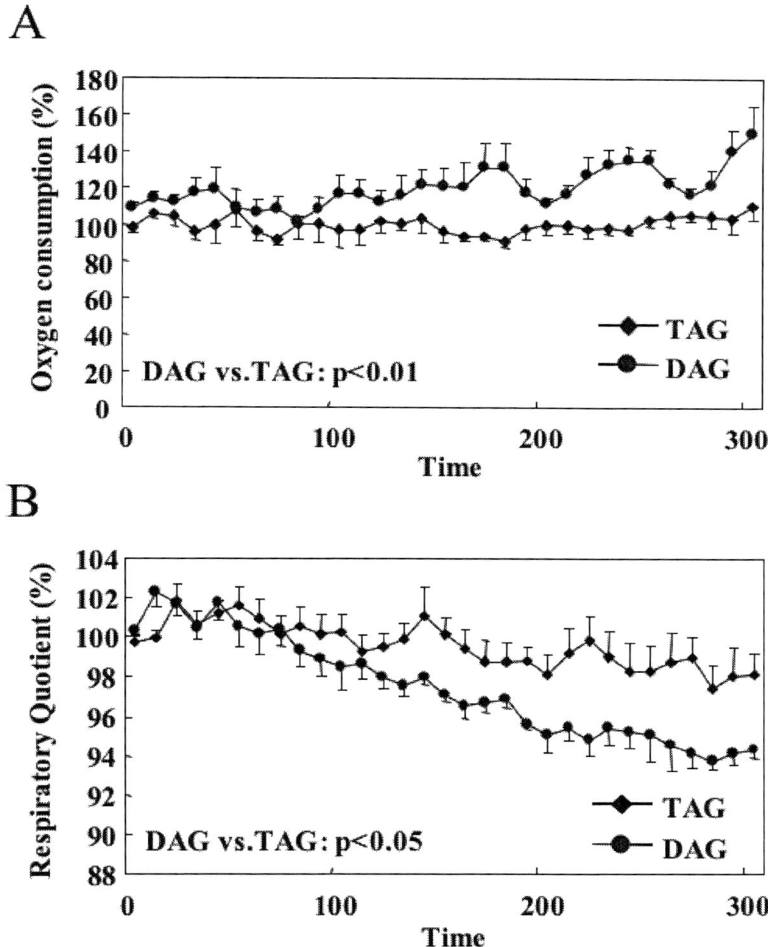

FIG. 4.9. Changes in oxygen consumption (A) and respiratory quotient (B) after lipid administration in rats. Data are means ± SD, (Kimura et al., 2006).

in Wistar rats after the administration of emulsions prepared with DAG or TAG by monitoring oxygen consumption and respiratory quotients (RQ).

Results are shown in Fig. 4.9. Oxygen consumption was enhanced after DAG infusion with increased combustion of lipids, as determined by the respiratory gas analysis. Oxygen consumption in the DAG group gradually increased after fat loading and reached a level significantly higher than that of the TAG group 3 h after fat loading. There was no such increase in the TAG group. Respiratory quotient values gradually decreased in both groups after fat loading. The RQ value reduction in the DAG group, however, was more pronounced than that in the TAG group. Kimura et al. concluded that these results were consistent with previous findings that DAG oil is less likely to become body fat, and may help to explain the antiobesity effect of DAG.

Summary and Conclusions

Continuous ingestion of DAG activates intestinal lipid metabolism (ß-oxidation) relatively early, followed by activation of the hepatic ß-oxidation system and a decrease in lipid synthesis. Kamphuis et al. (Kamphuis et al., 2003) reported that the RQ was significantly decreased and lipid oxidation was increased after DAG ingestion compared with TAG ingestion in an expiration analysis in humans, indicating that DAG itself is readily metabolized. The ease of DAG metabolism coupled with activation of the lipid-metabolizing system by continuous DAG ingestion may lead to a reduction in body fat accumulation in individuals over the long-term.

References

Cao, J., Lockwood, J., Burn, P., and Shi, Y. Cloning and Functional Characterization of a Mouse Intestinal Acyl-Coenzyme A:Monoacylglycerol Acyltransferase, MGAT2. *J. Biol. Chem.* **2003,** *278,* 13860-13866.

Cheng, D., Nelson, T.C., Chen, J., Walker, S.G., Wardwell-Swanson, J., Meegalla, R., Taub, R., Billheimer, J.T., Ramaker, M., and Feder, J.N. Identification of Acyl-Coenzyme A:Monoacylglycerol Acyltransferase 3, an Intestinal Specific Enzyme Implicated in Dietary Fat Absorption. *J. Biol. Chem.* **2003,** *278,* 13611-13614.

Hara, K., Onizawa, K., Honda, H., Otsuji, K., Ide, T., and Murata, M. Dietary Diacylglycerol-Dependent Reduction in Serum Triacylglycerol Concentration in Rats. *Ann. Nutr. Metab.* **1993,** *37,* 185-191.

Kamphuis, M.M., Mela, D.J., and Westerterp-Plantenga, M.S. Diacylglycerols Affect Substrate Oxidation and Appetite in Humans. *Am. J. Clin. Nutr.* **2003,** *77,* 1133-1139.

Kelly, D.P., Gordon, J.I., Alpers, R., and Strauss, A.W. The Tissue-Specific Expression and Developmental Regulation of Two Nuclear Genes Encoding Rat Mitochondrial Proteins. *J. Biol. Chem.* **1989,** *264,* 18921-18925.

Kimura, S., Tsuchiya, H., Inage, H., Meguro, S., Matsuo, N., Tokimitsu, I. Effects of dietary diacylglycerol on the energy metabolism. *Int. J. Vitam. Nutr. Res.* **2006,** *76,* 75-79.

Kondo, H., Hase, T., Murase, T., and Tokimitsu, I. Digestion and Assimilation Features of Dietary DAG in the Rat Small Intestine. *Lipids* **2003,** *38,* 25-30.

Kondo, H, Minegishi, Y, Komine, Y, Mori, T, Matsumoto, I, Abe, K, Tokimitsu, I, Hase, T, Murase, T. Differential regulation of intestinal lipid metabolism-related genes in obesity-resistant A/J vs. obesity-prone C57BL/6J mice. *Am. J. Physiol. Endocrinol. Metab.* **2006,** *291,* E1092-E1099.

Lehner, R., and Kuksis, A. Triacylglycerol Synthesis by sn-1,2(2,3)-Diacylglycerol Transacylase from Rat Intestinal Microsomes. *J. Biol. Chem.* **1993,** *268,* 8781-8786.

Li, A.C., Glass, C.K. Related PPAR- and LXR-dependent pathways controlling lipid metabolism and the development of atherosclerosis. *J. Lipid Res.* **2004,** *45,* 2161-2173.

Maki, K.C., Davidson, M.H., Tsushima, R., Matsuo, N., Tokimitsu, I., Umporowicz, D.M., Dicklin, M.R., Foster, G.S., Ingram, K.A., Anderson, B.D., Frost, S.D., and Bell, M. Consumption of Diacylglycerol Oil as Part of a Reduced-Energy Diet Enhances Loss of Body Weight and Fat in Comparison with Consumption of a Triacylglycerol Control Oil. *Am. J. Clin. Nutr.* **2002,** *76,* 1230-1236.

Meng, X., Zou, D., Shi, Z., Duan, Z., Mao, Z. Dietary diacylglycerol prevents high-fat diet-induced lipid accumulation in rat liver and abdominal adipose tissue. *Lipids.* **2004,** *39,* 37-41.

Murase, T., Mizuno, T., Omachi, T., Onizawa, K., Komine, Y., Kondo, H., Hase, T., and Tokimitsu, I. Dietary Diacylglycerol Suppresses High Fat and High Sucrose Diet-Induced Body Fat Accumulation in C57BL/6J Mice. *J. Lipid Res.* **2001,** *42,* 372-378.

Murase, T., Aoki, M., Wakisaka, T., Hase, T., and Tokimitsu, I. Anti-Obesity Effect of Dietary Diacylglycerol in C57BL/6J Mice: Dietary Diacylglycerol Stimulates Intestinal Lipid Metabolism. *J. Lipid Res.* **2002a,** *43,* 1312-1319.

Murase, T., Nagasawa, A., Suzuki, J., Wakisaka, T., Hase, T., and Tokimitsu, I. Dietary α-Linolenic Acid-Rich Diacylglycerols Reduce Body Weight Gain Accompanying the Stimulation of Intestinal β-Oxidation and Related Gene Expressions in C57BL/KsJ-db/db Mice. *J. Nutr.* **2002b,** *132,* 3018-3022.

Murase, T, Aoki, M, Tokimitsu, I. Supplementation with α-linolenic acid-rich diacylglycerol suppresses fatty liver formation accompanied by an up-regulation of β-oxidation in Zucker fatty rats. *Biochim. Biophys. Acta.* **2005,** *1733,* 224-231.

Murata, M., Hara, K., and Ide, T. Alteration by Diacylglycerols of the Transport and Fatty Acid Composition of Lymph Chylomicrons in Rats. *Biosci. Biotech. Biochem.* **1994,** *58,* 1416-1419.

Murata, M., Ide, T., and Hara, K. Reciprocal Responses to Dietary Diacylglycerol of Hepatic Enzymes of Fatty Acid Synthesis and Oxidation in the Rat. *Br. J. Nutr.* **1997,** *77,* 107-121.

Nagao, T., Watanabe, H., Goto, N., Onizawa, K., Taguchi, H., Matsuo, N., Yasukawa, T., Tsushima, R., Shimasaki, H., and Itakura, H. Dietary Diacylglycerol Suppresses Accumulation of Body Fat Compared to Triacylglycerol in Men in Double-Blind Controlled Trial. *J. Nutr.* **2000,** *130,* 792-797.

Nemali, M.R., Usuda, N., Reddy, M.K., Oyasu, K., Hashimoto, T., Osumi, T., Rao, M.S., and Reddy, J.K. Comparison of Constitutive and Inducible Levels of Expression of Peroxisomal Beta-Oxidation and Catalase Genes in Liver and Extrahepatic Tissues of Rats. *Cancer Res.* **1988,** *48,* 5316-5324.

Schoonjans, K., Staels, B., and Auwerx, J. Role of the Peroxisome Proliferator-Activated Receptor (PPAR) in Mediating the Effects of Fibrates and Fatty Acids on Gene Expression. *J. Lipid Res.* **1996,** *37,* 907-925.

Taguchi, H., Watanabe, H, Onizawa, K., Nagao, T., Gotoh, N., Yasukawa, T., Tsushima, R., Shimasaki, H., and Itakura, H. Double-Blind Controlled Study on the Effects of Dietary Diacylglycerol on Postprandial Serum and Chylomicron Triacylglycerol Responses in Healthy Humans. *J. Am. Coll. Nutr.* **2000,** *19,* 789-796.

Taguchi, H., Omachi, T., Nagao, T., Matsuo, N., Tokimitsu, I., and Itakura, H. Dietary Diacylglycerol Suppresses High Fat Diet-Induced Hepatic Fat Accumulation and Microsomal Triacylglycerol Transfer Protein Activity in Rats. *J. Nutr. Biochem.* **2002,** *13,* 678-683.

Watanabe, H., Onizawa, K., Taguchi, H., Kobori, M., Chiba, H., Naito, S., Matsuo, N., Yasukawa, T., Hattori, M., and Shimasaki, H. Nutritional Characterization of Diacylglycerols in Rats. *J. Jpn. Oil Chem. Soc.* **1997,** *46,* 301-308.

Watanabe, H., Yamaguti, T., Onizawa, K., Osaki, N., Harada, U., Tokimitsu, I., Shimasaki, H., and Itakura, H. Effects of α-Linolenic Acid-Rich Diacylglycerol on the Oxidation of Dietary Fats in Rats. *J. Oleo Sci.* **2001,** *50,* 839-842.

Yang, L.Y., and Kuksis, A. Apparent Convergence (at 2-Monoacylglycerol Level) of Phosphatidic Acid and 2-Monoacylglycerol Pathways of Synthesis of Chylomicron Triacylglycerols. *J. Lipid Res.* **1991,** *32,* 1173-1186.

Chapter 5

The Effect of Diacylglycerols on Energy Expenditure and Substrate Utilization in Humans

Brent D. Flickinger

Archer Daniels Midland Company, Decatur, IL

Introduction

The issue of obesity has become a primary focus of public health. Over the past 15 years, the rising numbers of obese adults and children around the world have led to prioritizing the fight against obesity as a primary objective by leading governmental and nongovernmental agencies. These agencies consistently agree that current and predicted levels of obesity have or will reach epidemic proportions.

Obesity is a complex issue with physical, behavioral, and cultural components. The physical aspects most pertinent for body weight control are energy expenditure and intake. The types and amounts of foods have clearly become an arena of major importance for understanding their role in obesity. In addition to the fundamental issue of energy balance, considerable effort is underway to find new food ingredients that affect energy expenditure or intake.

Diacylglycerol (DAG) oil was shown to affect body weight and body fat in both animals and humans when consumed regularly in the diet. The animal data suggest that the differences in body weight and body fat are primarily, if not solely, due to energy expenditure. Little is known concerning the effect of DAG oil on the relative role of changes in energy expenditure or energy intake as the force behind the observed differences of lower body weight.

Background

DAG oil is an edible oil containing at least 80% diglycerides. The dominant portion of diglycerides contains fatty acids on the ends of the glycerol backbone, which is described as an *sn*-1,3-diglyceride. Of the diglycerides in DAG oil, ~70% are 1,3-diglycerides, making them the major component of DAG oil.

DAG oil was shown to affect body weight and body fat in both animals and humans when consumed regularly in the diet compared with conventional oil (Murase et al., 2001; Murase et al., 2002; Nagao et al., 2000; Maki et al., 2002). Additionally, consumption of DAG oil was shown to lower serum triglyceride (TG) levels compared with conventional oil (Hara et al., 1993; Taguchi et al., 2000; Tada et al., 2001). The differences due to DAG oil were attributed to the content of 1,3-diacylglycerols in DAG oil.

The absence of a fatty acid in the *sn*-2 position of a 1,3-diglyceride creates a different monoglyceride (MG) after normal digestion. This digestive product known as a 1(3)-MG, as opposed to 2-MG formed during digestion of conventional (i.e., TG) oils, is not as efficient a substrate for TG resynthesis in the intestinal lumen. As a result, 1(3)-MG must follow the 2-MG pathway to a much slower degree or an alternative pathway for TG resynthesis. This alternative pathway is commonly described as the glycerol phosphate pathway. It involves the utilization of glycerol-3-phosphate as the starting point for TG synthesis. For 1-MG to follow this pathway, it must be hydrolyzed to glycerol, which releases a free fatty acid (FFA) that could be available for TG resynthesis or potentially for energy utilization. As a result, postprandial elevations in TG-rich lipoproteins are lowered significantly after consumption of DAG oil (Taguchi et al., 2000; Tada et al., 2001), and a certain quantity of fatty acids could be available for energy use.

Increasing fat for energy use, particularly during the postprandial phase, continues to be an evolving concept as a way of affecting energy balance by either stimulating energy expenditure or suppressing energy intake. The scientific literature has examples that describe the increased use of fat as an energy substrate. These studies used medium-chain fatty acids, conventional oils with medium-chain fatty acids, and certain preparations of fatty acids and resulted in increased fat oxidation, increased energy expenditure, or decreased food intake, thus helping to create a negative energy balance (Van Wymelbeke et al., 1998 and 2001; St-Onge et al., 2003; Jones et al., 1992; Cox et al., 2000). These examples demonstrate a common approach of attempting to provide fatty acids for energy use during the postprandial period. 1,3-Diglycerides present in DAG oil can serve as a potential source of fatty acids for energy utilization during the postprandial phase.

Effect of DAG Oil on Energy Utilization in Humans

To date two published studies described the acute effect of DAG oil on energy expenditure and energy substrate utilization in humans (Kamphuis et al., 2003; Saito et al., 2006). A metabolic chamber study examined the influence of DAG oil vs. conventional oil on energy expenditure, energy substrate utilization, and subjective appetite ratings of 12 healthy women during a 36-h stay in a metabolic chamber and a metabolic cart study examined the influence of DAG oil vs. conventional oil on energy expenditure and energy substrate utilization in the postprandial phase following a single test meal.

During the chamber experiment, 40% of the fat consumed was DAG oil or conventional oil with a similar fatty acid composition as part of a defined (55% en from carbohydrate, 15% en from protein, 40% en from fat), eucaloric diet. Before the chamber stay, the subjects consumed the defined, eucaloric diet for 3 d. The results indicated that DAG oil consumption increased fat oxidation but did not affect energy expenditure (Table 5.1). Fat oxidation was significantly increased during both d 1 and 2 (two 24-h segments from the total 36-h chamber stay) by 4.9 and 4.0 g/d, respectively. This agreed with the observation that serum levels or ketones, as measured by β-hydroxybutyrate, and FFA tended to be greater after DAG oil consumption. Serum ketone levels (d 2 at the first blood draw)

TABLE 5.1
Summary of Selected Energy Expenditure (EE) and Substrate Utilization Data Due to DAG Oil[a,b]

Variable	DAG oil	TAG oil
EE (MJ), 24-h		
Total	9.5	9.5
Sleeping MR	6.3	6.3
DIT	0.8	0.8
Activity-based EE	2.5	2.4
Fat oxidation (g/d)		
d 1	60.7**	55.8
d 2	64.6*	60.6
RQ		
Total, 36 h	0.849*	0.855
d 1	0.846*	0.853
d 2	0.851*	0.857

[a]Source: Kamphuis et al., 2003. Asterisks indicate different from TAG oil **P = 0.004; *P < 0.05.
[b]Abbreviations: MR, metabolic rate; DIT, diet-induced thermogenesis; RQ, respiratory quotient; DAG, diacylglycerol; TAG, triacylglycerol.

were greater after DAG oil consumption ($P < 0.05$ vs. TAG oil using post-hoc analysis). Neither overall energy expenditure nor individual components of energy expenditure (sleeping metabolic rate, diet-induced thermogenesis, and activity-based energy expenditure) differed after DAG oil or TAG oil consumption.

The subjective appetite results indicated differences in subjective appetite ratings due to DAG oil consumption. Although not observed after d 1, significant differences due to DAG oil consumption in cumulative d 2 ratings for hunger, appetite, estimate of prospective food intake, and desire to eat were observed. Relative to TAG oil, cumulative hunger scores were lower (less hunger), appetite lower (less appetite), estimate of prospective food intake lower (less estimated food intake), and desire to eat lower (less desire to eat) after DAG oil consumption. To summarize, these results indicate the ability of DAG oil consumption to increase fat oxidation and suppress subjective appetite ratings when substituted for conventional edible oil. A metabolic cart study examined the effect of DAG oil vs. TAG oil in healthy subjects during the postprandial phase (Saito et al., 2006). Following a meal containing 30 g of test oil, postprandial RQ was significantly lower at 2 and 5 hours while energy expenditure was significantly higher at 3 hours when DAG oil was consumed as compared to the control TAG oil.

Relating Energy Balance to Body Weight Regulation by DAG Oil

Changes in body fat and body weight loss were relatively mild after consumption of DAG oil over a period of several months, with an approximate difference in daily energy expenditure or intake between 50 and 100 kcal. In the study by Maki and colleagues (Maki et al., 2002), the increase in weight loss over 6 mo observed between DAG oil and conventional oil could be attributed to a daily energy deficit of ~48 kcal. In the study by Nagao and colleagues (Nagao et al., 2000), the difference in weight loss over 4 mo observed between DAG oil and conventional oil could be attributed to a daily energy deficit of ~90 kcal.

Correspondingly, the data from Kamphius and colleagues did not indicate a significant change in daily energy expenditure after DAG oil consumption compared with TAG oil (Kamphuis et al., 2003). A difference of 50–100 kcal over a 24-h period would be difficult to detect in a metabolic chamber setting under the best of conditions. These investigators also found a significant increase in fat oxidation of approximately 4–5 g/d. This degree of fat oxidation corresponds to a shift in using ~35–45 kcal from fat rather than other fuel sources, which may apparently affect energy intake (i.e., appetite) to a greater degree than energy expenditure.

Fat Metabolism and Appetite Regulation

Fatty acid oxidation and metabolism as controls for energy intake appear to be an important relation (Scharrer, 1999; Friedman, 1998; Tremblay, 1992; Ronnett et al., 2005) with new research observing that obese individuals tend to have decreased ability to oxidize dietary fat (Westerterp et al., 2008). Metabolites of fatty acid oxidation including ketones, glycerol, and FFA have been suggested to play a role in the control of food intake. Historically, ketones have received considerable attention. Ketogenic diets (severe carbohydrate restriction) have been utilized for weight loss and espoused with an underlying assumption that elevated serum ketone levels provide a degree of appetite suppression. However, the appetite-suppressing effect of elevated serum ketones remains to be conclusively documented.

Recent literature has focused on examining the effects of compounds that inhibit enzymes of fat metabolism. Using rodent models as well as human subjects, changes in fatty acid oxidation can be inversely correlated with changes in food intake. The administration of compounds that directly inhibit fatty acid oxidation or inhibit fatty acid synthesis significantly increases or decreases food intake, respectively, as well as changes in body weight over extended durations (Del Prete et al., 1998; Friedman et al., 1999; Rawson et al., 1996; Thupari et al., 2002; Loftus et al., 2000; Triscari & Sullivan, 1984). An example of a compound that decreases fatty acid oxidation is etomoxir, a carnitine palmitoyl transferase I inhibitor, which causes decreased mitochondrial oxidation of long-chain fatty acids and increases insulin resistance in mice due to intramyocellular lipid accumulation over prolonged administration (Dobbins et al., 2001). In a limited number of studies using human subjects, there is mixed evidence regarding the acute administration of etomoxir resulting from increased food intake (Hinderling et al., 2002; Kahler et al., 1999). An example of a compound that causes a shift toward decreased fatty acid synthesis is C75, a fatty acid synthase inhibitor, which results in decreased fatty acid synthesis and decreased food intake and body weight in mice through its proposed modulation of hypothalamic neuronal activity and neuropeptide Y levels (Shimokawa et al., 2002; Gao & Lane, 2003). To date, no study using C75 in human subjects has been published to date in the scientific literature.

The postprandial utilization of fatty acids for energy after DAG oil consumption was reported to occur via β-oxidation in the liver and/or small intestine based on rodent studies (Murase et al., 2001; Murase et al., 2002; Murata et al., 1997). Hepatic fat oxidation is believed to be under the control of the hepatic glucose metabolism, rather than fat intake, which has led to the argument that the status of glycogen stores and carbohydrate metabolism are more important parameters controlling total energy balance and fat mass than fat intake (Flatt, 1995; Flatt, 1996). Therefore, this mechanism would suggest that smaller glycogen stores, which were observed to correlate with lower body fat stores and decreased effort to replenish the loss of body fat, potentially result from increased postprandial fatty acid oxidation. Consequently, increased postprandial fatty acid oxidation in the liver would maintain ATP stores which are reported to be an important switch for food intake regulation (Friedman, 2007).

Summary

The utilization of DAG oil to provide 1,3-DAG results in differing metabolic effects compared with conventional edible oils. Body weight and body fat levels are significantly lowered using DAG oil in both humans and animals by improved energy balance *via* increased energy expenditure or decreased energy intake. In conclusion, the literature currently supports the hypothesis that DAG oil increases fat oxidation and may play a role in long-term weight maintenance *via* appetite suppression in humans.

References

Cox, J.E., Tyler, W.J., Randich, A., Kelm, G.R., Bharaj, S.S., Jandacek, R.J., and Meller, S.T. Suppression of Food Intake, Body Weight, and Body Fat by Jejunal Fatty Acid Infusions. *Am. J. Physiol.* **2000,** *278*, R604-R610.

Del Prete, E., Lutz, T.A., Althaus, J., and Scharrer, E. Inhibitors of Fatty Acid Oxidation (Mercaptoacetate, R-3-Amino-4-trimethylaminobutyric Acid) Stimulate Feeding in Mice. *Physiol. Behav.* **1998,** *63*, 751-754.

Dobbins, R.L., Szczepaniak, L.S., Bentley, B., Esser, V., Myhill, J., and McGarry, J.D. Prolonged Inhibition of Muscle Carnitine Palmitoyltransferase-1 Promotes Intramyocellular Lipid Accumulation and Insulin Resistance in Rats. *Diabetes* **2001,** *50*, 123-130.

Flatt, J.P. Use and Storage of Carbohydrate and Fat. *Am. J. Clin. Nutr.* **1995**, *61*, 952S-959S.

Flatt, J.P. Glycogen Levels and Obesity. *Int. J. Obes. Relat. Metab. Disord.* **1996**, *20* (Suppl. 2), S1-S11.

Friedman, M.I. Fuel Partitioning and Food Intake. *Am. J. Clin. Nutr.* **1998,** *67*, 513S-518S.

Friedman, M.I. Obesity and the Hepatic Control of Feeding Behavior. *Drug News Perspect.* **2007,** *20*, 573-578.

Friedman, M.I., Harris, R.B., Ji, H., Ramirez, I., and Tordoff, M.G. Fatty Acid Oxidation Affects Food Intake by Altering Hepatic Energy Status. *Am. J. Physiol.* **1999,** *276*, R1046-R1053.

Gao, S., and Lane, M.D. Effect of the Anorectic Fatty Acid Synthase Inhibitor C75 on Neuronal Activity in the Hypothalamus and Brainstem. *Proc. Natl. Acad. Sci. USA* **2003,** *100*, 5628-5633.

Hara, K., Onizawa, K., Honda, H., Otsuji, K., Ide, T., and Murata, M. Dietary Diacylglycerol-Dependent Reduction in Serum Triacylglycerol Concentration in Rats. *Ann. Nutr. Metab.* **1993,** *37*, 185-191.

Hinderling, V.B., Schrauwen, P., Langhans, W., and Westerterp-Plantenga, M.S. The Effect of Etomoxir on 24-h Substrate Oxidation and Satiety in Humans. *Am. J. Clin. Nutr.* **2002,** *76*, 141-147.

Jones, P.J., Ridgen, J.E., Phang, P.T., and Birmingham, C.L. Influence of Dietary Fat Polyunsaturated to Saturated Ratio on Energy Substrate Utilization in Obesity. *Metabolism* **1992,** *41*, 396-401.

Kahler, A., Zimmermann, M., and Langhans, W. Suppression of Hepatic Fatty Acid Oxidation and Food Intake in Men. *Nutrition* **1999,** *15*, 819-828.

Kamphuis, M.M., Mela, D.J., and Westerterp-Plantenga, M.S. Diacylglycerols Affect Substrate Oxidation and Appetite in Humans. *Am. J. Clin. Nutr.* **2003,** *77*, 1133-1139.

Loftus, T.M., Jaworsky, D.E., Frehywot, G.L., Townsend, C.A., Ronnett, G.V., Lane, M.D., and Kuhajda, F.P. Reduced Food Intake and Body Weight in Mice Treated with Fatty Acid Synthase Inhibitors. *Science* **2000,** *288*, 2379-2381.

Maki, K.C., Davidson, M.H., Tsushima, R., Matsuo, N., Tokimitsu, I., Umporowicz, D.M., Dicklin, M.R., Foster, G.S., Ingram, K.A., Anderson, B.D., Frost, S.D., and Bell, M. Consumption of Diacylglycerol Oil as Part of a Reduced-Energy Diet Enhances Loss of Body Weight and Fat in Comparison with Consumption of a Triacylglycerol Control Oil. *Am. J. Clin. Nutr.* **2002,** *76*, 1230-1236.

Murase, T., Mizuno, T., Omachi, T., Onizawa, K., Komine, Y., Kondo, H., Hase, T., and Tokimitsu, I. Dietary Diacylglycerol Suppresses High Fat and High Sucrose Diet-Induced Body Fat Accumulation in C57BL/6J Mice. *J. Lipid Res.* **2001,** *42*, 372-378.

Murase, T., Aoki, M., Wakisaka, T., Hase, T., and Tokimitsu, I. Anti-Obesity Effect of Dietary Diacylglycerol in C57BL/6J Mice: Dietary Diacylglycerol Stimulates Intestinal Lipid Metabolism. *J. Lipid Res.* **2002,** *43*, 1312-1319.

Murata, M., Ide, T., and Hara, K. Reciprocal Responses to Dietary Diacylglycerol of Hepatic Enzymes of Fatty Acid Synthesis and Oxidation in the Rat. *Br. J. Nutr.* **1997,** *77*, 107-121.

Nagao, T., Watanabe, H., Goto, N., Onizawa, K., Taguchi, H., Matsuo, N., Yasukawa, T., Tsushima, R., Shimasaki, H., and Itakura, H. Dietary Diacylglycerol Suppresses Accumulation of Body Fat Compared to Triacylglycerol in Men in a Double-Blind Controlled Trial. *J. Nutr.* **2000,** *130*, 792-797.

Rawson, N.E., Ulrich, P.M., and Friedman, M.I. Fatty Acid Oxidation Modulates the Eating Response to the Fructose Analogue 2,5-Anhydro-D-mannitol. *Am. J. Physiol.* **1996,** *271*, R144-R148.

Ronnett, G.V., Kim, E.K., Landree, L.E., and Tu, Y. Fatty Acid Metabolism as a Target for Obesity Treatment. *Physiol. Behav.* **2005,** *85*, 25-35.

Saito, S., Hase, K., and Tokimitsu, I. Effects of Diacylglycerol on Postprandial Energy Expenditure and Respiratory Quotient in Halthy Subjects. *Nutrition* **2006,** *22*, 30-35.

Scharrer, E. Control of Food Intake by Fatty Acid Oxidation and Ketogenesis. *Nutrition* **1999,** *15*, 704-714.

Shimokawa, T., Kumar, M.V., and Lane, M.D. (2002) Effect of a Fatty Acid Synthase Inhibitor on Food Intake and Expression of Hypothalamic Neuropeptides. *Proc. Natl. Acad. Sci. USA* **2002,** *99*, 66-71.

St-Onge, M.P., Ross, R., Parsons, W.D., and Jones, P.J. Medium-Chain Triglycerides Increase Energy Expenditure and Decrease Adiposity in Overweight Men. *Obes. Res.* **2003,** *11*, 395-402.

Tada, N., Watanabe, H., Matsuo, N., Tokimitsu, I., and Okazaki, M. Dynamics of Postprandial Remnant-Like Lipoprotein Particles (RLP) in Serum After Loading of Diacylglycerols. *Clin. Chem. Acta* **2001,** *311*, 109-117.

Taguchi, H., Watanabe, H., Onizawa, K., Nagao, T., Gotoh, N., Yasukawa, T., Tsushima, R., Shimasaki, H., and Itakura, H. Double-Blind Controlled Study on the Effects of Dietary Diacylglycerol on Postprandial Serum and Chylomicron Triacylglycerol Responses in Healthy Humans. *J. Am. Coll. Nutr.* **2000,** *19*, 789-796.

Thupari, J.N., Landree, L.E., Ronnett, G.V., and Kuhajda, F.P. C75 Increases Peripheral Energy Utilization and Fatty Acid Oxidation in Diet-Induced Obesity. *Proc. Natl. Acad. Sci. USA* **2002,** *99*, 9498-9502.

Tremblay, A. Human Obesity: A Defect in Lipid Oxidation or in Thermogenesis? *Int. J. Obes. Relat. Metab. Disord.* **1992,** *16*, 953-957.

Triscari, J., and Sullivan, A.C. Anti-Obesity Activity of a Novel Lipid Synthesis Inhibitor. *Int. J. Obes.* **1984,** *8* (Suppl. 1), 227-239.

Van Wymelbeke, V., Himaya, A., Louis-Sylvestre, J., and Fantino, M. Influence of Medium-Chain and Long-Chain Triacylglycerols on the Control of Food Intake in Men. *Am. J. Clin. Nutr.* **1998,** *68*, 226-234.

Van Wymelbeke, V., Louis-Sylvestre, J., and Fantino, M. Substrate Oxidation and Control of Food Intake in Men After a Fat-Substitute Meal Compared with Meals Supplemented with an Isoenergetic Load of Carbohydrate, Long-Chain Triacylglycerols, or Medium-Chain Triacylglycerols. *Am. J. Clin. Nutr.* **2001,** *74*, 620-630.

Westerterp, K.R., Smeets, A., Lejeune, M.P., Wouters-Adriaens, M.P.E. and Westerterp-Plantenga, M.S. Dietary Fat Oxidation as a Function of Body Fat. *Am. J. Clin. Nutr* **2008,** *87*, 132-135.

Part III

Health Benefits

Section A

Influence on Postprandial Lipid

Chapter 6

Fate of Hydrolysis Products Originating from Diacylglycerol and Triacylglycerol in the Intestine

Ikuo Ikeda[a] and Teruyoshi Yanagita[b]

[a]Laboratory of Food and Biomolecular Science, Department of Food Function and Health, Division of Bioscience and Biotechnology for Future Bioindustries, Graduate School of Agricultural Studies, Tohoku University, Sendai, 981-8555, Japan and [b]Laboratory of Nutrition Biochemistry, Department of Applied Biological Sciences, Saga University, Saga 840–8502, Japan

Introduction

After a meal, triacylglycerol is hydrolyzed by gastric and pancreatic lipases, and 2-monoacylglycerol and fatty acids are formed in the intestinal lumen. In contrast, it is thought that 1- or 3-monoacylglycerol and fatty acid are formed from digestion of 1, 3-diacylglycerol, whereas 2-monoacylglycerol and fatty acid are from 1, 2-diacylglycerol (Watanabe et al., 1997; Kondo et al., 2003). In this chapter, the fate of hydrolysis products from triacylglycerol (TAG) and diacylglycerol (DAG) in intestinal epithelial cells and their transport to lymph are discussed.

Triacylglycerol

After the incorporation of hydrolysis products from TAG, 2-monoacylglycerol and fatty acids, into intestinal epithelial cells, the fatty acids are bound to the 1- and 3positions of 2-monoacylglycerol; then, triacylglycerol (triglyceride) is resynthesized via the 2-monoacylglycerol pathway (Fig. 6.1) (Friedman & Nylund, 1980). The triglyceride is incorporated into chylomicrons, which are released into the intestinal lymph and poured into the bloodstream.

1,2-(2,3-)Diacylglycerol

Free fatty acid and 2-monoacylglycerol are formed from 1, 2(2, 3)-DAG in the intestinal lumen. After the incorporation into intestinal epithelial cells, these hydrolysis products are subjected to triglyceride synthesis as in the case of TAG absorption (Fig. 6.1). Because the amount of absorbed fatty acids are not sufficient to synthesize triglyceride, endogenous fatty acids are utilized. Alternatively, a part of 2-monoacylglycerol is hydrolyzed to glycerol and fatty acid, which may be utilized for triglyceride synthesis. The triglyceride is incorporated into chylomicrons and released into the intestinal lymph.

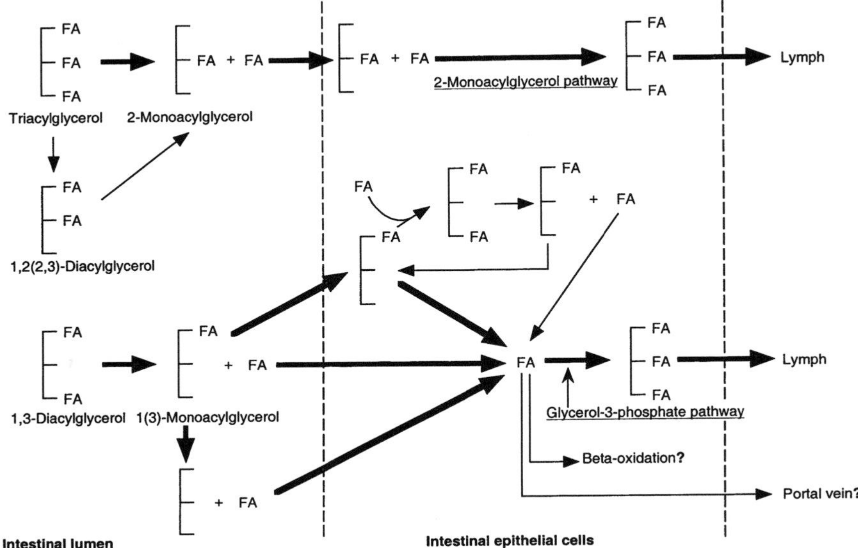

FIG. 6.1. A scheme of the intestinal digestion and absorption of TAG and DAG. The thick arrows indicate the main pathways of digestion and absorption of TAG and DAG. The fate of glycerol formed is not shown.

1,3-Diacylglycerol

Free fatty acid and 1- or 3-monoacylglycerol are formed from 1, 3-DAG in the intestinal lumen (Fig. 6.1) (Watanabe et al., 1997; Kondo et al., 2003). 1(3)-Monoacylglycerol is not generally formed in the intestinal lumen after TAG consumption. 1(3)-Monooleoylglycerol was detected in rat intestinal epithelial cells after an intraduodenal infusion of 1, 3-di[1-^{14}C]oleoylglycerol; therefore, it may be incorporated directly into intestinal epithelial cells (Kondo et al., 2003). Alternatively, 1(3)-monoacylglycerol may be further hydrolyzed to fatty acid and glycerol in the intestinal lumen (Kondo et al., 2003). 1(3)-Monoacylglycerol may be hydrolyzed to fatty acid and glycerol in intestinal cells, because the direct pathway for triglyceride synthesis from 1(3)-monoacylglycerol is not known.

When ^{14}C-linoleic acid and 1(3)-monoacylglycerol were incubated with intestinal epithelial cells, radioactivity was detected as 1, 3-DAG (Kondo et al., 2003). Therefore, at least a part of the 1(3)-monoacylglycerol incorporated into intestinal epithelial cells may be synthesized to 1, 3-DAG. However, because the rate of triglyceride synthesis from 1, 3-DAG is much lower than that from 1, 2-DAG (Kondo et al., 2003), 1, 3-DAG must be hydrolyzed by lipases in intestinal cells. A part of 1, 3-DAG may also be converted to 1, 2-DAG. Because >90% of chylomicron lipids are triglycerides after the ingestion of 1, 3-DAG in rats, and free fatty acid, monoacylglycerol, and DAG are not increased in lymph chylomicrons (Yanagita et al., 2004), it is thought that most of 1, 3-DAG finally become free fatty acids and glycerol and are then subjected to triglyceride synthesis. Because 2-monoacylglycerol cannot be sufficiently supplied, triglyceride synthesis from free fatty acids is possible via the glycerol-3-phosphate pathway, which is less active than the 2-monoacylglycerol pathway (Friedman & Nylund, 1980).

In our study, when rats with cannulated the thoracic duct were administered 1,3-di[1-^{14}C]oleoylglycerol or tri[1-^{14}C]oleoylglycerol intragastrically, lymphatic recovery of the radioactivity from dioleoylglycerol was ~50% of that from trioleoylglycerol 1 h after the treatment (Fig. 6.2). The total lymphatic recovery of radioactivity from ^{14}C-dioleoylglycerol after 24 h was slightly but significantly lower than that from ^{14}C-trioleoylglycerol, 81.3±1.0 vs. 86.5±1.2 %, respectively (Yanagita et al., 2004). Although the recovery from DAG was slightly lower than that from TAG, our results showed that a large part of the fatty acids from 1,3-DAG ingested by rats was transported to the lymph as in the case of TAG. More than 90% of the radioactivity from ^{14}C-dioleoylglycerol and ^{14}C-trioleoylglycerol was recovered in the triglyceride fraction in the lymph, indicating that fatty acids from DAG are slowly absorbed and transported to the lymph, probably because most of fatty acids from DAG are slowly synthesized to triglyceride via the glycerol-3-phosphate pathway. Taguchi et al. (Taguchi et al., 2001) observed that fecal excretion of fat after the feeding of DAG was almost the same as in the case of TAG, suggesting that intestinal absorption of DAG is comparable to that of TAG.

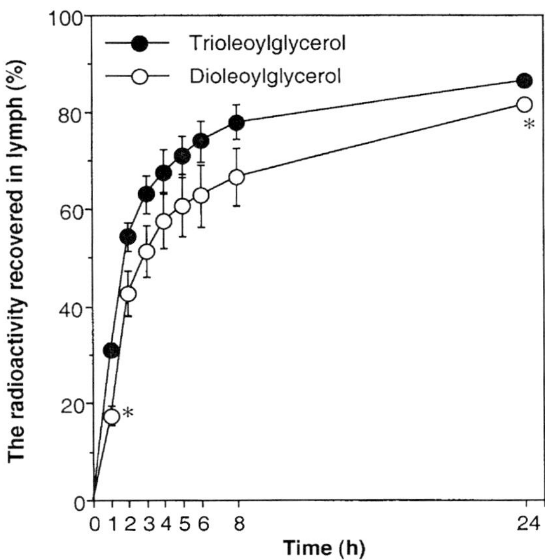

FIG. 6.2. Lymphatic absorption of 1,3-dioleoylglycerol and trioleoylglycerol in rats. Rats cannulated in the thoracic duct were given an emulsion containing 200 mg of 1,3-di[1-^{14}C]oleoylglycerol or tri[1-^{14}C]oleoylglycerol via a stomach tube and lymph was collected for 24 h. *Different from trioleoylglycerol, $P < 0.05$.

We showed that total 24-h recovery of di[1-^{14}C]oleoylglycerol was ~5% lower than that of tri[1-^{14}C]oleoylglycerol (Yanagita et al., 2004). There are three possible causes for the slightly lower recovery of fatty acids originated from 1, 3-DAG in lymph. First, a part of the fatty acids from DAG might be transferred to the portal vein. Watanabe et al. (Watanabe et al., 1997) observed an increase in portal venous fatty acids after the ingestion of DAG compared with TAG. They suggested that fatty acids transported via the portal vein might be preferentially ß-oxidized in the liver (Watanabe et al., 1997). Medium-chain fatty acids are transported via the portal vein to the liver and preferentially ß-oxidized. Although portal venous transport of long-chain fatty acids had been reported (Kiyasu et al., 1952), most long-chain fatty acids are almost quantitatively absorbed as chylomicron triglyceride *via* the lymph (Ikeda et al., 1997). Therefore, it is likely that the amount of fatty acids transported *via* the portal vein after DAG feeding is small, if any.

Second, there is a possibility that a part of the fatty acids from DAG was oxidized in intestinal cells. Murase et al. (Murase et al., 2002) reported that when DAG were fed to mice for 10 d, intestinal ß-oxidation was stimulated. They also observed increases in mRNA expression of acyl-CoA oxidase, medium-chain acyl-CoA dehydrogenase, and uncoupling protein-2 in the intestine. We observed that when 1, 3-di[1-^{14}C]oleoylglycerol or tri[1-^{14}C]oleoylglycerol was administered intragastrically to rats cannulated in the thoracic duct, a small but significantly higher radioactivity originated from 1, 3-DAG was detected in the free cholesterol fraction of lymph lipids collected for 24 h after the administration (Yanagita et al., 2004). Therefore, we cannot deny the possibility that a small part of [1-^{14}C]oleic acid hydrolyzed from 1, 3-DAG was more effectively ß-oxidized to acetyl CoA than that from TAG and then utilized as a substrate for cholesterol synthesis.

Third, intestinal absorption of DAG may be less effective than that of TAG under our experimental conditions because a large amount (200 mg) of DAG or TAG was given as a bolus. Therefore, lymphatic absorption of dietary DAG might be the same as that of TAG under ad libitum feeding conditions. In that case, it is not likely that fatty acids from DAG are more effectively ß-oxidized in intestinal cells.

Proposed Mechanisms of the Antiobesity Effect of Diacylglycerol

It was reported that dietary DAG had antiobesity activity and prevented postprandial hypertriacylglycerolemia in experimental animals and humans (Watanabe et al., 1997; Murase et al., 2002; Murase et al., 2001; Taguchi et al., 2000; Nagao et al., 2000). Some mechanisms have been suggested for the antiobesity activity of DAG. Murata et al. (Murata et al., 1997) and Murase et al. (Murase et al., 2001) showed in rats and mice that hepatic ß-oxidation may be stimulated by the feeding of DAG compared with TAG. They observed that the feeding of TAG reduced hepatic triglyceride concentration in rats and mice. Recently, Murase et al. (Murase et al., 2002) showed that intestinal ß-oxidation was enhanced after the feeding of DAG in mice. Watanabe et al. (Watanabe et al., 1997) showed that postprandial oxygen consumption was increased in rats fed DAG for 7 d compared with TAG. These observations suggest that dietary DAG may activate hepatic and intestinal ß-oxidation and increase energy expenditure.

We propose another mechanism for the antiobesity activity of DAG. As described, the lymphatic transport of chylomicron triglyceride in rats given DAG is delayed (Yanagita et al., 2004). This may be the reason why DAG suppressed postprandial hypertriacylglycerolemia in rats and humans (Watanabe et al., 1997; Taguchi et al., 2000). We suggest that delayed absorption of DAG, compared with TAG, may be the important determinant in preventing body fat accumulation. After a meal, the increase in blood glucose is followed by insulin secretion. Insulin activates peripheral lipoprotein lipase, which hydrolyzes chylomicron triglycerides; a major part of the free fatty acids formed is incorporated into peripheral tissues, in particular, adipocytes. Although a part of the fatty acids is incorporated into muscle cells and utilized as energy, glucose, rather than fatty acids, is preferentially utilized as an energy source in muscle cells. During the period of high insulin levels in blood after a meal, the amount of chylomicron triglycerides entering the bloodstream after DAG consumption is less than that for TAG because lymphatic transport of triglyceride is delayed after the feeding of DAG (Fig. 6.2). With the reduction in the concentration of blood insulin, the deposition of fatty acids from chylomicron triglycerides decreases and their utilization as energy increases in the body. Therefore, fatty acids from chylomicron triglycerides after DAG feeding may be utilized preferentially for fatty acid oxidation rather than for body fat accumulation. It seems that a balance between fat deposition and energy consumption shifts in favor of energy consumption in the feeding of DAG vs. TAG. The difference may be extremely small in one meal; however, the reduction in body fat deposition in the feeding of DAG, compared with that of TAG, is caused by the long-term accumulation of slightly increased energy consumption over lipogenesis.

Some studies have pointed out that slower absorption of dietary fat prevents body fat accumulation in mice (Han et al., 1999a; Han et al., 1999b). It has been also reported that the nibbling meal pattern, compared with the gorging meal pattern, suppresses body fat accumulation in humans and experimental animals (Fabry et al., 1964; Bellisle et al., 1997; Baker & Huebotter, 1973). Fabry et al. (Fabry et al., 1964) demonstrated that mean skinfold thickness was inversely related to meal frequency in Czechoslovakian men aged 60-64 y. Baker and Huebotter (Baker & Huebotter, 1973) showed that lipogenesis was activated in mice restricted to a single large meal once a day rather than small meals twice a day. When larger amounts of glucose and triglycerides enter the bloodstream in the gorging meal pattern, compared with the nibbling meal pattern, insulin secretion is greater and therefore, peripheral lipoprotein lipase increases. The delayed absorption of the fatty acids from dietary DAG seems to resemble the lower flow of triglyceride into the bloodstream observed in the nibbling feeding pattern.

However, several studies also showed that total daily energy expenditure was not changed by meal patterns, gorging vs. nibbling, as reviewed by Bellisle et al. (Bellisle et al., 1997). Those authors concluded that body mass reduction caused by the nibbling meal patterns might be an artifact. The difference in energy expenditure between the nibbling and gorging feeding patterns may be small when measured over 1 d, making it difficult to observe significant differences. If it were possible to measure energy expenditure for a long-term period, e.g., 1 wk or 1 mo, a significant difference in energy expenditure between the nibbling and gorging feeding patterns might be observed.

We proposed here a new mechanism to explain the antiobesity activity of DAG. However, the relation between the acceleration of ß-oxidation by DAG feeding shown by Murata et al. (Murata et al., 1997) and Murase et al. (Murase et al., 2001) and our proposed mechanism is not clear. Researches are now working to show that delayed absorption of fatty acids from dietary DAG causes the suppression of body fat accumulation.

References

Baker, N., and Huebotter, R.J. Lipogenic Activation After Nibbling and Gorging in Mice. *J. Lipid Res.* **1973,** *14,* 87-4.

Bellisle, F., McDevitt, R., and Prentice, A.M. Meal Frequency and Energy Balance. *Br. J. Nutr.* **1997,** *77 (Suppl. 1),* S57-S70.

Fabry, P., Hejl, Z., Fodor, J., Braun, T. and Zvolankova, K. The Frequency of Meals: Its Relation to Overweight, Hypercholesterolaemia, and Decreased Glucose-Tolerance. *Lancet ii,* **1964,** *614*-15.

Friedman, H.I., and Nylund, B. Intestinal Fat Digestion, Absorption, and Transport. *Am. J. Clin. Nutr.* **1980,** *33,* 1108-139.

Han, L.-K., Takaku, T., Li, J., Kimura, Y. and Okuda, H. Anti-Obesity Action of Oolong Tea. *Int. J. Obes.* **1999a,** *23,* 98-05.

Han, L.-K., Kimura, Y., and Okuda, H. Reduction in Fat Storage During Chitin-Chitosan Treatment in Mice Fed a High-Fat Diet. *Int. J. Obes.* **1999b,** *23,* 174-79.

Ikeda, I., Yoshida, H., and K. Imaizumi, Effects of Trioelin or Oleic Acid on Lymphatic Recovery of Docosahexaenoic Acid Given as Ethyl Ester and Their Intramolecular Distribution in Lymph Triglyceride of Rats. *Lipids* **1997,** *32,* 949-52.

Kiyasu, J.Y., Bloom, B., and Chaikoff, I.L. The Portal Transport of Absorbed Fatty Acids. *J. Biol. Chem.* **1952,** *199,* 415-19.

Kondo, H., Hase, T., Murase, T., and Tokimitsu, I. Digestion and Assimilation Features of Dietary DAG in the Rat Small Intestine. *Lipids* **2003,** *38,* 25-0.

Murase, T., Mizuno, T., Omachi, T., Onizawa, K., Komine, Y., Kondo, H., Hase, T., and Tokimitsu, I. Dietary Diacylglycerol Suppresses High Fat and High Sucrose Diet-Induced Body Fat Accumulation in C57BL/6J Mice. *J. Lipid Res.* **2001,** *42.* 372-78.

Murase, T., Aoki, M., Wakisaka, T., Hase, T., and Tokimitsu, I. Anti-Obesity Effect of Dietary Diacylglycerol in C57BL/6J Mice: Dietary Diacylglycerol Stimulates Intestinal Lipid Metabolism. *J. Lipid Res.* **2002,** *43,* 1312-319.

Murata, M., Ide, T., and Hara, K. Reciprocal Responses of Dietary Diacylglycerol of Hepatic Enzymes of Fatty Acid Synthesis and Oxidation in the Rat. *Br. J. Nutr.* **1997,** *77,* 107-21.

Nagao, T., Watanabe, H., Goto, N., Onizawa, K., Taguchi, H., Matsuo, N., Yasukawa, T., Tsushima, R., Shimasaki, H., and Itakura, H. Dietary Diacylglycerol Suppresses Accumulation of Body Fat Compared to Triacylglycerol in Men in a Double-Blind Controlled Trial. *J. Nutr.* **2000,** *130,* 792-97.

Taguchi, H., Watanabe, H., Onizawa, K., Nagao, T., Gotoh, N., Yasukawa, T., Tsushima, R., Shimasaki, H., and Itakura, H. Double-Blind Controlled Study on the Effects of Dietary

Diacylglycerol on Postprandial Serum and Chylomicron Triacylglycerol Responses in Healthy Humans. *J. Am. Coll. Nutr.* **2000,** *19,* 789-96.

Taguchi, H., Nagao, T., Watanabe, H., Onizawa, K., Matsuo N., Tokimitsu, I., and Itakura, H. Energy Value and Digestibility of Dietary Oil Containing Mainly 1,3 Diacylglycerol Are Similar to Those of Triacylglycerol. *Lipids* **2001,** *36,* 379-82.

Watanabe, H., Onizawa, K., Taguchi, H., Kobori, M., Chiba, H., Naito, S., Matsuo, N., Yasukawa, T., Hattori, M., and Shimasaki, H. Nutritional Characterization of Diacylglycerols in Rats (in Japanese). *J. Jpn. Oil Chem. Soc.* **1997,** *46,* 301-07.

Yanagita, T., Ikeda, I., Wang, Y.M., Nakagiri, H. Comparison of the lymphatic transport of radio-labeled 1,3-dioleoylglycerol and trioleoylglycerol in rats. *Lipids* **2004,** *39,* 827-32. Erratum in: *Lipids.* **2005,** *40,* 533.

Chapter 7

Suppressive Effects of Diacylglycerol Oil on Postprandial Serum Triglyceride Elevation in Animals

Naohiro Gotoh[a] and Hiroyuki Shimasaki[b]

[a]Department of Food Science and Technology, Tokyo University of Marine Science and Technology, Minato-Ku, Tokyo 108-8477, Japan and [b]Department of Health and Nutrition, University of Human Arts and Sciences, Saitama, 339-8539, Japan

Introduction

Diacylglycerol (DAG) is a component of edible oils and fats that has been shown to have nutritional functions distinct from ordinary triacylglycerol (TAG) oil. Studies conducted to date suggest that the primary effect of DAG oil is on the plasma triglyceride concentrations after ingestion of DAG oil, compared with TAG oil. In 1993, it was reported that dietary DAG oil has the ability to reduce serum triglyceride concentrations in rats that consumed an ad libitum diet containing 10% test oil (Hara et al., 1993). Serum triglyceride levels in rats were decreased with increasing amounts of DAG oil at the expense of TAG in the test oil. This phenomenon has also been observed in pigs (Kristensen et al., 2006), mice (Yasunaga et al., 2007), and dogs (Bauer et al., 2006). However, the mechanism of this phenomenon has not yet been clarified. In this chapter, we summarize animal studies focusing on the mechanism of the suppressive effects of DAG oil on postprandial blood TAG elevation. In addition, some preliminary results on differences in the clearance of lipoprotein formed after DAG oil vs. TAG oil ingestion are discussed.

Reduced Chylomicron Secretion by DAG Oil Ingestion

In 1994, Murata and co-workers (Murata et al., 1994) reported that DAG oil in the diet reduced postprandial serum TAG levels in rat lymphatic fluid. These findings indicated that DAG oil was capable of decreasing serum TAG levels in rats, not only after chronic administration, but also postprandially. In this study, DAG and TAG oils were prepared with almost the same fatty acid composition. The test oil, consisting of 1, 3-DAG:1, 2-DAG (65.2:32.6), and TAG oil (control oil) were emulsified and infused directly into the stomach of anesthetized rats via a polymer tube. Chylomicrons in rat lymph were collected at 1-hour intervals up to 5 hours after the infusion, and the triglyceride and cholesterol levels in the lymph were measured. The transport rate and the cumulative amounts of TAG in the lymph of rats caused by DAG oil infusion were significantly lower than those caused by TAG oil (Fig. 7.1). Similar results were obtained for cholesterol levels in the lymph (Fig. 7.2). These results indicate that the amount of chylomicrons excreted from the intestinal mucosal cells to the lymph was significantly lower in rats that had been fed dietary DAG oil than in rats that were fed TAG oil.

To explain these results, Murata and co-workers (Murata et al., 1994) considered a pathway in which triglyceride is resynthesized in intestinal mucosal cells, with a focus on the structural differences between the two types of oil. TAG oil is generally hydrolyzed by pancreatic lipase to free fatty acids (FFA) and 2-monoacylglycerol in the small intestine lumen (Fig. 7.3A). These form a micelle together with bile salts and other lipids, such as cholesterol and phospholipids, and are absorbed by intestinal cells. In intestinal cells, triglyceride is resynthesized from 2-monoacylglycerol and two FFA (Fig. 7.4A), and combined with a protein to form a chylomicron, which is released from the intestinal cell into the lymphatic system by fusion of the secretory vacuole with the cell membrane. In the case of DAG oil, the metabolic pathways in the intestinal cells are considerably different from those of TAG oil. Dietary DAG oil is mainly in the form of 1, 3-DAG. When it is hydrolyzed by pancreatic lipase in the lumen of the small intestine, 2-monoacylglycerol is not generated during digestion because no fatty acid is present at the 2-position of the 1, 3-DAG (Fig. 7.3B). The dominant structures of DAG and monoacylglycerol are 1, 3-DAG and 1-monoacylglycerol, respectively. Consequently, it can be assumed that no conversion occurs from 1, 3-DAG to 1, 2-DAG and from 1-monoacylglycerol to 2-monoacylglycerol during digestion in the small intestine lumen. 1, 3-DAG would be hydrolyzed initially to 1-monoacylglycerol and then, at least in part, to glycerol and FFA. These products are absorbed into the intestinal mucosa cells. However, triglyceride cannot be synthesized from 1-monoacylglycerol in the intestinal cells because it would be a slow process in comparison with the 2-monoacylglycerol pathway (Fig. 7.4B). In one experiment, triglyceride could be synthesized via the glycerol-3-phosphate pathway. These findings suggest that the effect of DAG oil in

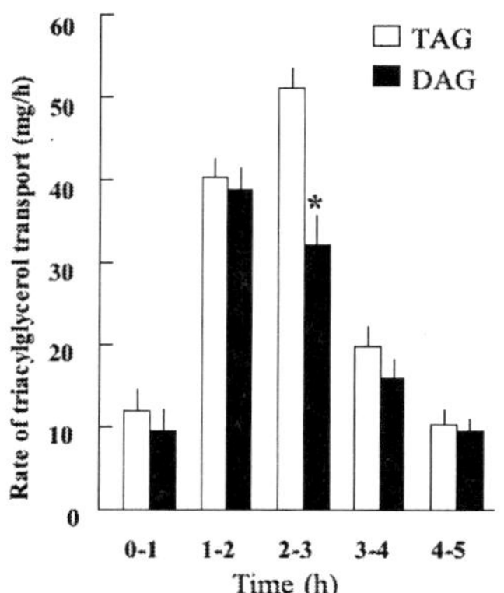

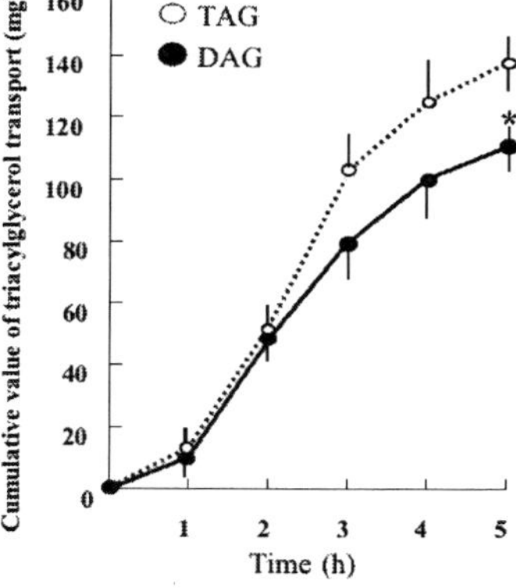

FIG. 7.1. Rate (A) and cumulative values (B) of triglyceride transport by chylomicrons in thoracic duct lymph of rats infused with emulsions containing TAG oil or DAG oil (Murata et al., 1994). Each value represents the mean ± SE, n = 6. *Different from the corresponding values for rats infused with the TAG emulsion, $P < 0.05$.

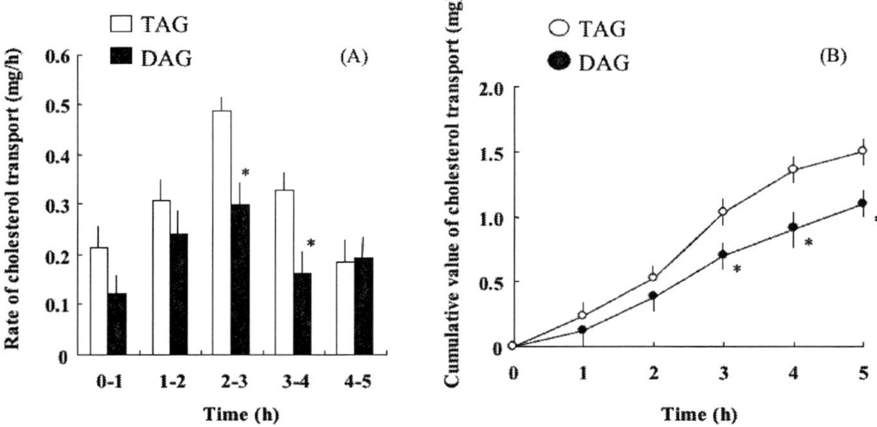

FIG. 7.2. Rate (A) and cumulative values (B) of cholesterol transport by chylomicrons in thoracic duct lymph of rats infused with emulsions containing TAG oil or DAG oil (Murata et al., 1994). Each value represents the mean ± SE, n = 6. *Different from the corresponding values for rats infused with the TAG emulsion, $P < 0.05$.

decreasing serum triglyceride levels in rats is to reduce postprandial triglyceride concentration in the lymph due to a different metabolic pathway of DAG oil compared with TAG oil.

The same phenomenon was also observed and the same mechanism was proposed

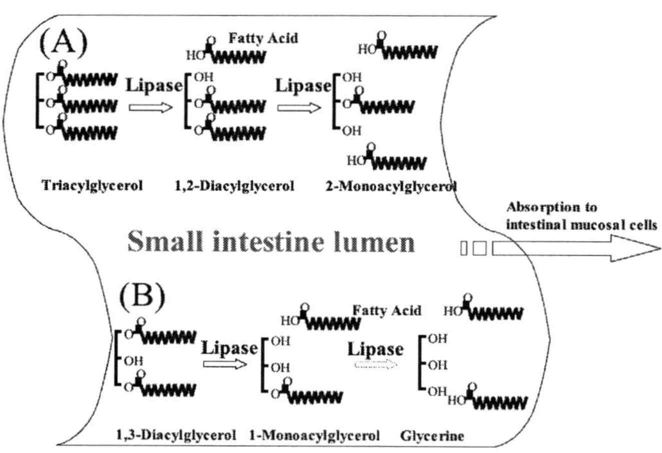

FIG. 7.3. Digestion pathway of (A) TAG and (B) 1,3-DAG in the small intestine.

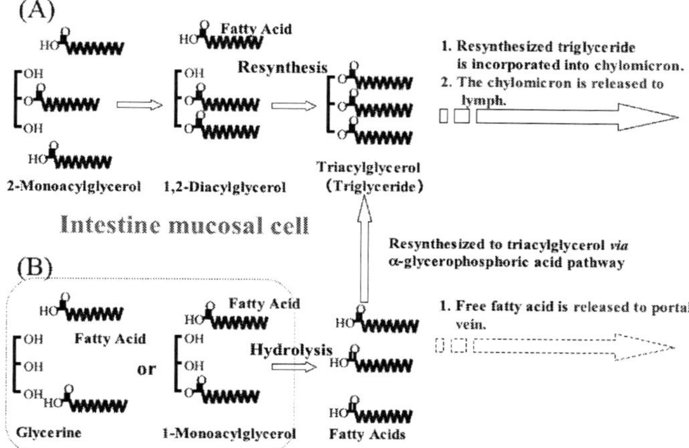

FIG. 7.4. Metabolism of hydrolysates formed from (A) TAG and (B) 1,3-DAG in intestinal mucosal cells.

by Yanagita and co-workers in rats (Yanagita et al., 2004). They compared lymphatic transport of ^{14}C-labeled oleic acid, by measuring the hourly totals of recovered radioactivity incorporated into several kinds of glycerides for each of the first 8 hours after direct infusion of ^{14}C-labeled DAG or TAG into the stomach. Almost all of the ^{14}C-labeled oleic acid was found in resynthesized TAG, and the recovery of the ^{14}C-labeled oleic acid after ^{14}C-labeled DAG infusion was almost half of that after ^{14}C-labeled TAG infusion ($P < 0.05$) in the first hour (Fig. 7.5). However, no significant difference between the two groups in the amount of ^{14}C-labeled oleic acid incorporated into resynthesized TAG was detected after that. Based on these findings, they also hypothesized that the slowing of resynthesis of TAG from infused DAG could be a key cause of postprandial serum TAG suppression. Their results also supported the mechanism proposed by Murata and co-workers, that resynthesis of TAG occurs in the intestinal mucosa cells via the glycerol-3-phosphate pathway after intake of DAG.

Watanabe and co-workers (Watanabe et al., 1997) also found similar results in rats that had been fed DAG. However, they proposed a different mechanism for the postprandial suppression of serum TAG levels. They administered diolein or triolein to rats in the form of emulsions and observed changes in TAG and FFA concentrations in the jugular and portal veins. A significantly lower level of TAG was found in the jugular veins of rats that had been fed the diolein test oil compared with the triolein. This observation was consistent with the previous study (Murata et al., 1994).

On the other hand, the concentration of FFA in the portal vein increased rapidly 30 min after the administration of diolein. In contrast, no significant change in FFA levels was observed in the control rats that had been fed triolein (Fig. 7.5). This finding sug-

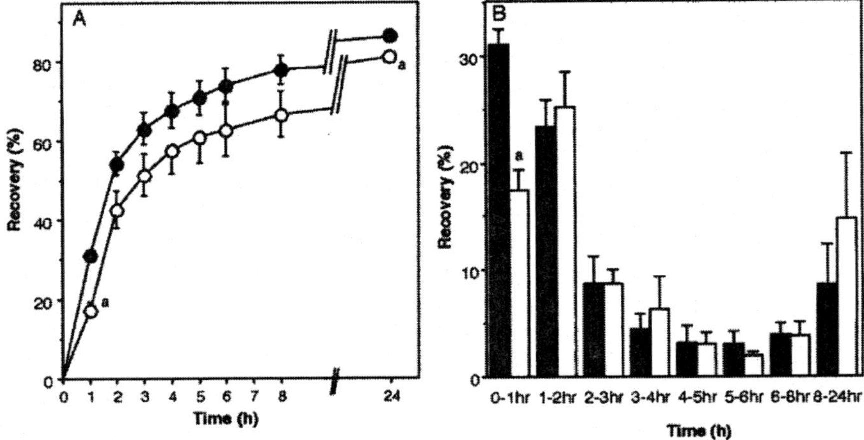

FIG. 7.5. Cumultative recovery (A) and secretion rate for each interval (B) for ^{14}C-radioactivity in the thoracic duct lymph of rats infused with an emulsion containing 1,3-di[^{14}C]oleoylglycerol or tri[^{14}C]oleoylglycerol. Values are expressed as the percent recovery of administrated labeled dioleoylglycerol or trioleoylglycerol. Closed circles and bars denote the trioleoylglycerol group, and open circles and open bars, the 1,3-dioleoylglycerol group. Each value represents the mean ± SE for seven rats. Value with a superscript "a" are significantly different from the corresponding values for the rats infused with ^{14}C-labeled trioleoylglycerol at $P < 0.05$. Source: Yanagita et al., 2004

gested that the FFA resulting from the digestion of 1, 3-DAG share a route of absorption via the hepatic portal vein.

On the other hand, the concentration of FFA in the portal vein increased rapidly 30 min after the administration of diolein. In contrast, no significant change in FFA levels was observed in the control rats that had been fed triolein. This finding suggested that the FFA resulting from the digestion of 1, 3-DAG share a route of absorption via the hepatic portal vein. As a consequence, Watanabe and co-workers concluded that the fatty acid derived from 1-monoacylglycerol by digestion is released directly from intestinal mucosal cells to the portal vein rather than being esterified to TAG via the α-glycerophosphoric acid pathway to reduce the serum TAG level (Fig. 7.4).

Kristensen and co-workers also observed FFA transportation in the portal vein and plasma TAG level in the mesenteric artery in pigs (Kristensen et al., 2006). However, they did not find any increase in the FFA level in the portal vein after intake of DAG, compared with that after TAG intake. Furthermore, significant suppressive effects of DAG (compared with TAG) on plasma TAG level were not found in the first 4 hours, but were found at 5 and 6 hours after intake of feed containing test oils. These results do not support the two kinds of proposed mechanisms mentioned above. The difference in animal species might be the main cause of this finding. Yasunaga and co-workers also carried out the same kind of study with mice (Yasunaga et al., 2007). They infused DAG and

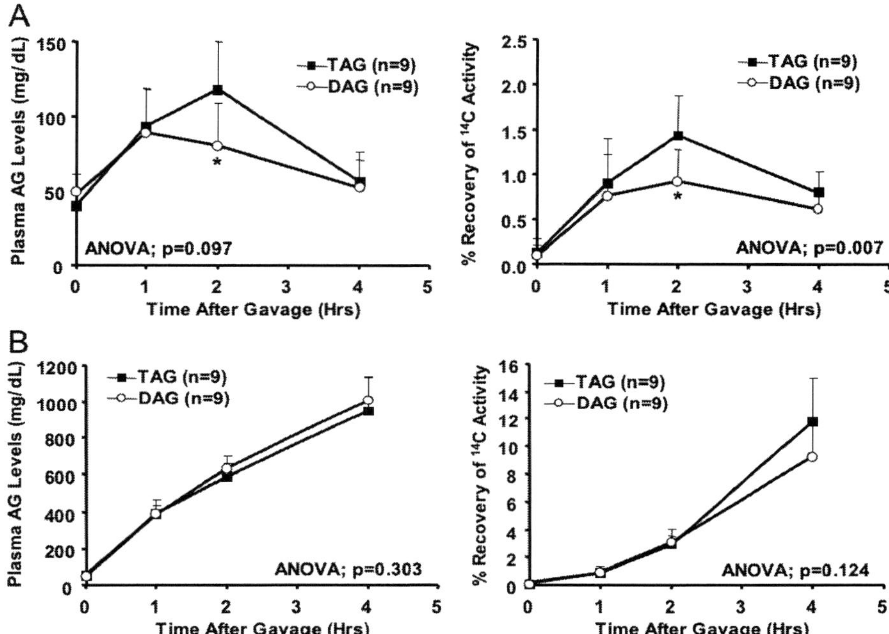

FIG. 7.6. Plasma levels of acylglycerols (AG) after gavage of radiolabeled DAG and TAG emulsions with and without Triton WR1339. A: Plasma levels of acylglycerols (left panel; means ± SD, mg/dL) and ^{14}C-radiolabeled plasma lipids (right panel; as means ± SD, percent of radioactivity recovered) over 4 hr after gavage of either TAG or DAG ^{14}C-radiolabeled emulsions without concomitant Triton WR1339 injection. Nine mice in each group. B: Plasma levels of acylglycerols and ^{14}C-radiolabeled plasma lipids over 4 hours after administration of either TAG or DAG emulsion with concomitant Triton WR1339 injection. Data represent means ± SD, and the asterisks denote statistically significant differences between TAG and DAG groups at each individual time point ($P < 0.05$). Repeated two-way ANOVA was performed on the overall data set, and *P* values are indicated. Source: Yasunaga et al., 2007.

TAG consisting of ^{14}C-labeled oleic acid into mice to follow the administered fatty acids in blood. They found that significantly higher amounts of ^{14}C-labeled fatty acid were distributed in 1, 3-DAG ($P < 0.001$) and monoacylglycerol ($P < 0.01$) and a lower amount was detected in TAG ($P < 0.01$) in plasma lipids after gavage of ^{14}C-labeled DAG, compared with that of ^{14}C-labeled TAG. These results were consistent with those obtained by Yanagita and co-workers (Yanagita et al., 2004). In addition, acylglycerol levels in the plasma of mice that were previously treated with or without Triton WR1339, a blocker against the catabolism of lipoproteins by lipoprotein lipase (LPL) in plasma, were compared between ^{14}C-labeled DAG and TAG infused groups. The acylglycerol levels in the plasma of ^{14}C-labeled DAG-administered mice were significantly suppressed compared with those of ^{14}C-labeled TAG-administered mice in the groups not treated with Triton WR1339 (Fig. 7.6A). On the contrary, no significant difference on the acylglycerol levels in plasma was observed between ^{14}C-labeled DAG and TAG-administered mice in the groups treated with Triton WR1339 (Fig. 7.6B). These results suggest that clearance of chylomicron formed after the DAG infusion by LPL was more rapid than that after the TAG administration. Based on these results, they proposed that the suppression of the postprandial TAG level in plasma by DAG oil could be due to the more efficient lipolysis of chylomicron formed after DAG intake, compared with that after TAG intake.

Lipoprotein Subclass Analysis After the Ingestion of DAG Oil in Rats

In spite of the fact that the study of lipoprotein metabolism after the ingestion of DAG oil is one of the important issues in mechanistic studies of DAG functions, few reports exist on the detailed analysis of these lipoproteins. Nakagiri and co-workers (Nakagiri et al., 2002) reported the results of experiments in which variations of triglyceride in lipoproteins collected from the peripheral blood and from the lymph were examined in detail. They used high-performance liquid chromatography (HPLC) based on the gel filtration method established by Okazaki et al. (Okazaki et al., 2000). In their experiment, emulsions containing test oils (6.67 g/kg body weight) were administered to SD male rats (n = 10) after 14 hours of food deprivation. Blood samples collected before and at 1, 2, 3, 4, and 6 hours after the loading were analyzed by HPLC. Lymphatic fluid was collected simultaneously during the first 3 hours and analyzed similarly. Increases in serum triglyceride, chylomicron-triglyceride, and VLDL-triglyceride after DAG loading were significantly suppressed compared with TAG loading. The suppression of triglyceride secretion was also observed in the lymphatic fluid. Interestingly, the triglyceride content in the chylomicron fraction obtained from serum in the DAG loading group was significantly smaller than that in the TAG loading group, whereas this difference was not observed in the lymphatic fluid. This indicates that the catabolism of lipoproteins in the blood circulation might be faster for the lipoproteins generated after DAG oil intake than after TAG oil intake. These results suggest that the subclass of chylomicron (or chylomicron-remnant) after DAG ingestion might differ from that after TAG ingestion in its structure or composition, which was not detected in the lymphatic fluid by the method employed. More experiments would help confirm this phenomenon regarding the rate of catabolism of lipoproteins formed after DAG ingestion.

Summary

Lipid levels in the lymph and serum after the administration of DAG oil to animals are lower than those when TAG oil is administered. Mainly three kinds of mechanism on this phenomenon have been proposed already; however, the detailed mechanism is still unclear. Further research will be required to elucidate this problem.

References

Bauer, J.E., Nagaoka, D., Porterpan, B., Bigley, K., Umeda, T., and Otsuji, K. Postprandial Lipolytic Activities, Lipids, and Carbohydrate Metabolism Are Altered in Dogs Fed Diacylglycerol Meals Containing High- and Low-Glycemic-Index Starches. *J. Nutr.* **2006**, *136*, 1955S-1957S.

Hara, K., Onizawa, K., Honda, H., Otsuji, K., Ide, T., and Murata, M. Dietary Diacylglycerol-Dependent Reduction in Serum Triaclyglycerol Concentration in Rats. *Ann. Nutr. Metab.* **1993**, *37*, 185-191.

Kristensen, J.B., Jorgensen, H., and Mu, H. Diacylglycerol Oil Does Not Affect Portal Vein Transport of Nonesterified Fatty Acids but Decreases the Postprandial Plasma Lipid Response in Catheterized Pigs. *J. Nutr.* **2006**, *136*, 1800-1805.

Murata, M., Hara, K., and Ide, T. Alternation by Diacylglycerols of the Transport and Fatty Acid Composition of Lymph Chylomicrons in Rats. *Biosci. Biotech. Biochem.* **1994**, *58*, 1416-1419.

Nakagiri, H., Watanabe, H., Otsuka, A., Hase, T., Tokimitsu, I.. Okazaki, M., and Tada, N. Dynamics of Lipoprotein Profiles in Serum After Loading of Diacylglycerols in Rats, The 18th International Congress of Clinical Chemistry and Laboratory Medicine. 2002, October 20-25, Kyoto.

Okazaki, M., Usui, S., Hosaki, S. in *Handbook of Lipoprotein Testing, 2nd edn.* 2000, pp. 647-669, American Association of Clinical Chemistry Press, Washington.

Watanabe, H., Onizawa, K., Taguchi, H., Kobori, M., Chiba, H., Naito, S., Matsuo, N., Yasukawa, T., Hattori, M., and Shimasaki H. Nutritional Characterization of Diacylglycerols in Rats. *J. Jpn. Oil Chem. Sci.* **1997**, *46*, 301-307.

Yanagita, T., Ikeda, I., Wang, Y.M., and Nagagiri, H. Comparison of the Lymphatic Transport of Radiolabeled 1,3-Dioleoylglycerol and Trioleoylglycerol in Rats. *Lipids* **2004**, *39*, 827-832.

Yasunaga, K., Saito, S., Zhang, Y-L., Hernandez-Ono, A., and Ginsberg, H.N. Effects of triacylglycerol and Diacylglycerol Oils on Blood Clearance, Tissue Uptake, and Hepatic Apolipoprotein B Secretion in Mice. *J. Lipid Res.* **2007**, *48*, 1108-1121.

Chapter 8

Effects of Diacylglycerol Oil on Postprandial Increase in Serum Triglyceride and Remnant Lipoproteins in Humans

Norio Tada

Division of General Medicine, Department of Internal Medicine, Kashiwa Hospital, The Jikei University School of Medicine, Chiba, 277-8567 Japan

Introduction

In healthy humans, ingested fats are digested and almost completely absorbed (Carey et al., 1983), entering the blood circulation largely in the form of chylomicrons. Serum triglyceride levels thus increase after the ingestion of a meal containing oil and fat. Serum triglyceride levels decrease as triglyceride in the form of chylomicrons is cleared from the circulation by enzymatic hydrolysis (Goldberg, 1996) and receptor-mediated uptake, mainly by hepatocytes (Eriksson et al., 1991). It takes more than 8 h for the increased serum triglyceride from ingestion of fatty meals to return to preprandial levels. Therefore, humans typically spend a great deal of time in a postprandial hypertriglyceridemic state. Because chylomicron remnants, metabolites of chylomicron in the circulation, are implicated in atherogenic lipoproteins (Zilversmit, 1979; Patsch, 1987; Gianturco & Bradley,1991; Patsch et al., 1992; Cohn, 1998), the suppression of the postprandial increase in chylomicron remnants will lead to a reduction in CAD risks.

Factors that influence the magnitude of postprandial lipemia have been investigated in humans (Cohen et al., 1988; Cohn et al., 1988; Jeppesen et al., 1995; Dubois et al., 1998). Food containing diacylglycerol (DAG) oil was reported to reduce fasting serum triglyceride levels, to suppress postprandial increases in serum triglyceride levels, and to prevent the accumulation of body fat in animals and humans (Hara et al., 1993; Murase et al., 2001; Taguchi et al., 2000; Tada et al., 2001; Nagao et al., 2000). The therapeutic application of DAG oil has been discussed elsewhere (Tada, 2004; Yanai et al., 2007). In this chapter, the results of studies on the effects of DAG on the postprandial response of serum lipids and lipoproteins under various conditions are summarized.

Effects of DAG on Serum Lipids Studied in Humans

Animal experiments demonstrated that the continuous feeding of DAG oil containing mainly 1,3-DAG led to a decrease in serum triglyceride levels in a dose-dependent manner (Hara et al., 1993; Murata et al., 1997). These authors showed that DAG oil decreased fasting serum triglyceride levels in rats after 17 d of feeding compared with feeding a triacylglycerol oil (TAG oil) with a similar fatty acid composition. Experiments with lymph fistula-rats revealed that the assembly of chylomicrons was retarded and the lymphatic secretions of chylomicron-cholesterol and chylomicron-triglyceride were significantly decreased in DAG-infused rats compared with TAG-infused rats (Murata et al., 1994).

In agreement with these results, Taguchi et al. (Taguchi et al., 2000) conducted a human study in which 40 normolipidemic men were enrolled and were fed fat emulsions containing either DAG oil or TAG oil at different doses: 10 g (n = 13), 20 g (n = 10), and 44 g (n = 17). The characteristics of the subjects are shown in Table 8.1. The subjects ingested each amount (44, 20, and 10g) of two test emulsified oils differing only in the type of fat (either DAG or TAG oil) with an interval of 7 d. The emulsion contained 10 g/100 g of fat and 13 g/100 g of reduced starch hydrolysate and additives for stabilization and flavoring.

The postprandial serum triglyceride response increased with the intake of each fat emulsion (Fig. 8.1). At 6 h after the ingestion of 44 g of fat, serum triglyceride concentrations in the DAG group were significantly lower than those in the TAG group (Fig. 8.1A). The extent of postprandial lipemia induced by both oils at a dose of 44 g was compared by calculating the area bounded by the curve above the fasting levels (AUC). The postprandial lipemia (mean ± SD) induced by the DAG emulsion [6.54 ± 5.12 (mmol·h)/L] was significantly ($P < 0.05$) lower (23%) than that induced by the TAG emulsion [8.45 ± 7.54 (mmol·h)/L]. Similar effects were reported to be detectable with smaller doses (Fig. 8.1B, C).

Taguchi et al. (Taguchi et al., 2000) found almost no differences in serum cholesterol concentrations between the DAG and TAG emulsions. Serum phospholipid levels were significantly increased after fat ingestion at all doses, and the levels in the DAG group were significantly lower than those in the TAG group. Although serum nonesterified fatty acid levels tended to decrease in both groups, they returned to near baseline levels 4–8 h postprandially, and no difference was observed between the DAG and TAG emulsion at any of the doses used. Postprandial changes in plasma glucose and serum insulin levels also did not differ between the groups.

For a detailed analysis of the effects of DAG on postprandial lipemia in humans, they separated and analyzed lipoproteins after the ingestion of 20 g fat. Changes in plasma chylomicron lipid levels after the ingestion of 20 g fat are shown in Figure 8.2.

TABLE 8.1
Characteristics of the Subjects[a]

	44-g fat group	20-g fat group	10-g fat group
n	17	10	13
Age (y)	33 ± 8	36 ± 6	34 ± 4
Height (m)	1.72 ± 0.08	1.69 ± 0.06	1.71 ± 0.04
Weight (kg)	70.6 ± 13.2	67.9 ± 10.8	66.3 ± 9.4
Body mass index (kg/m^2)	23.8 ± 3.3	23.6 ± 3.5	22.6 ± 2.2
Fasting serum triglyceride (mmol/L)	1.12 ± 0.53	1.17 ± 0.44	0.94 ± 0.40
Fasting serum cholesterol (mmol/L)	4.73 ± 0.87	4.87 ± 0.79	4.60 ± 0.69

[a]Values are means ± SD. Source: Taguchi et al., 2000.

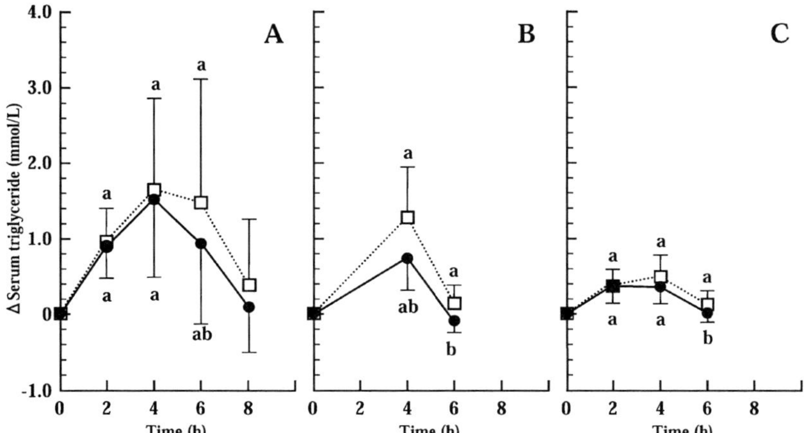

FIG. 8.1. Mean (± SD) changes (△) from baseline in serum triglyceride levels after the ingestion of DAG (●) or TAG emulsion (□). (A) 44-g fat ingestion (n = 17); (B) 20-g fat ingestion (n = 10); (C) 10-g fat ingestion (n = 13); [a]different from fasting levels, $P < 0.05$ (Student's t-test for paired values); [b]different from the corresponding values for the TAG emulsion; $P < 0.05$ (Student's t-test for paired values). Source: Taguchi et al., 2000.

The increments in the postprandial concentrations of chylomicron-triglyceride, -cholesterol and -phospholipid after DAG ingestion were smaller than those after TAG ingestion. The difference in the levels of chylomicron-triglyceride at 4 h was significant (Fig. 8.2A).

Fasting and the postprandial lipid levels of plasma VLDL, LDL, and HDL after ingestion of the fat emulsions are also reported (Table 8.2). Plasma VLDL triglyceride concentrations at 4 and 6 h in the DAG group were significantly lower than those in the

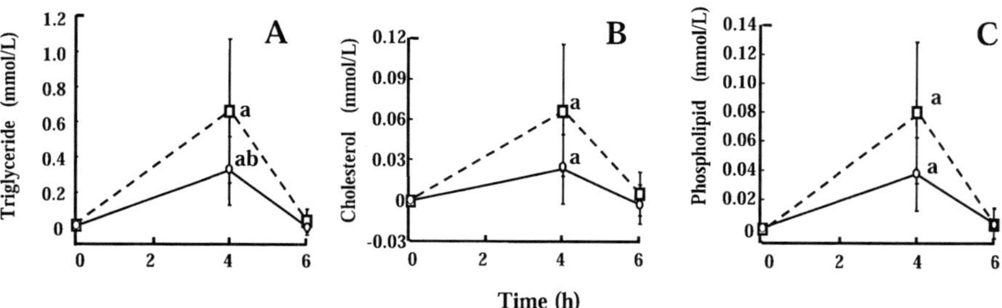

FIG. 8.2. Mean (± SD) changes (△) from baseline in plasma chylomicron lipid concentrations after ingestion of DAG emulsion (○) or TAG emulsion (□) at a dose of 20 g fat (n = 10). (A) Chylomicron triglyceride concentrations; (B) Chylomicron cholesterol concentrations; (C) Chylomicron phospholipid concentrations; [a]different from fasting levels, $P < 0.05$ (Student's t-test for paired values); [b]different from the corresponding values for the TAG emulsion, $P < 0.05$ (Student's t-test for paired values). Source: Taguchi et al., 2000.

TABLE 8.2
Fasting and Postprandial (4 and 6 h) Concentrations of Plasma Lipoprotein Lipid After Ingestion of 20-g Fat Emulsions[a]

Lipoprotein lipid		Total cholesterol		Triglyceride		Phospholipid	
Fat emulsion		DAG	TAG	DAG	TAG	DAG	TAG
				(mmol/L)			
VLDL							
Fasting		0.35 ± 0.22	0.36 ± 0.19	0.49 ± 0.32	0.65 ± 0.43	0.22 ± 0.14	0.24 ± 0.12
Postprandial	4 h	0.45 ± 0.25[b]	0.51 ± 0.21[b]	0.78 ± 0.44[bc]	1.01 ± 0.53[b]	0.32 ± 0.21[b]	0.35 ± 0.15[b]
	6 h	0.35 ± 0.23	0.41 ± 0.20	0.42 ± 0.27[c]	0.70 ± 0.44	0.21 ± 0.18	0.25 ± 0.13
LDL							
Fasting		2.88 ± 0.68	3.08 ± 0.75	0.24 ± 0.07	0.26 ± 0.05	0.79 ± 0.17	0.89 ± 0.18
Postprandial	4 h	2.86 ± 0.70	3.03 ± 0.80	0.24 ± 0.05[c]	0.27 ± 0.05	0.79 ± 0.19	0.88 ± 0.24
	6 h	2.90 ± 0.59	3.10 ± 0.75	0.24 ± 0.05	0.25 ± 0.04	0.78 ± 0.15	0.89 ± 0.18
HDL							
Fasting		1.46 ± 0.26	1.31 ± 0.17	0.10 ± 0.03	0.09 ± 0.02	1.25 ± 0.20	1.17 ± 0.21
Postprandial	4 h	1.43 ± 0.18[c]	1.26 ± 0.19	0.10 ± 0.02	0.11 ± 0.02[b]	1.26 ± 0.21	1.16 ± 0.24[c]
	6 h	1.50 ± 0.15[c]	1.34 ± 0.19	0.10 ± 0.03	0.10 ± 0.02	1.26 ± 0.21	1.18 ± 0.23

[a]Values are means ± SD, n = 10. Source: Taguchi et al., 2000.
[b]Different from fasting levels, $P < 0.05$ (paired *t*-test).
[c]Different from the TAG-based emulsion, $P < 0.05$ (paired *t*-test).

TAG group. Changes in LDL and HDL lipid levels after the ingestion of the fat emulsion were either small or not observed at all. There were no marked changes in HDL lipid levels after fat ingestion, but there was a tendency for HDL cholesterol levels in the DAG group to be slightly higher than those in the TAG group. Thus, the postprandial increments in serum triglyceride and chylomicron-triglyceride after ingestion of 10–44 g of DAG emulsion was suppressed in humans compared with the same amount of TAG emulsion.

Postprandial Elevation of Remnant Lipoproteins Is Smaller After DAG Ingestion

An increase in remnant lipoproteins (remnants) caused by postprandial hyperlipidemia was recognized recently as a potent atherogenic phenomenon (Karpe & Hamsten, 1995). Remnant lipoproteins are metabolites of triglyceride-rich lipoproteins such as chylomicrons and VLDL in plasma. Remnant-cholesterol is incorporated by macrophages and accumulates in the arterial walls, where primary lesions of atherosclerosis then develop (Ross & Ziversmit, 1997; Goldstein et al., 1980; Mahley et al., 1980; Havel, 1994; Simons et al., 1987). A reduction in remnant lipoproteins is, therefore, an effective strategy for preventing and treating postprandial hyperlipidemia.

To investigate the effects of DAG on postprandial hyperlipidemia, we examined the differences in the dynamics of remnants after loading DAG or TAG oils with the same fatty acid composition (Tada et al., 2001). As an indicator of the level of remnants, we used the concentration of cholesterol (RLP-cholesterol) and triglyceride (RLP-triglyceride) in remnant-like lipoprotein particles (RLP) in the serum (Nakajima et al.,

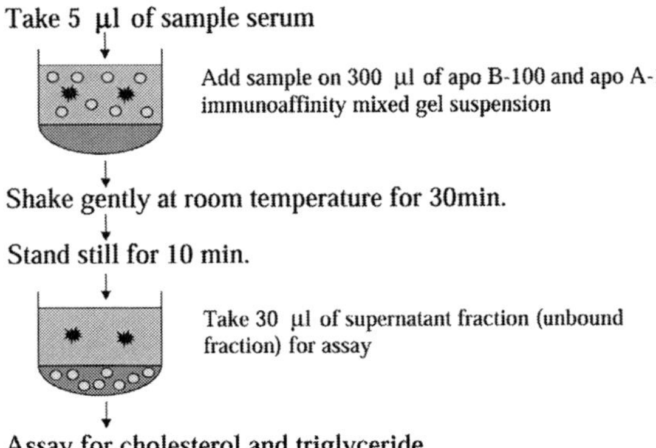

FIG. 8.3. Assay method for unbound lipoproteins (RLP): ●, unbound lipoproteins; ○, bound lipoproteins.

1993; Nakajima et al., 1996; Tanaka et al., 1995). Quantification of RLP was carried out by the method of Nakajima et al. (Nakajima et al., 1993; Nakajima et al., 1996). The RLP fraction was isolated directly from each serum sample by using an immunoaffinity-mixed gel conjugated with anti-human monoclonal antibody against apolipoprotein (apo) B-100 and apo AI (Fig. 8.3).

Male volunteers (n = 6) (Table 8.3) ingested emulsified test oils (Table 8.4) at a dose of 30 g fat/m^2 of body surface area early in the morning after fasting for at least 12 h. This was designed as a randomized, double-blind, crossover study with a crossover interval of 1 mo. Changes in serum lipid parameters were examined by collecting blood samples before fat loading, and after 2, 3, 4, 6, and 8 h.

Serum cholesterol did not change but serum triglyceride increased, peaking at 4 h after loading of either fat, and decreasing below initial levels at 8 h (Fig. 8.4). In a comparison of serum triglyceride changes between DAG and TAG loading, significant differences at several time points after fat loading were observed. Overall, the elevation in

TABLE 8.3
Characteristics of Subjects

	Age (y)	Height (m)	Body weight (kg)	BMI (kg/m^2)	Triglyceride (mg/dL)	Cholesterol (mg/dL)	
						Total	HDL
	31	1.73	82.7	27.6	141	186	33
	40	1.78	74.8	23.6	119	258	43
	38	1.80	76.5	23.6	450	217	34
	33	1.68	71.2	25.2	105	198	41
	32	1.74	78.7	26.0	102	102	40
	39	1.69	66.0	23.1	201	233	59
Mean	35.5	1.74	75.0	24.9	136	199	42

Source: Tada et al., 2001

TABLE 8.4
Composition of Creamed Test Oils

Ingredient	(%)
Oil (DAG or TAG)	35
Casein sodium	2
Fatty acid sucrose polyester	0.5
Lecithin from soybeans	0.36
Aspartame	0.2
Water	61.94

Source: Tada et al., 2001

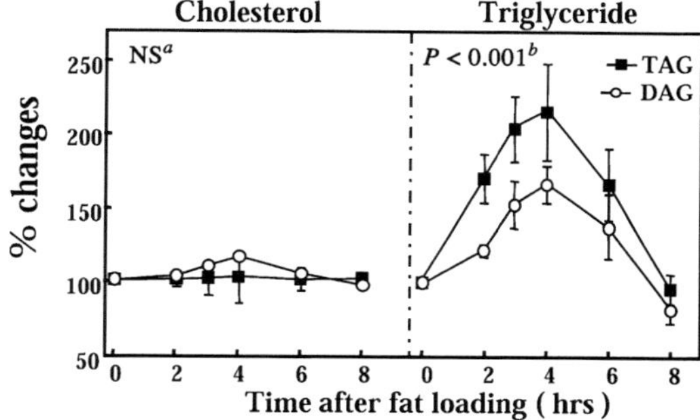

FIG. 8.4. The percentage change in serum levels of total cholesterol and triglyceride after ingestion of a fat emulsion (30 g/m^2 body surface area). [a]Not significant; [b]*P*-value for significant main effect by two-way ANOVA between TAG and DAG. Values are means ± SE, n = 6. Source: Tada et al., 2001.

serum triglyceride concentrations was significantly smaller after loading with DAG oil than with TAG oil, as determined by two-way ANOVA ($P < 0.05$).

Relative changes in serum RLP-lipids levels are shown in Figure 8.5. Differences between DAG and TAG loading observed in the serum triglyceride levels seemed to be more pronounced in the RLP-lipid fraction. Increases in RLP-triglyceride and RLP-cholesterol concentrations during fat loading were significantly smaller (two-way ANOVA, $P < 0.001$) with DAG loading than with TAG loading (Fig. 8.5). The initial levels of RLP-cholesterol for DAG and TAG group were 0.14 ± 0.08 and 0.13 ± 0.05 mg/dL, respectively, and 0.24 ± 0.21 and 0.22 ± 0.12 mg/dL, respectively for RLP-triglyceride.

Intergroup differences in the postprandial response of serum triglyceride and RLP-lipids levels were also evaluated by calculation of the areas under the curve (AUC) of each parameter. Table 8.5 shows differences in the AUC for serum triglyceride, RLP-cholesterol, and RLP- triglyceride after DAG and TAG loading. The AUC for serum triglyceride and RLP-cholesterol tended to decrease and that for RLP- triglyceride decreased significantly during DAG loading compared with TAG loading. The differences in the concentrations of triglyceride-rich lipoproteins during the ingestion of DAG

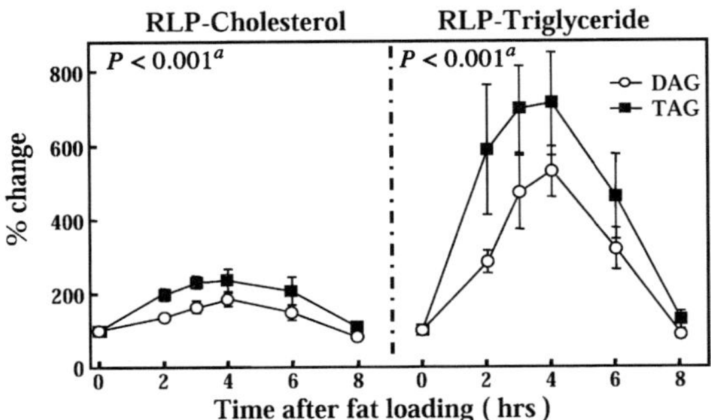

FIG. 8.5. The percentage change in cholesterol and triglyceride in the RLP fraction. [a]*P*-value for significant main effect by two-way ANOVA between DAG and TAG. Values are means ± SE, n = 6. Source: Tada et al., 2001.

TABLE 8.5
Differences in Area Under the Curve for Triglyceride and Remnant-Like Particles Between DAG and TAG Ingestion

Variable Area under the curve	DAG	 [mmol·h/L]	TAG
Triglyceride	16.20 ± 7.98		18.59 ± 6.85
Cholesterol in RLP	1.63 ± 1.09		1.97 ± 0.95
Triglyceride in RLP	0.76 ± 0.64*		1.06 ± 0.63

[a]Values are means ± SD, n = 6. Source: Tada et al., 2001.
*Different from TAG ingestion, $P < 0.05$.

and TAG might be due to the process of digestion and absorption in the intestine or of clearance in the circulation especially by lipoprotein lipase (LPL) after absorption. However, no differences in changes in serum LPL mass, apo CII, or apo CIII levels, or in free fatty acid concentrations at every point up to 8 h postprandially were observed between DAG and TAG loading in this study.

Postprandial Lipid Responses After DAG Oil Consumption in Pre-diabetics and Diabetics

The impact of DAG oil consumption on postprandial serum lipid responses was examined in subjects diagnosed as pre-diabetic or type 2 diabetic. A single ingestion study in subjects with or without insulin resistance demonstrated a significant reduction in the increase of postprandial serum triglyceride levels after consumption of a test meal containing DAG oil (Takase et al., 2005). In this study, analysis revealed that subjects with insulin resistance had a markedly lower increase in postprandial serum TG level after DAG oil consumption than the subjects without insulin resistance. Another set of clinical studies in subjects with type 2 diabetes (Tada et al., 2005) and impaired glucose tolerance (Ai et al., 2007) showed this reduced increase in postprandial serum TG level to be consistent with healthy and pre-diabetic subjects.

Suppressive effects of DAG oil on postprandial hyperlipidemia in diabetes. Therapeutic approaches to reduce remnant levels in the postprandial phase are believed to be important for the management of patients with diabetes and those with metabolic syndrome (Cottrell et al., 2003). As shown in the previous section, substitution of DAG oil intake for TAG oil significantly suppressed postprandial increases in serum triglyceride and lipids in remnants (Taguchi et al., 2000; Tada et al., 2001). The intake of DAG oil has been reported to reduce fasting serum triglyceride and hemoglobin A1c (HbA1c) levels in type 2 diabetics and to prevent the accumulation of body fat in experimental animals and in humans (Yamamoto et al., 2001; Tada & Yoshida, 2003). We were therefore interested in whether this favorable effect of DAG intake on postprandial hyperlipidemia could be applied to diabetic subjects without any serious adverse phenomena. We investigated the effects of oral DAG loading on postprandial changes in serum lipids and related parameters in moderately controlled diabetes (Tada et al., 2005).

The subjects were 6 patients with type 2 diabetes mellitus (5 females and 1 male, aged 46–70 years) who had moderately controlled HbA1c levels that were below 8%. Emulsified DAG oil or TAG oil with a fatty acid composition similar to DAG oil was orally administered (30g fat/m^2 of body surface) to the subjects after fasting for at least 12 hours. Serum cholesterol and triglyceride, lipids in RLP, and other lipid parameters, including serum ketone bodies, were measured prior to and 2 h, 4 h, and 6 h after fat loading.

As shown in Table 8.6, DAG oil loading significantly suppressed postprandial increases in serum triglyceride and lipids in RLP, compared with TAG oil. The incremental area under the curve (IAUC) for serum triglyceride and lipids in RLP after fat loading were also significantly smaller for DAG oil compared with TAG oil. However, the responses of serum insulin, free fatty acids and ketone bodies were essentially the same for DAG and TAG oil ingestion.

Although the number of subjects in this study was small, the findings again showed suppressed postprandial elevation in serum triglyceride, RLP-triglyceride, and RLP-cholesterol with DAG oil intake, compared with TAG oil. These results suggested that substituting DAG oil intake for TAG oil may be beneficial to moderately controlled diabetics, due to its effect in reducing postprandial hyperlipidemia.

Suppressive effects of DAG oil on postprandial hyperlipidemia in non-diabetic subjects with and without insulin resistance. A strong relationship between postprandial hyperlipidemia, insulin resistance and impaired glucose tolerance has been reported (Ai et al., 2001). As the magnitude of postprandial hyperlipidemia is accentuated in insulin-resistant and hyperinsulinemic individuals, it has been speculated that DAG oil may have a beneficial effect on these conditions. From the reports described in the previous sections, DAG oil appears to be effective in suppression of postprandial hyper-

TABLE 8.6
Changes in Serum Concentrations of Lipids

		Time (h) after fat loading				
	Test oil	0	2	4	6	*p* value of ANOVA
T-Chol. (mmol/L)	DAG	5.25±0.16	5.35±0.13	5.22±0.10	5.28±0.13	
		(0.00)	(0.08±0.08)	(-0.05±0.08)	(0.03±0.08)	*p*=0.330
	TAG	5.38±0.16	5.35±0.13	5.33±0.13	5.28±0.13	
		(0.00)	(-0.03±0.05)	(-0.05±0.05)	(-0.10±0.05)	
LDL-C (mmol/L)	DAG	3.28±0.23	3.34±0.26	3.23±0.23	3.18±0.23	
		(0.00)	(0.08±0.08)	(-0.05±0.05)	(-0.10±0.08)	*p*=0.617
	TAG	3.26±0.28	3.26±0.26	3.18±0.26	3.18±0.28	
		(0.00)	(0.00±0.05)	(-0.08±0.05)	(-0.08±0.05)	
HDL-C (mmol/L)	DAG	1.66±0.21	1.63±0.18	1.60±0.18	1.63±0.21	
		(0.00)	(-0.03±0.03)	(-0.08±0.05)	(0.00±0.03)	*p*=0.567
	TAG	1.63±0.18	1.66±0.18	1.58±0.18	1.60±0.18	
		(0.00)	(0.00±0.05)	(-0.08±0.03)	(-0.05±0.00)	
TAG (mmol/L)	DAG	0.94±0.19	1.46±0.33	1.59±0.42	1.41±0.26	
		(0.00)	(0.52±0.23)	(0.64±0.23)	(0.14±0.11)	*p*=0.005
	TAG	1.11±0.20	1.77±0.32	1.80±0.34	1.41±0.36	
		(0.00)	(0.68±0.27)	(0.69±0.18)	(0.32±0.17)	
FFA (mmol/L)	DAG	0.51±0.06	0.63±0.07	0.87±0.09	0.82±0.11	
		(0.00)	(0.11±0.04)	(0.36±0.07)	(0.31±0.09)	*p*=0.567
	TAG	0.59±0.08	0.75±0.11	0.79±0.06	0.85±0.13	
		(0.00)	(0.16±0.16)	(0.21±0.11)	(0.27±0.13)	
RLP-C (mmol/L)	DAG	0.11±0.01	0.17±0.03	0.17±0.04	0.13±0.02	
		(0.00)	(0.06±0.03)	(0.06±0.02)	(0.03±0.01)	*p*=0.021
	TAG	0.11±0.02	0.21±0.02	0.20±0.03	0.15±0.03	
		(0.00)	(0.10±0.02)	(0.09±0.02)	(0.05±0.02)	
RLP-TAG (mmol/L)	DAG	0.14±0.04	0.48±0.15	0.52±0.19	0.29±0.08	
		(0.00)	(0.34±0.14)	(0.38±0.14)	(0.15±0.05)	*p*=0.004
	TAG	0.15±0.04	0.67±0.19	0.59±0.13	0.41±0.13	
		(0.00)	(0.52±0.18)	(0.44±0.10)	(0.26±0.09)	

Values are Mean ± SE. Mean ± SE changes from baseline are shown in parentheses (DELTA). *Significantly different from TAG ingesition at the same time points by paired t-test: $p<0.05$. *P* values are calculated by repeated-measures two-way ANOVA(DELTA). DAG: Diacylglycerol, TAG: Triacylglycerol, T-Chol.: Total cholesterol, FFA: Free fatty acids. Source:Tada et al., 2005

lipidemia in healthy subjects and also in diabetic subjects. However, systematic and extensive investigation of the effect of DAG in relation to impaired glucose tolerance was still required.

Takase, et al. investigated the possible relationship between the effect of DAG oil ingestion on postprandial lipidemia and insulin resistance (Takase et al., 2005). They conducted a single dose ingestion study in a double-blind crossover manner. Eighteen subjects were recruited from male volunteers without any history of diabetes, dyslipidemia or daily medication that would interfere with lipid metabolism. Insulin resistance was assessed by the homeostasis model assessment (HOMA-R) formula as: HOMA-R= fasting insulin (uU/mL) x fasting glucose (mg/dL)/405. The subjects were classified into insulin sensitive (IS, n=10, HOMA-R$<$2.0) and insulin resistant (IR, n=8, HOMA-R$\geq$2.0) groups. The amount of test oil ingested was 10g/60kg body weight. Blood samples were collected at 0, 2, 3, and 4h after ingestion. After a 1-week washout period, the same procedure was repeated, except that the test food was reversed.

Correlations between BMI, age, HOMA-R and lipid profiles were analyzed for the baseline values. Only fasting serum triglycerides and RLP-cholesterol correlated with HOMA-R and were significantly higher in the IR as compared with the IS group. Postprandial increments of triglyceride and RLP-cholesterol after DAG ingestion were significantly lower compared with those after TAG oil ingestion. In the case of TAG oil ingestion, increments of serum triglycerides positively correlated with HOMA-R and were significantly higher in the IR as compared with the IS group. In contrast, these increments remained constant after DAG ingestion in both groups. In the IR group, the postprandial lipidemia (assessed by the maximum incremental concentration from baseline following ingestion of test oil, designated as Cmax) was reduced after DAG ingestion to about half of that after TAG ingestion, as shown in Table 8.7. A remarkable reduction in postprandial hyperlipidemia after DAG ingestion was observed only in the IR group. In conclusion, DAG reduced postprandial lipidemia, especially in subjects with insulin resistance, and may be beneficial in preventing atherosclerosis and related disorders.

Suppressive effects of DAG oil on postprandial hyperlipidemia in insulin resistance and glucose intolerance. Ai et al. investigated the effect of DAG ingestion on postprandial hyperlipidemia in insulin resistance and glucose intolerance more systematically (Ai et al., 2007). In their study, all subjects received a 75g oral glucose tolerance test and evaluated their glucose intolerance and insulin resistance. Then impaired glucose tolerance subjects (IGT group, n=14) and normal glucose tolerance subjects (NGT group, n=11) participated in this study. This was designed in a double-blind crossover style with a 1-week interval. All subjects ingested emulsified test oil (TAG or DAG oil) containing 57% water, 33% lipid, 3% protein and 7% carbohydrate at a dose of 17g fat/m^2 of body surface area. Blood samples were collected before loading, and at 2h and 4h after loading. Plasma glucose and serum insulin, total cholesterol, triglyceride, high density lipoprotein cholesterol, low density lipoprotein cholesterol were determined. Results are shown in Table. 8.8. In the IGT subjects, after the DAG and TAG load, the serum concentrations of triglyceride, RLP-triglyceride, and RLP-cholesterol increased throughout the 4h study period. The responses of these variables above baseline after the DAG load were significantly smaller than those after the TAG load ($p<0.05$). In contrast, in the NGT subjects, changes in these parameters were much smaller than those observed for IGT subjects. The difference in the integrated responses for serum RLP-cholesterol concentrations during oral fat tolerance tests between DAG and TAG in all subjects can be easily explained by the integrated response of insulin rather than glucose during oral glucose tolerance test ($r=0.7$, $p<0.01$). DAG was more effective in insulin resistant and

TABLE 8.7

Maximum Incremental Concentrations from Baseline Following Ingestion of Test Oil (Cmax) of Lipids in Insulin Sensitive (IS) and Insulin Resistant (IR) Groups.

			IS		IR		
			(mg/dL)	*p* vs TAG[1)]	(mg/dL)	*p* vs TAG[2)]	p vs IS[3)]
DAG ingestion	Triglyceride	Total	32 ± 7	0.582	36 ± 5	0.003	0.664
		RLP	19.7 ± 4.0	0.378	20.5 ± 5.0	0.012	0.904
		Chyromicron	18 ± 4	0.550	15 ± 2	0.010	0.505
		VLDL	13 ± 3	0.911	17 ± 3	0.009	0.427
	Cholesterol	Total	4 ± 1	0.589	6 ± 1	0.375	0.410
		RLP	1.8 ± .4	0.392	2.2 ± .4	0.074	0.461
		Chyromicron	1 ± 0	1.000	1 ± 0	0.007	0.214
		VLDL	2 ± 1	0.922	3 ± 1	0.069	0.725
TAG ingestion	Triglyceride	Total	35 ± 6	-	69 ± 10	-	0.037
		RLP	23.5 ± 4.1	-	47.2 ± 8.7	-	0.063
		Chyromicron	17 ± 3	-	32 ± 6	-	0.171
		VLDL	13 ± 3	-	29 ± 4	-	0.012
	Cholesterol	Total	5 ± 2	-	3 ± 1	-	0.700
		RLP	2.2 ± .5	-	4.5 ± 1.0	-	0.043
		Chyromicron	1 ± 0	-	2 ± 0	-	0.503
		VLDL	2 ± 1	-	4 ± 1	-	0.141

Values represent mean ± SEM. 1) Paired t-test between DAG and TAG in IS group. 2) Paired t-test between DAG and TAG in IR group. 3) Student's t-test between IS and IR group. Source: Takase et al., 2005.

TABLE 8.8
Serum Concentrations of Triglyceride (TG), RLP-Cholesterol (RLP-C), and RLP-Triglyceride (RLP-TG) During Oral Fat Tolerance Test in Subjects with Normal Glucose Tolerance (NGT) and Impaired Glucose Tolerance (IGT).

	Time (h) after fat loading		
	0	2	4
TG (mmol/L)			
NGT			
TAG	0.87 ± 0.12	1.21 ± 0.19	1.06 ± 0.19
DAG	0.80 ± 0.08	1.03 ± 0.11	1.00 ± 0.21
IGT			
TAG	1.42 ± 0.17	2.02 ± 0.22	2.31 ± 0.25
DAG	1.67 ± 0.24	2.16 ± 0.25	2.41 ± 0.29
RLP-C (mmol/L)			
NGT			
TAG	0.09 ± 0.02	0.13 ± 0.22	0.10 ± 0.02
DAG	0.08 ± 0.01	0.10 ± 0.01	0.09 ± 0.02
IGT			
TAG	0.16 ± 0.03	0.26 ± 0.03	0.29 ± 0.04
DAG	0.19 ± 0.04	0.27 ± 0.04	0.29 ± 0.04
RLP-TG (mmol/L)			
NGT			
TAG	0.19 ± 0.08	0.47 ± 0.12	0.31 ± 0.10
DAG	0.10 ± 0.03	0.32 ± 0.07	0.32 ± 0.13
IGT			
TAG	0.17 ± 0.04	0.55 ± 0.09	0.68 ± 0.10
DAG	0.23 ± 0.08	0.54 ± 0.10	0.63 ± 0.10

Data are mean ± S.E.M. Source: Ai et el., 2007.

hyperinsulinemic participants regardless of glucose intolerance, and may be beneficial in reducing the extent of CAD risk in such individuals.

Summary

The present review summarizes the short-term effects of DAG on lipid metabolism, with a focus on the postprandial state of serum lipids and lipoproteins. In sum, substituting DAG for TAG appears to improve postprandial hyperlipidemia. Clinical studies in subjects with type 2 diabetes or impaired glucose tolerance also showed this reduced elevation in postprandial serum triglyceride levels to be consistent with healthy and prediabetes subjects. Developments in molecular biology have led to the recognition of food-derived lipids and their metabolites, including cholesterol, fatty acids, bile acids, and lithocholic acids, as ligands of each corresponding nuclear receptor and regulators of key physiologic events (Tada, 2004). Further analysis in this field may provide new insights for elucidating the chemical and biological properties of DAG, especially the effects of 1,3-DAG, on digestion, absorption, and metabolic processes, and for managing life-style related chronic disease such as the metabolic syndrome. Recently, data indicating a smaller postprandial elevation of serum triglyceride levels were presented for the DAG oil incorporated into practical meals (Tomonobu et al., 2006) in addition to the studies using fat emulsions. These observations encourage further investigations such as the above-mentioned mechanistic studies.

References

Ai, M., Tanaka, A., Ogita, K., Sekinc, M., Numano, F., Numano, F., and Reaven, G.M. Relationship between plasma insulin concentration and plasma remnant lipoprotein response to an oral fat load in patients with type 2 diabetes. *J. Am. Coll. Cardiol.* **2001**, *38*, 1628-32.

Ai, M., Tanaka, A., Shoji, K., Ogita, K., Hase, T., Tokimitsu, I., and Shimokado, K. Suppressive effects of diacylglycerol oil on postprandial hyperlipidemia in insulin resistance and glucose intolerance. *Atherosclerosis*. **2007**, 195, 398-403.

Carey, M.C., Small, D.M., and Bliss, C.M. Lipid Digestion and Absorption, *Annu. Rev. Physiol.* **1983,** *45,* 651–677.

Cohen, J.C., Noakes, T.D., and Spinnler-Benade, A.J. Serum Triglyceride Responses to Fatty Meals: Effects of Meal Fat Content, *Am. J. Clin. Nutr.* **1988,** *47,* 825–827.

Cohn, J.S., McNamara, J.R., Cohn, S.D., Ordovas, J.M., and Schaefer, E.J. Postprandial Plasma Lipoprotein Changes in Human Subjects of Different Ages, *J. Lipid Res.* **1988,** *29,* 469–479.

Cohn, J.S. Postprandial Lipemia: Emerging Evidence for Atherogenicity of Remnant Lipoproteins, *Can. J. Cardiol.* **1998,** *14,* 18B–27B.

Cottrell, D.A., Marshall, B.J., and Falko, J.M. Therapeutic approaches to dyslipidemia in diabetes mellitus and metabolic syndrome. *Curr Opin Cardiol* **2003,** *18,* 301-308.

Dubois, C., Beaumier, G., Juhel, C., Armand, M., Portugal, H., Pauli, A.M., Borel, P., Latgé, C., and Lairon, D. Effects of Graded Amounts (0–50 g) of Dietary Fat on Postprandial Lipemia and Lipoproteins in Normolipidemic Adults, *Am. J. Clin. Nutr.* **1998,** *67,* 31–38.

Eriksson, M., Angelin, B., Henriksson, P., Ericsson, S., Vitols, S., and Berglund, L. Metabolism of Lipoprotein Remnants in Humans: Studies During Intestinal Infusion of Fat and Cholesterol in Subjects with Varying Expression of the Low Density Lipoprotein Receptor, *Arterioscler. Thromb.* **1991,** *11,* 827–837.

Gianturco, S.H., and Bradley, W.A. A Cellular Basis for the Atherogenicity of Triglyceride-Rich Lipoproteins, *Atheroscler. Rev.* **1991,** *22,* 9–14.

Goldberg, I.J. Lipoprotein Lipase and Lipolysis: Central Roles in Lipoprotein Metabolism and Atherogenesis, *J. Lipid Res.* **1996,** *37,* 693–707.

Goldstein, J.L., Ho, Y.K., Brown, M.S., Innerarity, T.L., and Mahley, R.W. Choesteryl Ester Accumulation in Macrophages Resulting from Receptor-Mediated Uptake and Degradation of Hypercholesterolemic Canine Beta-Very Low Density Lipoproteins, *J. Biol. Chem.* **1980,** *255,* 1839–1848.

Hara, K., Onizawa, K., Honda, H., Otsuji, K., Ide, T., and Murata, M. Dietary Diacylglycerol-Dependent Reduction in Serum Triacylglycerol Concentration in Rats, *Ann. Nutr. Metab.* **1993,** *37,* 185–191.

Havel, R.J. Postprandial Hyperlipidemia and Remnant Lipoproteins, *Curr. Opin. Lipidol.* **1994,** *5,* 102–109.

Jeppesen, J., Chen, Y.D., Zhou, M.Y., Wang, T., and Reaven, G.M. Effect of Variations in Oral Fat and Carbohydrate Load on Postprandial Lipemia, *Am. J. Clin. Nutr.* **1995,** *62,* 1201–1205.

Karpe, F., and Hamsten A. Postprandial Lipoprotein Metabolism and Atherosclerosis, *Curr. Opin. Lipidol.* **1995,** *6,* 123–129.

Mahley, R.W., Innerarity, T.L., Brown, M.S., Ho, Y.K., and Goldstein, J.L. Cholesteryl Ester Synthesis in Macrophages: Stimulation by ß-Very Low Density Lipoproteins from Cholesterol-Fed Animals of Several Species, *J. Lipid Res.* **1980,** *21,* 970–980.

Murase, T., Mizuno, T., Omachi, T., Onizawa, K., Komine, Y., Kondo, H., Hase, T., and Tokimitsu, I. Dietary Diacylglycerol Suppresses High Fat and High Sucrose Diet-Induced Body Fat Accumulation in C57BL/6J Mice, *J. Lipid Res.* **2001,** *42,* 372–378.

Murata, M., Hara, K., and Ide, T. Alteration by Diacylglycerol of the Transport and Fatty Acid Composition of Lymph Chylomicron in Rats, *Biosci. Biotech. Biochem.* **1994,** *58,* 1416–1419.

Murata, M., Ide, T., and Hara, K. Reciprocal Responses to Dietary Diacylglycerol of Hepatic Enzymes of Fatty Acid Synthesis and Oxidation in the Rat, *Br. J. Nutr.* **1997,** *77,* 107–121.

Nagao, T., Watanabe, H., Goto, N., Onizawa, K., Taguchi, H., Matsuo, N., Yasukawa, T., Tsushima, R., Shimasaki, H., and Itakura, H. Dietary Diacylglycerol Suppresses Accumulation of Body Fat Compared to Triacylglycerol in Men in a Double-Blind Controlled Trial, *J. Nutr.* **2000,** *130,* 792–792.

Nakajima, K., Saito, T., Tamura, A., Suzuki, M., Nakano, T., Adachi, M., Tanaka, A., Tada, N., Nakamura, H., Campos, E., and Havel, R.J. Cholesterol in Remnant-Like Lipoproteins in Human Serum Using Monoclonal Anti Apo B-100 and Anti Apo A-I Immunoaffinity Mixed Gels, *Clin. Chim. Acta* **1993,** *223,* 53–71.

Nakajima, K., Okazaki, M., Tanaka, A., Pullinger, C.R., Wang, T., Nakano, T., Adachi, M., and Havel, R.J. Separation and Determination of Remnant-Like Particles in Human Serum Using Monoclonal Antibodies to Apo B-100 and Apo A-I, *J. Clin. Ligand Assay* **1996,** *19,* 177–183.

Patsch, J.R. Postprandial Lipaemia, *Baillieres Clin. Endocrinol. Metab.* **1987,** *1,* 551–580.

Patsch, J.R., Miesenbock, G., Hopferwieser, T., Muhlberger, V., Knapp, E., Dunn, J.K., Gotto, A.M., and Patsch, W., Jr. Relation of Triglyceride Metabolism and Coronary Artery Disease: Studies in the Postprandial State, *Arterioscler. Thromb.* **1992,** *12,* 1336–1345.

Ross, A.C., and Ziversmit, D.B. Chylomicron Remnant Cholesteryl Esters as the Major Constituent of Low Density Lipoproteins in Plasma of Cholesterol Fed Rabbits, *J. Lipid Res.* **1997,** *18,* 169–181.

Simons, L.A., Dwyer, T., Simons, J., Bernstein, L., Mock, P., Poonia, N.S., Balasbramaniam, S., Baron, D., Brason, J., Morgan, J., and Roy, P. Chylomicrons and Chylomicron Remnants in Coronary Artery Disease: A Case-Control Study, *Atherosclerosis* **1987,** *65,* 181–189.

Tada, N., Watanabe, H., Matsuo, N., Tokimitsu, I., and Okazaki, M. Dynamics of Postprandial Remnant-Lipoprotein Particles in Serum After Loading of Diacylglycerols, *Clin. Chem. Acta* **2001,** *311,* 109–117.

Tada, N., and Yoshida, H. Diacylglycerol on lipid metabolism. *Curr Opin Lipidol.* **2003,** *14,* 29-33.

Tada, N. Physiological Actions of Diacylglycerol Outcome, *Curr. Opin. Clin. Nutr. Metab. Care* **2004,** *7,* 145–150.

Tada, N., Shoji, K., Takeshita, M., Watanabe, H., Yoshida, H., Hase, T., Matsuo, N., and Tokimitsu, I. Effects of diacylglycerol ingestion on postprandial hyperlipidemia in diabetes. *Clin Chim Acta.* **2005,** *353(1-2),* 87-94.

Taguchi, H., Watanabe, H., Onizawa, K., Nagao, T., Gotoh, N., Yasukawa, T., Tsushima, R., Shimasaki, H., and Itakura, H. Double-Blind Controlled Study on the Effects of Dietary Diacylglycerol on Postprandial Serum and Chylomicron Triacylglycerol Responses in Healthy Humans, *J. Am. Coll. Nutr.* **2000,** *19,* 789–796.

Takase, H., Shoji, K., Hase, T., and Tokimitsu, I. Effect of diacylglycerol on postprandial lipid metabolism in non-diabetic subjects with and without insulin resistance. *Atherosclerosis.* **2005,** *180(1),* 197-204.

Tanaka, A., Ejiri, N., Fujinuma, Y., Yui, K., Tamura, M., Nakajima, K., Morohoshi, M., Fujisawa, K., Uchimura, I., and Nurano, F. Remnant-Like Particles and Restenosis of Coronary Arteries After PTCA, *Ann. N.Y. Acad. Sci.* **1995,** *748,* 595–598.

Tomonobu, K., Hase, T., and Tokimitsu, I. Dietary Diacylglycerol in a Typical Meal Suppresses Postprandial Increases in Serum Lipid Llevels Compared with Dietary Triacylglycerol, *Nutrition.* **2006,** *22,* 128-35.

Yamamoto K, Asakawa H., and Tokunaga K., Watanabe, H., Matsao, N., Tokimitsu, I., and Yagi, N. Long-term Ingestion of Dietary Diacylglycerol Lower Serum Triacylglycerol in Type II Diabetic Patients with Hypertriglyceridemia. *J Nutr* **2001,** *131,* 3204-3207.

Yanai, H., Tomono, Y., Ito, K., Furutani, N., Yoshida, H., and Tada, N. Diacylglycerol Oil for the Metabolic Syndrome. Nutr. J. **2007**, *6*: 43-49 (http://www.nutritionj.com/content/6/2/43)

Zilversmit, D.B. Atherogenesis: A Postprandial Phenomenon, *Circulation* **1979,** *60,* 473–485.

Section B

Prevention of Obesity

Chapter 9

Antiobesity Effect and Prevention of Insulin Resistance by a Long-Term Consumption of Dietary Diacylglycerol in Experimental Animal Models

Tadashi Hase[a] and Hiroshige Itakura[b]

[a]Biological Science Laboratories, Kao Corporation, Ichikai-machi, Haga-gun, Tochigi 321-3497, Japan, and [b]Department of Food Sciences, College of Life Sciences, Ibaraki Christian University, Ibaraki 319-1221, Japan

Introduction

Obesity is a metabolic disorder resulting from disequilibrium between energy uptake and expenditure. It is a major risk factor for a variety of lifestyle-related diseases, including diabetes, hypertension, hyperlipidemia, and arteriosclerosis (Matsuzawa et al., 1995a; Rimm et al., 1995; Zimmet et al., 2001). Diacylglycerol (DAG) oil has energy values and absorption rates that are similar to triacylglycerol (TAG) oil (Taguchi et al., 2001). However, studies with animal models clearly show that a long-term ingestion of DAG oil prevents visceral fat accumulation as compared to ingestion of TAG oil. As shown in Chapter 4, the ingestion of DAG oil activates lipid metabolism in the small intestine and liver, and enhances whole-body energy expenditures (Murase et al., 2002a; Murata et al., 1997). Recent studies have shown that DAG improves glucose metabolism in addition its antiobesity effect. This chapter summarizes the effects of DAG oil consumption on body fat accumulation and on the insulin sensitivity in experimental animal models.

Antiobesity Effect of Diacylglycerol

A series of studies were conducted comparing the long-term effects of DAG and TAG in animals. In the studies described below, DAG was prepared by esterifying glycerol with fatty acids from rapeseed oil and soybean oil by the method of Watanabe et al. (Watanabe et al. 2003) and was comprised of the structural isomers 1,2-DAG and 1,3-DAG in a ratio of 3:7. The TAG oil was prepared by mixing rapeseed oil, soybean oil, and safflower oil to give a fatty acid composition that was similar to that of the DAG.

Watanabe et al. (1997) analyzed the body fat accumulation of male Sprague-Dawley (SD) rats fed a diet containing either 10% (w/w) DAG oil or TAG oil. The percentage of body fat, measured by bioimpedance (EM-SCAN SA-2), was monitored periodically at 3 and 4 weeks. No differences in food consumption were noted in the two groups, but in the DAG diet group, the percentage of body fat was significantly lower than that in the TAG diet group at both 3 and 4 weeks. Meng et al. reported the inhibitory effects of 1,3-DAG on diet-induced lipid accumulation in liver and abdominal fat of rats (Meng et al., 2004). Male SD rats were given free access to diets containing 7 wt% TAG (low TAG), 20 wt% TAG (high TAG), or 20 wt% DAG (high DAG), respectively, for 8 wk. The body weight of rats in the 20% high-TAG group increased significantly, and the weights of their abdominal adipose tissue and liver also showed a significant increase compared with rats in the low-TAG group. However, the high-DAG diet resulted in both a significant reduction in body weight gain and an increase in the ratio of abdominal fat to body weight compared with the high-TAG diet. Sugimoto et al. investigated the efficacies of DAG and TAG oil on body fat accumulation in male Wistar rats (Sugimoto et al. 2003a) and female Wistar fatty rats (Sugimoto et al., 2003b). There were no significant differences of body fat accumulation between the DAG diet group and the TAG diet group. Thus, in at least this one instance, the efficacy of DAG in Wistar rats is debatable as the difference in body fat between the treatment groups was relatively small compared with the obese animal model described below.

Additional research demonstrated that dietary DAG suppresses the accumulation of high-fat and high-sucrose diet-induced body fat in mice (Murase et al., 2001). This study was designed to examine the effects of dietary DAG on the development of obesity and the resulting hyperinsulinemia and hyperleptinemia in C57BL/6J mice, known as a model of diet-induced obesity and diabetes mellitus. Male C57BL/6J mice (7 wk old) were randomly assigned to 1 of 3 study groups in which they were allowed free access to 1 of 3 synthetic diets: a control (low-TAG) diet containing 5% (w/w) TAG oil, 20% protein, and 66.5% starch; a high-TAG diet containing 30% (w/w) TAG oil, 20% protein, 28.5% starch and 13% sucrose; or a high-DAG diet containing 30% (w/w) DAG oil instead of TAG oil. The mice were maintained on these diets for 5 consecutive months and monitored for several variables including body weight, food intake, feed efficiency [body weight gain per cage (g)/kcal of food consumed per cage by day], abdominal fat pad weights, liver triacylglycerol levels, liver cholesterol level, and serum insulin and leptin levels. Feces were collected 3 times on a per-cage basis for 24 h during the final week of feeding.

Mice that were fed the high-TAG diet for 5 months had a significantly greater weight gain compared with those that were fed the control diet. However, mice that were fed a high-DAG diet had a significant reduction in body weight compared with the high-TAG diet group (Fig. 9.1, Table 9.1). High-DAG and high-TAG groups took more energy on average than the control group; however, no difference was observed between the high-TAG and high-DAG groups. Consistent with this, mice fed high DAG had a significantly lower feed efficiency than mice fed high TAG. When fecal lipid quantity was examined, high-DAG and high-TAG groups had increases in fecal lipids compared with the control group. The amounts of fecal lipid in the high DAG-group and high TAG-group mice were not significant difference, indicating that there were no significant difference of absorption rate between the DAG-group and TAG-group (Table 9.1).

After 5 months of high-TAG diet consumption, the weights of the 4 individual abdominal fats (epididymal white adipose tissue (WAT), mesenteric WAT, retroperitoneal WAT, perirenal WAT) were significantly increased. In contrast, a high-DAG group

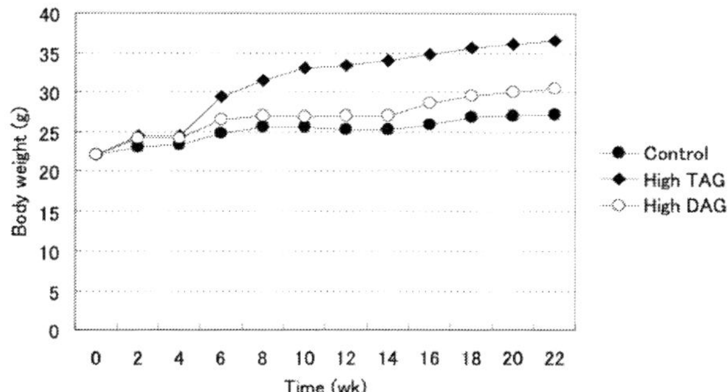

FIG. 9.1. Comparison of body weight in C57BL/6J mice fed the 3 diets for 5 mo. Each point represents the mean body weight of 5 mice. *Source*: Murase et al., 2001.

TABLE 9.1
Final Body Weight, Feed Efficiency, Energy Intake, and Fecal Lipid[a]

	Control	High TAG	High DAG
Final body weight[b] (g)	27.6 ± 1.7	36.4 ± 2.3**	30.2 ± 2.4††
Feed efficiency[c]	0.519	1.052	0.654
Energy intake [kcal/(cage·d)]	57.2 ± 3.9	69.7 ± 8.1**	65.2 ± 8.4**
Fecal lipid (mg/g dried feces)[d]	19.1 ± 2.9	64.3 ± 1.1**	60.5 ± 7.9**

[a]Values are means ± SD, n = 5. Source: Murase et al., 2001.
[b]Mice were killed after 5 mo of consuming each diet, and final body weight was determined.
[c]Feed efficiency was calculated as follows: [body weight gain per cage (g)]/[kcal of food consumed per cage per day].
[d]Feces were collected 3 times on a per-cage basis for 24 h during the final week of feeding.
**Different from controls, $P < 0.01$; ††different from the High-TAG group, $P < 0.01$.

significantly suppressed fat accumulation in all 4 areas (Fig. 9.2). These findings parallel the results for body weight gain.

The diet had a significant effect on the liver. Mice in the high-TAG diet group had an increased liver weight, whereas the increase in liver weight was suppressed by the high-DAG diet. Although there was no significant difference in liver triacylglycerol concentrations (mg/g liver) between the control and high TAG-fed mice, the high DAG-fed mice had a significantly lower triacylglycerol concentration than that of the high TAG-fed mice. The total liver triacylglycerol (mg) of TAG-fed mice was significantly higher than that of the controls. However, the total liver triacylglycerol of DAG-fed mice was significantly lower than that of the TAG-fed mice. Although no difference in cholesterol concentrations (mg/g liver) was observed between the groups, the liver total cholesterol (mg) of DAG-fed mice was significantly lower than that of TAG-fed mice (Table 9.2). These findings suggest that a DAG diet might be useful in preventing triacylglycerol and cholesterol accumulation in the liver.

The diet had significant effects on serum insulin and leptin levels. Elevated serum insulin levels in the high TAG-fed mice indicated the development of hyperinsulinemia

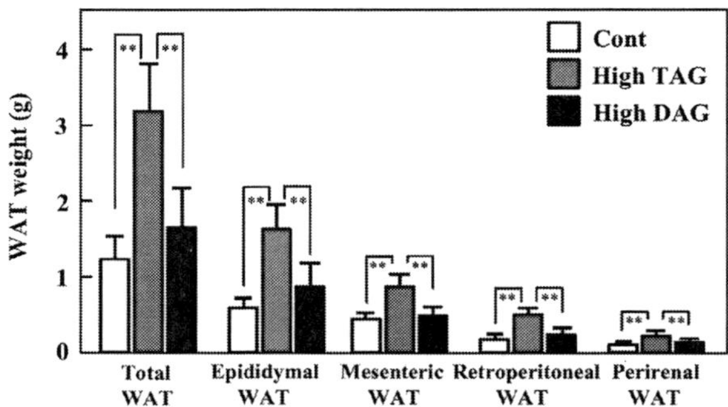

FIG. 9.2. White adipose tissue (WAT) weight in C57BL/6J mice after 5 mo of consuming the 3 diets. Values are means ± SD, n = 5. **Different from control and high-TAG, $P <$ 0.01. Source: Murase et al., 2001.

TABLE 9.2
Liver Weight, Triglyceride, and Cholesterol[a]

	Control	High TAG	High DAG
Weight (g)	0.907 ± 0.057	1.246 ± 0.106**	1.104 ± 0.050†
Triacylglycerol (mg/g liver)	79.4 ± 11.7	81.6 ± 13.7	60.4 ± 5.5**,†
Cholesterol (mg/g liver)	3.81 ± 0.41	3.69 ± 0.48	3.46 ± 0.35
Total triacylglycerol (mg)	71.9 ± 11.4	101.6 ± 18.2**	66.7 ± 7.0††
Total cholesterol (mg)	3.46 ± 0.42	4.59 ± 0.61**	3.81 ± 0.33†

[a]Values are means ± SD, n = 5. Source: Murase et al., 2001.
**Different from controls, $P < 0.01$; †different from the High-TAG group, $P < 0.05$; ††different from High-TAG group, $P < 0.01$.

during the 5 mo of feeding. This hyperinsulinemia was not observed in mice fed the DAG diet. Leptin levels in the high TAG-fed group were 10-fold higher than those in the control group, whereas the high-DAG diet significantly suppressed the elevation of leptin levels (data not shown).

Furthermore, Murase et al. (2002) examined the effects of dietary DAG and TAG with similar fatty acid compositions on the development of obesity in C57BL/6J mice during 8 mo of feeding. In this study, the antiobesity effect of a lower content of DAG in the feed was examined. Male C57BL/6J mice (7 wk old) were randomly assigned to 1 of 3 study groups in which they were allowed free access to 1 of 3 synthetic diets: a control (low TAG) containing 5% (w/w) TAG oil, 20% protein, and 66.5% starch; a high TAG containing 25% (w/w) TAG oil, 20% protein, 28.5% starch and 13% sucrose; or a high DAG containing 15% (w/w) DAG oil and 10% (w/w) TAG oil. The mice were fed these diets for 8 consecutive months. The results are as follows. Dietary DAG consistently reduced body fat accumulation induced by a high-fat diet. Compared with the low-TAG diet, mice that were fed the high-TAG diet for 8 mo had significant increases in body weight and adipose tissue weight. Conversely, in mice fed the high-DAG diet, body weight gain, epididymal WAT, perirenal WAT, and interscapular BAT weight were reduced by 43, 52, 85, and 81%, respectively, compared with the high-TAG diet (Table 9.3). In addition, under conditions of food deprivation, DAG-fed mice had significantly lower glucose, insulin, and leptin concentrations than those of high TAG-fed (Table 9.3).

No differences in mean energy intake were observed between the high-DAG and high-TAG groups. It was demonstrated previously that lipid absorption does not differ between these groups (Taguchi et al., 2001; Murase et al., 2001), suggesting that the reduced energy intake in the DAG group does not account for the reduced accumulation of body fat.

These results stimulated our interest in the mechanisms of the antiobesity effects of DAG. The effects of DAG on the mRNA expression of genes involved in lipid metabolism in various tissues were also analyzed. The details of these studies are given in Chapter 4. In the above studies, DAG was prepared by esterifying glycerol with fatty acids from rapeseed oil and soybean oil, which contain mainly oleic acid and linoleic acid as fatty acid constituents. These results suggest that DAG, which contains mainly oleic acid and linoleic acid as fatty acids, is very effective in preventing and altering obesity compared with TAG with a similar fatty acid composition.

Antiobesity Effect of α-Linolenic Acid–Rich Diacylglycerol

The above results suggest that the structure of the acylglycerol, but not the fatty acid composition, markedly affects the nutritional behavior of this class of lipids. On the other hand, n-3 polyunsaturated fatty acids (n-3 PUFA), such as eicosapentaenoic acid (EPA), docosahexaenoic acid (DHA), and α-linolenic acid (ALA) were reported to have a number of physiologic functions, such as anti-arteriosclerosis, anti-hypertension, and immunity control (Eritsland et al., 1995; Dehmer et al., 1988; Bonaa et al., 1990;

TABLE 9.3
Body Weight, Energy Intake, Fat Weight and Plasma Analysis[a]

	Control	High TAG	High DAG
n	10	20	10
Body weight[b] (g)	30.8 ± 2.1***	41.9 ± 5.4	37.1 ± 4.1*
Energy intake[c] [kcal/(cage·d)]	54.8 ± 8.9***	65.3 ± 3.8	62.2 ± 8.7
Epididymal WAT (g)	0.98 ± 0.23***	1.81 ± 0.47	1.38 ± 0.56*
Perirenal WAT (g)	0.11 ± 0.03***	0.31 ± 0.16	0.14 ± 0.09**
Interscapular WAT (g)	0.24 ± 0.08***	0.62 ± 0.4	0.38 ± 0.14
Interscapular BAT (g)	0.13 ± 0.02***	0.29 ± 0.11	0.16 ± 0.11***
Glucose (mg/dL)[d]	67.64 ± 12.65***	111.65 ± 25.18	83.28 ± 13.20**
Insulin (ng/mL)[d]	0.60 ± 0.13*	1.45 ± 1.04	0.64 ± 0.57*
Leptin (ng/mL)[d]	3.47 ± 1.23***	24.00 ± 18.67	7.40 ± 8.62**

[a]Values are means ± SD, n = 5. Source: Murase et al., 2001.
[b]Mice were killed after 8 mo of consuming the respective diets and body fat weight was then determined.
[c]Energy intake was measured on a per-cage basis over the course of 24 h 1 d/wk.
Different from the High-TAG group: *$P < 0.05$; **$P < 0.01$; ***$P < 0.001$.
[d]Blood was obtained from mice in the fasting condition (12 hr) 10 days prior to the end of the experimant (8 mo).

Robinson et al., 1993; Neuringer et al., 1986). The n-3 PUFA affect lipid metabolism, resulting in a decreased lipid concentration in the blood and a reduction in body weight and body fat weight (Westphal et al., 2000; Harris, 1989; Hun et al., 1999; Ikemoto et al., 1996). In the three studies below, α-linolenic acid-rich DAG (ALA-DAG), which has a 1,3-DAG structure and fatty acids containing >60% ALA, was the focus of interest, and the antiobesity effect of ALA-DAG was examined in C57/BL6J mice. ALA-DAG was prepared from perilla oil in the presence of immobilized lipase. The DAG content in the ALA-DAG oil was >80%, and the ratio of 1,2-DAG and 1,3-DAG was 3:7.

In the first study, male C57BL/6J mice (7 wk old) were randomly assigned to 1 of 5 study groups (Hun et al., 1999). This experiment included a low-fat (LF) diet, a high-fat (HF) diet, and a HF diet with the addition of incremental percentages of ALA-DAG oil (1, 2, and 4%, w/w). After a 7-d acclimatization, the mice were allowed free access to food and water. During the 4 wk in which the mice were fed these diets, they were monitored weekly for body weight. The amount of food consumed over a 24-h period was measured and the energy intake was calculated once a week. At the end of the experiment, the visceral fats [epididymal white adipose tissue (WAT), mesenteric WAT, retroperitoneal WAT, and perinephric WAT] were dissected and weighed.

No significant differences in the initial body weights or mean energy intakes during the test period were observed within each group. All of the groups showed gains in body weight by wk 4. The HF group had a significantly greater weight gain than the LF group. Conversely, the ALA-DAG-added group had a smaller body weight gain than the HF group. At the end of the 4 wk, the ALA-DAG-4%-added group had the lowest body weight and weight gain. The visceral fat weight of the HF group was signifi-

TABLE 9.4
Effects of α-Linolenic Acid-Rich Diacylglycerol (ALA-DAG) Diet on Body Weight at 4 Weeks[a]

	Dietary treatment				
			High fat + ALA-DAG		
	Low fat	High fat	ALA-DAG 1%	ALA-DAG 2%	ALA-DAG 4%
Body weight (g)					
Initial	21.6 ± 1.3	21.6 ± 1.3	21.6 ± 0.9	21.6 ± 1.2	21.6 ± 1.2
Final	26.3 ± 1.2*	29.0 ± 1.9	26.4 ± 0.3*	26.6 ± 1.3*	25.9 ± 2.0**
Gain (4 wk)	4.8 ± 0.4***	7.4 ± 1.5	4.9 ± 0.7***	5.0 ± 0.7***	4.3 ± 0.8***
Visceral-fat weight (g)					
Total	0.95 ± 0.10**	1.46 ± 0.37	1.11 ± 0.18*	1.00 ± 0.15**	0.94 ± 0.16**
Epididymal	0.49 ± 0.06**	0.76 ± 0.21	0.56 ± 0.12*	0.51 ± 0.08**	0.49 ± 0.10**
Mesenteric	0.32 ± 0.04*	0.41 ± 0.08	0.34 ± 0.03*	0.33 ± 0.05*	0.30 ± 0.03**
Retroperitoneal	0.10 ± 0.03***	0.24 ± 0.07	0.16 ± 0.05*	0.12 ± 0.02***	0.11 ± 0.03***
Perinephric	0.04 ± 0.00	0.05 ± 0.02	0.04 ± 0.01	0.04 ± 0.01	0.04 ± 0.01
Energy intake [kcal/(cage·d)]	55.8 ± 7.1	62.9 ± 4.2	59.4 ± 5.5	57.2 ± 4.7	57.9 ± 5.1

[a]Values are means ± SD, n = 5. Source: Hase et al., 2001.
Different from High Fat: $^*P < 0.05$; $^{**}P < 0.01$; $^{***}P < 0.001$.

cantly higher than that of the LF group. The ALA-DAG-added groups had significantly lower values than the HF group for total visceral fat weight, epididymal WAT, mesenteric WAT, retroperitoneal WAT, and perinephric WAT (Table 9.4). The visceral fat weight tended to decrease with linearly increasing ALA-DAG content.

In the second study, male C57/BL6J mice (7 wk old) were fed a low-fat (LF) diet, a high-fat (HF) diet, or a HF diet with the addition of 3% ALA-DAG for 20 wk (Hase et al., 2001). As above, C57/BL6J mice were used as a model for diet-induced obesity and diabetes mellitus. In this study, serum insulin and leptin levels in addition to body weight and the weight of the visceral fats were also examined. The effects on body weight gain and visceral fat weight were similar to the results in the first study (Hase et al., 2001) (Table 9.5). Serum insulin and leptin concentrations for the HF group were higher than those for the LF group. Serum insulin and leptin concentrations of the 3% ALA-DAG group were significantly lower than those of the HF group (Table 9.5). These results suggest the possibility that ALA-DAG may be effective against diabetes mellitus.

In addition to the two studies discussed above, Murase et al. (Murase et al., 2002b) examined the effect of ALA-DAG in genetically obese C57BL/KsJ-db/db mice, which develop severe obesity because of their leptin signaling deficiency (Chen et al., 1996). Three oils were used in this study, i.e., a TAG oil mixture that was prepared by combining soybean and rapeseed oils, an α-linolenic acid-rich TAG oil (ALA-TAG oil), and an α-linolenic acid-rich DAG oil (ALA-DAG oil) that was prepared by esterifying glycerol with fatty acids from the linseed oil. The ALA-DAG and ALA-TAG oils had similar fatty acid compositions. Female C57BL/KsJ-*db/db* mice (5 wk old) were randomly assigned

TABLE 9.5
Effects of ALA-DAG on Body Weight, Liver Weight, Visceral-Fat Weight, and Serum Insulin and Leptin at 20 wk[a]

	Dietary treatment		
	Low fat	High fat	High fat + ALA-DAG
Body weight (g)			
Initial	22.0 ± 0.6	22.5 ± 0.9	22.0 ± 0.7
Final	27.6 ± 1.7***	36.4 ± 2.3	32.0 ± 3.7*
Gain (20 wk)	5.6 ± 1.2***	13.8 ± 1.9	10.0 ± 3.1*
Visceral-fat weight (g)			
Total	1.264 ± 0.380***	3.180 ± 0.623	2.284 ± 0.873
Epididymal	0.578 ± 0.194***	1.620 ± 0.328	1.188 ± 0.516
Mesenteric	0.454 ± 0.101***	0.874 ± 0.161	0.634 ± 0.156*
Retroperitoneal	0.162 ± 0.069***	0.488 ± 0.094	0.340 ± 0.151*
Perinephric	0.070 ± 0.022**	0.198 ± 0.068	0.122 ± 0.056*
Energy intake [kcal/(cage·d)]	57.4 ± 2.8**	68.6 ± 1.3	66.5 ± 4.2
Insulin (pg/mL)	176.40 ± 73.95**	897.60 ± 505.26	287.30 ± 72.05**
Leptin (ng/mL)	1.59 ± 1.07***	16.72 ± 7.66	6.75 ± 5.27*

[a]Values are means ± SD, n = 5. Source: Hase et al., 2001.
Different from High Fat: *$P < 0.05$; **$P < 0.01$; ***$P < 0.001$.

TABLE 9.6
Body Weight, Weight Gain, and Energy Intake in Mice Fed TAG, ALA-DAG, or ALA-TAG Diets for 1 mo[a]

	TAG	ALA-DAG	ALA-TAG
Body weight (g)	42.3 ± 1.6[d]	38.6 ± 2.7[c]	40 ± 2.5[c,d]
Body weight gain (g/mo)	12.3 ± 1.7[d]	8.6 ± 2.4[c]	10.2 ± 1.7[c,d]
Energy intake[b] (kJ/d)	296.9 ± 28.4	296.9 ± 31.7	288.6 ± 38.4

[a]Values are means ± SD, n = 8; means without a common letter differ, $P < 0.05$. Source: Murase et al., 2002b.
[b]Energy intake: 4 mice/cage, n = 2.

to 1 of 3 study groups in which they were had free access to water and 1 of the 3 synthetic diets: containing 14% (w/w) TAG oil, 4% (w/w) ALA-TAG oil with 10% TAG oil, or 4% (w/w) ALA-DAG oil with 10% TAG oil. The mice were fed these diets for 1 mo and monitored for variables of body weight and food intake. There were no significant differences in energy intake during the test period among the three groups. Compared with the mice fed the diet supplemented with the ALA-TAG, the ALA-DAG diet reduced body weight gain, indicating that dietary ALA-DAG more effectively suppresses increases in body weight than ALA-TAG (Table 9.6).

As shown in the second study (Hase et al., 2001), the intake of 3% ALA-DAG oil (in 30% fat in diet) significantly suppressed body weight gain and hyperinsulinemia induced by high-fat and high-sucrose diets; however, the effects are not directly comparable to the effects of 30% DAG oil (in 30% fat in diet). ALA-DAG may be more effective in suppressing body fat accumulation than DAG. Although further studies will be required, these data suggest that ALA-DAG might be an excellent oil having the characteristics of both a DAG structure and a n-3 PUFA.

Improvement of Impaired Glucose Tolerance by DAG Ingestion

The accumulation of visceral fat is associated with the development of hyperinsulinemia and insulin resistance (Matsuzawa et al., 1995b). Several animal studies have shown that the ingestion of DAG oil prevents the accumulation of visceral fat and suppresses the development of abnormal carbohydrate metabolism (Murase et al., 2001, 2002a). Therefore, studies targeted specifically to investigating the effects of a DAG diet on the development of insulin resistance were conducted.

Meguro et al. investigated the effects of long-term ingestion of DAG oil, in comparison with TAG oil, on impaired glucose tolerance as well as on visceral fat accumulation in young (8 wk) and old (32 wk) rats (Meguro et al., 2006) fed a high-sucrose-diet. Male Wistar rats were fed either high-sucrose (57.5% sucrose w/w) or control diets containing either 10% (w/w) DAG or TAG oil with a similar fatty acid composition for 48 wk in 8-wk-old rats and for 24 wk in 32-wk-old rats. Researchers measured plasma lipids, the size of the islets of Langerhans, and levels for insulin, glu-

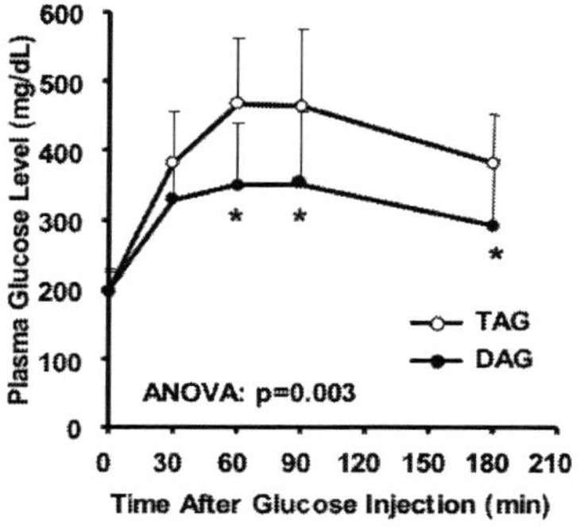

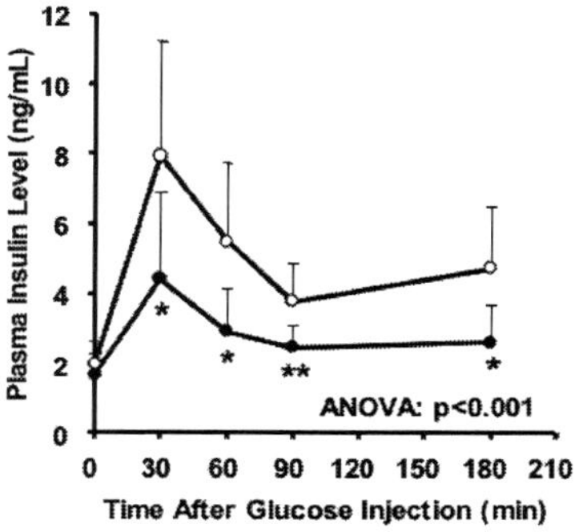

FIG. 9. 3. Consumption of the DAG diet was associated with better glucose tolerance in comparison with the TAG diet. Changes in plasma glucose (left panel) and insulin (right panel) levels during OGTT conducted after 10 wks of consuming TAG (n=9) or DAG (n=8) diets in BATless mice. Values are the mean ± SD. Repeated 2-way ANOVA was performed. The asterisks denote statistically significant differences between the TAG and DAG groups at individual time points and for the AUC (assessed by Student t test; *P<0.05). Source: Saito et al. 2007.

cose, and adipocytokine. Also an oral glucose tolerance test (OGTT) was performed during the study period. For rats in both age groups that were fed a high-sucrose diet, the DAG oil group had lower plasma glucose and a lower insulin response in the OGTT and lower homeostasis model assessment-R (HOMA-R) levels compared with the TAG oil group. HOMA-R levels are a known index of insulin resistance. Furthermore, in 8-wk-old rats that were fed a high-sucrose diet, the DAG oil group accumulated less visceral fat and showed decreases of plasma adiponectin and suppressed increases of plasma insulin, leptin, and the size of islet of Langerhans compared with the TAG oil group. No difference in the OGTT was found between the DAG and TAG oil groups in either age group of rats fed a control diet. These results suggest that, in comparison with TAG oil ingestion, DAG oil ingestion prevents the high-sucrose-diet-induced development of impaired glucose tolerance.

Saito et al. investigated the effect of dietary 1,3-DAG on the development of insulin resistance and obesity using brown adipose tissue-deficient (BATless) mice, a model of high-fat diet induced insulin resistance and obesity. The mice (8 wk old) were fed Western-type diets (WTD) containing either DAG oil (n = 8) or standard TAG oil (n = 9) for 15 wks. Although BATless mice became obese on both TAG- and DAG-enriched WTD (TAG-WTD and DAG-WTD), the mice eating DAG-WTD gained less weight and had less body fat accumulation. The results of glucose tolerance tests conducted after 5 weeks of each WTD were similar. However, after 10 weeks of each WTD, impaired glucose tolerance developed in the TAG-WTD group but was prevented in the DAG-WTD group (Fig. 9.3.). Exploratory analyses of gene expression suggested that consumption of DAG-WTD was associated with reduced phosphoenolpyruvate carboxykinase gene expression in liver and increased expression of the genes for peroxisome proliferator-activated receptor alpha, lipoprotein lipase, and uncoupling proteins 2 and 3 in skeletal muscle. There were no effects of the DAG-WTD on either fasting and postprandial plasma triglyceride levels, hepatic triglyceride content, or the rate of secretion of triglyceride from the liver. These findings suggest that diets enriched in 1,3-DAG oil may reduce WTD-induced insulin resistance and body fat accumulation by suppressing gluconeogenesis in the liver and by stimulating fat oxidation in skeletal muscle.

Moreover, Otsuka Long-Evand Tokushiua Fatty rats (OLETF rats), an animal model with some characteristics of human type 2 diabetes, were used to evaluate the long-term effect of DAG oil. Mori et al. studied the effects of dietary DAG by long-term feeding on the progression of diabetic features using this animal model (Mori et al., 2005). After 22 wk of DAG feeding, an OGTT revealed enhanced glucose disposal in the DAG-fed rats compared with the TAG-fed rats. Serum levels of adiponectin, an important insulin-sensitizing adipocytokine, were also higher in the DAG-fed rats than in the TAG-fed rats ($P<0.05$).

Summary

The results in this chapter show that the continuous consumption of DAG suppresses body fat accumulation in experimental animal models. As a result of the long-term consumption of DAG, an antiobesity effect was clearly observed in obese animals models, compared with the normal experimental animals. In Chapter 10 and 11, these effects are also examined in humans. Recent animal studies showed that the ingestion of DAG oil not only prevents the accumulation of visceral fat but also suppresses the development of abnormal carbohydrate metabolism. The results of this collective research indicate the possibility that dietary intervention with DAG may contribute to the prevention and management of obesity in humans, and also may help prevent various lifestyle related diseases such as diabetes, hyperlipidemia, and arteriosclerosis.

Further research into the nutritional characteristics of dietary DAG and its molecular mechanisms may provide new insights for the management of obesity and lipid nutrition.

References

Bonaa, K.H., Bjerve, K.S., Straume, B., Gram, I.T., and Thelle, D. Effect of Eicosapentaenoic and Docosahexaenoic Acids on Blood Pressure in Hypertension. A Population-Based Intervention Trial from the Tromso Study. *N. Engl. J. Med.* **1990,** *322,* 795–801.

Chen, H., Charlat, O., Tartaglia, L.A., Woolf, E.A., Weng, X., Ellis, S.J., Lakey, N.D., Culpepper, J., Moore, K.J., Breitbart, R.E., Duyk, G.M., Tepper, R.I., and Morgenstern, J.P. Evidence That the Diabetes Gene Encodes the Leptin Receptor: Identification of a Mutation in the Leptin Receptor Gene in db/db Mice. *Cell* **1996,** *84,* 491–495.

Dehmer, G.J., Popma, J.J., van den Berg, E.K., Eichhorn, E.J., Prewitt, J.B., Campbell, W.B., Jennings, L., Willerson, J.T., and Schmitz, J.M. Reduction in the Rate of Early Restenosis After Coronary Angioplasty by a Diet Supplemented with n-3 Fatty Acids. *N. Engl. J. Med.* **1988,** *319,* 733–740.

Eritsland, J., Arnesen, H., Seljeflot, I., and Hostmark, A.T. Long-Term Metabolic Effects of n-3 Polyunsaturated Fatty Acids in Patients with Coronary Artery Disease. *Am. J. Clin. Nutr.* **1995,** *61,* 831–836.

Harris, W.S. Fish Oils and Plasma Lipid and Lipoprotein Metabolism in Humans: A Critical Review. *J. Lipid Res.* **1989,** *30,* 785–807.

Hase, T., Mizuno, T., Onizawa, K., Kawasaki, K., Nakagiri, H., Komine, Y., Murase, T., Meguro, S., Tokimitsu, I. Shimasaki, H., and Itakura, H. Effects of α-Linolenic Acid-Rich Diacylglycerol on Diet-Induced Obesity in Mice. *J. Oleo Sci.* **2001,** *50,* 701–710.

Ikemoto, S., Takahashi, M., Tsunoda, N., Maruyama, K., Itakura, H., and Ezaki, O. High-Fat Diet-Induced Hyperglycemia and Obesity in Mice: Differential Effects of Dietary Oils. *Metabolism* **1996,** *45,* 1539–1546.

Matsuzawa, Y., Nakamura, T., Shimomura, I., and Kotani, K. Visceral Fat Accumulation and Cardiovascular Disease. *Obes. Res.* **1995a ,** *3,* 645S–647S.

Matsuzawa, Y., Shimomura, I., Nakamura, T., Keno, Y., Kotani, K., and Tokunaga, K. Pathophysiology and Pathogenesis of Visceral Fat Obesity. *Obes. Res.* **1995b,** *2(Suppl.),* 187S-194S.

Meguro, S., Osaki, N., Matsuo, N., Tokimitsu, I. Effect of diacylglycerol on the development of impaired glucose tolerance in sucrose-fed rats. *Lipids.* **2006,** 347-55. Erratum in: *Lipids.* **2006,** 627-8. *Lipids.* **2006,** 881.

Meng, X., Zou, D,. Shi, Z., Duan, Z., Mao, Z. Dietary diacylglycerol prevents high-fat diet-induced lipid accumulation in rat liver and abdominal adipose tissue. *Lipids.* **2004,** *39,* 37-41.

Mori, Y., Nakagiri, H, Kondo, H., Murase, T., Tokimitsu, I., Tajima, N. Dietary diacylglycerol reduces postprandial hyperlipidemia and ameliorates glucose intolerance in Otsuka Long-Evans Tokushima Fatty (OLETF) rats. *Nutrition.* **2005,** *21,* 933-9.

Murase, T., Mizuno, T., Omachi, T., Onizawa, K., Komine, Y., Kondo, H., Hase, T., and Tokimitsu, I. Dietary Diacylglycerol Suppresses High Fat and High Sucrose Diet-Induced Body Fat Accumulation in C57BL/6J Mice. *J. Lipid Res.* **2001,** *42,* 372–378.

Murase, T., Aoki, M., Wakisaka, T., Hase, T., and Tokimitsu, I. Anti-Obesity Effect of Dietary Diacylglycerol in C57BL/6J Mice: Dietary Diacylglycerol Stimulates Intestinal Lipid Metabolism. *J. Lipid Res.* **2002a,** *43,* 1312–1319.

Murase, T., Nagasawa, A., Suzuki, J., Wakisaka, T., Hase, T., and Tokimitsu, I. Dietary α-Linolenic Acid–Rich Diacylglycerols Reduce Body Weight Gain Accompanying the Stimulation of Intestinal ß-Oxidation and Related Gene Expressions in C57BL/KsJ-db/db Mice. *J. Nutr.* **2002b,** *132,* 3018–3022.

Murata, M., Ide, T., and Hara, K. Reciprocal Responses to Dietary Diacylglycerol of Hepatic Enzymes of Fatty Acid Synthesis and Oxidation in the Rat. *Br. J. Nutr.* **1997,** *77,* 107–121.

Neuringer, M., Connor, W.E., Lin, D.S., Barstad, L., and Luck, S. Biochemical and Functional Effects of Prenatal and Postnatal Omega 3 Fatty Acid Deficiency on Retina and Brain in Rhesus Monkeys. *Proc. Natl. Acad. Sci. USA* **1986,** *83,* 4021–4025.

Rimm, E.B., Stampfer, M.J., Giovannucci, E., Ascherio, A., Spiegelmen, D., Colditz, G.A., and Willett, W.C. Body Size and Fat Distribution as Predictors of Coronary Heart Disease Among Middle-Aged and Older US Men. *Am. J. Epidemiol.* **1995,** *141,* 1117–1127.

Robinson, D.R., Xu, L.L., Tateno, S., Guo, M., and Colvin, R.B. Suppression of Autoimmune Disease by Dietary n-3 Fatty Acids. *J. Lipid Res.* **1993,** *34,* 1435–1444.

Saito, S., Hernandez-Ono A., Ginsberg, H. N. Dietary 1,3-diacylglycerol protects against diet-induced obesity and insulin resistance. *Metabolism.* **2007,** *56,* 1566-75.

Sugimoto, T., Kimura, T., Fukuda, H., and Iritani, N. Comparisons of Glucose and Lipid Metabolism in Rats Fed Diacylglycerol and Triacylglycerol Oils, *J. Nutr. Sci Vitaminol.* **2003a,** *49,* 47–55.

Sugimoto, T., Fukuda, H., Kimura, T., and Iritani, N. Dietary Diacylglycerol-rich Oil Stimulation of Glucose Intolerance in Genetically Obese Rats. *J. Nutr. Sci. Vitaminol.* **2003b** , *49,* 139–144.

Taguchi, H., Nagao, T., Watanabe, H., Onizawa, K., Matsuo, N., Tokimitsu, I., and Itakura, H. Energy Value and Digestibility of Dietary Oil Containing Mainly 1,3-Diacylglycerol Are Similar to Those of Triacylglycerol. *Lipids* **2001,** *36,* 379–382.

Umeda, T., Bauer, J.E., and Otsuji, K. Weight loss effect of dietary diacylglycerol in obese dogs. *J. Amin. Physiol. Anim. Nutr. (Berl).* **2006,** *90,* 208-215.

Watanabe, H., Onizawa, K., Taguchi, H., Kobori, M., Chiba, H., Naito, S., Matsuo, N., Yasukawa, T., Hattori, M. and Shimasaki, H. Nutritional Characterization of Diacylglycerols in Rats. *J. Jpn. Oil Chem. Soc.* **1997,** *46,* 301–307.

Watanabe, T., Shimizu, M., Sugiura, M., Sato, M., Kohori, J., Yamada, N., Nakanishi, K. Optimization of Reaction Conditions for the Production of DAG Using Immnobilized 1,3-Regiospecific Lipase Lipozyme RM IM. *JAOCS.* **2003,** *80,* 1201-1207.

Westphal, S., Orth, M., Ambrosch, A., Osmundsen, K., and Luley, C. Postprandial Chylomicrons and VLDLs in Severe Hypertriacylglycerolemia Are Lowered More Effectively than Are Chylomicron Remnants After Treatment with n-3 Fatty Acids. *Am. J. Clin. Nutr.* **2000,** *71,* 914–920.

Zimmet, P., Alberti, K.G., and Shaw, J. Global and Societal Implications of the Diabetes Epidemic. *Nature* **2001,** *414,* 782–787.

Chapter 10

Clinical Studies Evaluating the Benefits of Diacylglycerol for Managing Excess Adiposity

Kevin C. Maki[a], Noboru Matsuo[b], and Mary R. Dicklin[a]

[a]Provident Clinical Research & Consulting, Inc, Glen Ellyn, IL, and [b]Kao Corporation, Health Care Food Research Laboratories, Tokyo, Japan

Introduction

Obesity is a growing problem worldwide (Flegal et al., 2002; World Health Organization, 1998). In the United States, 65% of adults are overweight or obese, defined as an excess of body weight compared with set standards (body mass index [BMI] = 25 kg/m^2), and 31% are obese (BMI = 30 kg/m^2) (Flegal et al., 2002). Overweight and obesity, particularly abdominal or visceral obesity, are known risk factors for a variety of diseases including diabetes, heart disease, stroke, hypertension, gallbladder disease, osteoarthritis, sleep apnea, and some forms of cancer (National Institute of Health, National Heart, Lung, and Blood Institute, 1998).

Energy restriction and increased physical activity are the first lines of therapy for promoting fat loss (National Institute of Health, National Heart, Lung, and Blood Institute, 1998). The effects of changes in the mixture of macronutrients contained in the diet, as well as changes in the specific type of fat consumed, on energy metabolism and body weight regulation are not well understood. Diacylglycerol (DAG) has the same fatty acid composition as triacylglycerol (TAG), but different metabolic characteristics that may be beneficial in preventing and managing obesity (Watanabe & Matsuo, 2003). DAG has been shown to prevent the accumulation of body weight and fat in animals (Murase et al., 2001) and humans (Maki et al., 2000; Nagao et al., 2000; Maki et al., 2002; Takei et al., 2001). The use of DAG in conjunction with an energy-restricted diet shows promise as a means by which weight loss and fat reduction may be enhanced. Rudkowska et al. (2005) recently reviewed the effects of DAG on ß-oxidation and body composition. In the present chapter, data from long-term clinical trials examining the beneficial effects of DAG consumption for the prevention and management of excess adiposity will be presented, along with a discussion of the possible mechanisms for these effects.

Clinical Studies

Body Weight and Body Composition Changes After 16 wk of Treatment with 10 g DAG Oil- or TAG Oil-Containing Foods

Nagao et al. (2000) conducted a double-blind study to investigate the effects of long-term ingestion of DAG on body fat accumulation in humans. This trial involved 38 healthy men ranging in age from 27 to 49 y with an average BMI of ~24 kg/m^2 and who had no history of diabetes or lipemia. During a 4-wk control lead-in period, the subjects were asked to consume ~50 g of total fat daily. After this control period and completion of baseline anthropometric measurements, subjects were randomly assigned to receive either DAG- (n = 19) or TAG-containing (n = 19) food products for a 16-wk treatment period.

The study food products included bread, mayonnaise, and shortbread, each containing 5 g of 1,3-DAG-rich or TAG-rich oil. Subjects selected two study foods each day at breakfast throughout the 16-wk treatment period so that the daily DAG or TAG oil consumption (from study foods) was 10 g. In addition, subjects consumed ~5 g of fat from other foods of their own choosing at breakfast. Self-selected meals at lunch and dinner were aimed at providing 15 and 20 g of total fat, respectively. Subjects kept records of daily food consumption throughout the control period and during the last 4 wk of treatment. According to these dietary records, mean daily energy intakes (~7950 kJ), fat intakes (~43 g/d), and percentages of energy from fat (~21%) did not differ between the DAG and TAG groups.

Changes from baseline in body weight, BMI, anthropometrics, and body composition (assessed by computed tomography) after 16 wk of treatment are summarized in Table 10.1. Decreases in body weight (–2.6 and –1.1 kg; $P < 0.01$) BMI (–0.9 and –0.4 kg/m^2; $P <0.01$), waist circumference (–4.4 and –2.5 cm; $P < 0.05$), and the total fat area (–38 and –17 cm^2; $P < 0.05$) for DAG and TAG, respectively, were significant in the two treatment groups; however, the reduction from baseline in these variables was significantly greater in the DAG group compared with the TAG group ($P < 0.05$ for all comparisons, a value considered significant in this chapter). Further, reductions in visceral fat area (–16 and –5 cm^2), subcutaneous fat area (–22 and –8 cm^2), and the visceral-to-subcutaneous fat area ratio (–0.05 and –0.00) for DAG and TAG, respectively, were each significantly different in the DAG group ($P < 0.05$ for all), but not in the TAG group.

It should be noted that although the subjects were randomly assigned to receive either DAG- or TAG-containing food products, baseline values for total fat area (227 and 182 cm^2) and visceral fat area (79 and 56 cm^2) for DAG and TAG, respectively, differed between groups ($P < 0.05$). This is important because studies have reported that a larger initial visceral fat area is significantly related to a greater reduction in visceral fat with body weight loss (Leenen et al., 1992). Because of this finding, the authors performed an analysis of covariance procedure to clarify the effect of DAG on body fat. These analyses, in which changes from baseline were evaluated using baseline values as a

TABLE 10.1
Anthropometric and Body Composition Values of Men Who Consumed Bread, Mayonnaise, and Shortbread Containing Either Diacylglycerol (DAG) Oil or Triacylglycerol (TAG) Oil for 16 wk[a,b]

		DAG oil	TAG oil
Body weight (kg)	Baseline[c]	72.1 ± 1.8	68.1 ± 1.3
	Change[c]	−2.6 ± 0.3**	−1.1 ± 0.4
Body mass index (kg/m^2)	Baseline	24.1 ± 0.4	23.5 ± 0.3
	Change	−0.9 ± 0.1**	−0.4 ± 0.1
Waist circumference (cm)	Baseline	85.0 ± 1.4	82.0 ± 1.0
	Change	−4.4 ± 0.6*	−2.5 ± 0.6
Hip circumference (cm)	Baseline	97.1 ± 0.9	96.1 ± 0.7
	Change	−1.1 ± 0.5	−0.7 ± 0.5
Waist/Hip ratio	Baseline	0.87 ± 0.01	0.85 ± 0.01
	Change	−0.04 ± 0.01	−0.02 ± 0.01
Body fat (g/100 g)	Baseline	21.6 ± 1.2	20.3 ± 1.0
	Change	−1.1 ± 0.3	−1.5 ± 0.7
Total fat area (cm^2)	Baseline	227 ± 16*	182 ± 15
	Change	−38 ± 3*†	−17 ± 8
Visceral fat area (cm^2)	Baseline	79 ± 7*	56 ± 6
	Change	−16 ± 2**†	−5 ± 3
Subcutaneous fat area (cm^2)	Baseline	148 ± 11	126 ± 10
	Change	−22 ± 3**	−8 ± 4
Visceral/Subcutaneous fat ratio	Baseline	0.55 ± 0.04	0.46 ± 0.04
	Change	−0.05 ± 0.02	−0.00 ± 0.02
Liver/Spleen ratio[d]	Baseline	1.24 ± 0.04	1.23 ± 0.03
	Change	0.06 ± 0.03	0.01 ± 0.02

[a]Values are means ± SEM, n = 19/group. Asterisks indicate significantly different from the TAG oil group, *P < 0.05, **P < 0.01. †Significantly different from TAG oil group by analysis of covariance, P < 0.05.
[b]Source: Nagao et al., 2000.
[c]Baseline, value at wk 0; change = wk 16 value − wk 0 value.
[d]Ratio of Hounsfield units of computed tomography.

covariate, confirmed that the decreases in total fat and visceral fat were significantly greater in the DAG group than in the TAG group ($P < 0.05$).

Hepatic fat content decreased during DAG treatment, as indicated by a slight increase of 0.06 in the liver-to-spleen density ratio ($P < 0.05$). There was no significant change in hepatic fat content in the TAG group, and no significant difference between the TAG and DAG groups.

These results indicated that DAG, in contrast to TAG, suppressed body weight and regional fat deposition, including visceral and hepatic fat, in healthy but near- obese men. The authors suggested that the differing responses to DAG and TAG may be attributable to differences in their metabolism at the level of the small intestine (Hara et al., 1993; Murata et al., 1994; Naito et al., 1997). Impaired postprandial triglyceride clearance was shown to be associated with visceral obesity (Couillard et al., 1998; Mekki et al., 1999). Postprandial serum triglyceride levels increase less after DAG ingestion than after TAG ingestion, which may explain in part the observed reduction in visceral fat after repeated ingestion of the DAG oil.

Body Weight and Body Composition Changes After 16 wk of Treatment with 15 g DAG-Mayonnaise vs. TAG-Mayonnaise

In a long-term, double-blind, parallel study reported by Takei et al. (2001), the effects on body fat accumulation of consuming DAG-containing mayonnaise (DAG-M) compared with a control mayonnaise prepared with TAG (TAG-M) with the same fatty acid composition were examined. Subjects were healthy men (n = 43) with a mean age of ~39 y and a mean BMI of ~24 kg/m^2. After a 4-wk lead-in period during which all participants consumed TAG-M, the subjects were randomly assigned to receive either DAG-M (n = 23) or TAG-M (n = 20) for a 16-wk treatment period. Randomization was performed using the visceral fat area at the initiation of the control period as an index to ensure that there would be no significant difference between treatment groups in baseline visceral fat area.

Subjects ingested one pack (15 g) of DAG-M or TAG-M/d. The DAG oil in DAG-M contained 80–83% DAG and was prepared in the presence of immobilized lipase. Subjects were instructed to maintain a daily fat intake, including the fat from the DAG or TAG mayonnaise, in the range of 45–55 g/d. The subjects recorded everything consumed each day in a dietary record that was analyzed for seven consecutive days at 4-wk intervals. These records indicated no significant differences in mean energy, carbohydrate, protein, or lipid intakes between the DAG-M and TAG-M treatment groups. Mean energy intakes during treatment were 1927 and 1878 kcal/d for DAG-M and TAG-M

TABLE 10.2

Changes in the Rate of Variation from Baseline to wk 16 in Anthropometric and Body Composition Indices of Men Who Consumed Either DAG- or TAG-Containing Mayonnaise (M)[a,b,c]

	DAG-M Group	TAG-M Group
Body weight	97.6 ± 0.5***	97.9 ± 0.4***
Waist	96.3 ± 0.5***	97.2 ± 0.7***
Hip	97.6 ± 0.6*	97.3 ± 0.4***
Waist/Hip ratio	98.7 ± 0.7††	99.9 ± 0.7
Skinfold thickness	88.3 ± 2.6***	85.7 ± 2.2***
Total fat	90.4 ± 3.2†	98.3 ± 3.1
Subcutaneous fat	89.9 ± 3.2**	97.6 ± 3.2
Visceral fat	91.9 ± 4.8**†	101.3 ± 3.7
Liver/Spleen ratio	104.5 ± 3.3†	99.0 ± 2.3

[a]Values are means ± SEM, n = 23 (DAG-M) or 20 (TAG-M). Asterisks indicate different from the initial level, *$P < 0.05$, **$P < 0.01$, ***$P < 0.001$. Daggers indicate different from TAG, †$P < 0.05$, ††$P < 0.01$, by two way ANOVA using all measurement data (4, 8, 12, 16 wk).

[b]Source: Takei et al., 2001.

[c]Initial levels were regarded as 100.

groups, respectively, and mean lipid intakes were ~49 and ~51 g/d for DAG-M and TAG-M groups, respectively.

Treatment groups did not differ significantly in baseline body weight, waist and hip circumferences, waist-to-hip ratio, skinfold thickness, total fat, visceral fat, subcutaneous fat, and the liver-to-spleen ratio. These indices were measured every 4 wk. Changes in some of these indices at the final point of measurement (wk 16) are shown in Table 10.2, expressed as a percentage of the initial values. Decreases in body weight (~98% of initial for both) waist circumference (~96 and 97% of initial), hip circumference (~98 and 97% of initial), and skinfold thickness (~88 and 86% of initial) for DAG and TAG, respectively, were significantly different in both treatment groups ($P < 0.05$ for all). Further, total fat area and subcutaneous fat area were significantly decreased from baseline at 8 and 16 weeks of treatment in the DAG-M group ($P < 0.05$ for all). Although the changes in visceral fat from baseline to 12 and 16 wk were not significant for either treatment group, visceral fat in the DAG-M group tended to be lower than at baseline (12 wk $P = 0.057$, 16 wk $P = 0.083$).

Throughout the treatment period, body weight, waist and hip circumferences, and skinfold thickness did not differ significantly between the DAG-M and TAG-M groups. Compared with the TAG-M treatment group, the decrease in waist circumference tended to be larger in the DAG-M group ($P = 0.060$), whereas the reduction in hip circumference tended to be smaller ($P = 0.096$). The waist-to-hip ratio, total fat area, and visceral fat area were each significantly decreased in the DAG-M group compared with the TAG-M group ($P < 0.05$ for all), and the decrease in subcutaneous fat area tended to be larger in the DAG-M group than in the TAG-M group ($P = 0.084$). The liver-to-spleen ratio, an index of accumulated fat in the liver, was significantly increased in the DAG-M group compared with the TAG-M group ($P = 0.039$), which indicated a decrease in the amount of fat in the liver.

Results of this study showed several trends suggestive of greater reductions in body weight and body fat with DAG compared with TAG, but there were fewer significant differences between treatments than in the study of Nagao et al. (2000). The decrease in total fat area of 6.7% after 16 wk of DAG-M ingestion (Takei et al., 2001) was similar to the 6.8% reduction in total fat area after 16 weeks of DAG-containing oil consumption reported by Nagao et al. (2000). However, compared with Nagao and colleagues (2000), Takei et al. (2001) did not find a significant difference in BMI between DAG and TAG treatment groups. The more marked effects on BMI seen with the DAG-containing oil were suggested to be related to a lower lipid intake (~43 g/d), and a higher ratio of DAG intake to total lipid intake (Nagao et al., 2000).

Body Weight and Body Composition Changes After 24 wk of Treatment with DAG Oil- vs. TAG Oil-Containing Foods in Obese Men and Women

In a randomized, double-blind, controlled, parallel trial, Maki et al. (2002) assessed the efficacy of DAG oil (1,2-DAG:1,3-DAG ratio of 3:7) for reducing body weight and fat mass when incorporated into a reduced-energy diet. Men and women (n = 131) with a mean BMI of ~34 kg/m^2 and a mean abdominal circumference of ~106 cm participated in the study. Subjects ranged in age from 19–71 y, were in good health (no history of diabetes mellitus, elevated fasting glucose concentration, or elevated blood levels of triglyceride or cholesterol) and had a normal activity level. After baseline assessments, subjects were randomly assigned to receive either DAG oil (n = 65) or TAG oil (control, n = 66) food products as part of a reduced-energy diet (~2100–3350 kJ/d, based on an estimate of initial maintenance energy needs) for a 24-wk treatment period. Study food products included muffins, crackers, instant soup mix, sugar cookies, and granola bars, which subjects substituted for other foods in their diets with a goal of achieving ~15% of total energy from either DAG or TAG oil. The DAG oil was ~90% DAG (by weight) and was prepared from a mixture of rapeseed and soybean oils under the presence of lipase. The TAG oil was prepared from a mixture of rapeseed, soybean, and safflower oils.

According to individually calculated energy requirements, 2–5 servings of study foods were supposed to be incorporated into the diet each day. However, because the food products were relatively high in energy and sucrose, some subjects reported that they found it difficult to consume the number of servings of study products required while maintaining a reduced-energy diet. According to 3-d dietary records kept by the subjects during the trial, no differences between groups in the changes from baseline in energy or percentage of energy from carbohydrate were observed; however the percentage of energy from fat decreased to a greater degree in the TAG group (-4.4% for TAG

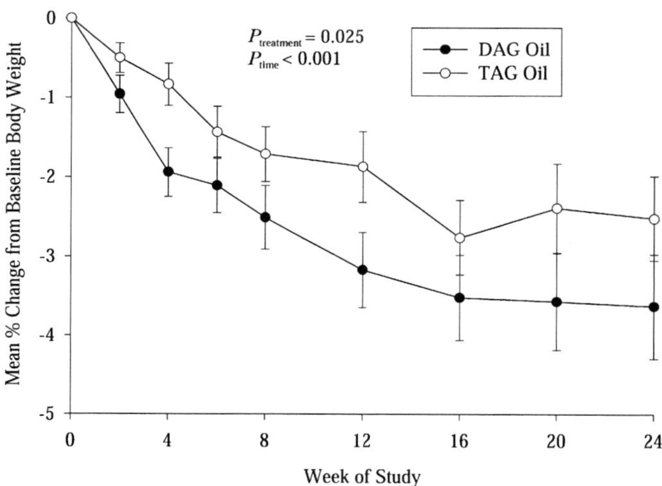

FIG. 10.1. Change in body weight (%) from baseline to wk 2, 4, 6, 8, 12, 16, 20, and 24 among subjects assigned to diacylglycerol (DAG) oil or triacylglycerol (TAG) oil treatment groups. Values are means ± SEM. Body weights at baseline were 98.0 ± 1.6 and 97.6 ± 1.8 kg in the DAG and TAG oil groups, respectively. Source: Maki et al., 2002.

vs. –1.0% for DAG, $P = 0.043$). Physical activity, assessed by a 7-d physical activity recall questionnaire administered periodically during the trial, increased by ~11 and ~13% in DAG and TAG groups, respectively, which was not significant.

Changes in body weight from baseline throughout the treatment period are shown in Figure 10.1. Body weight decreased in both the DAG and TAG groups, but from wk 2 of the study onward, reductions in body weight were greater in the DAG oil group than in the TAG oil group. Reductions in body weight at wk 12 (DAG –3.2% vs. TAG –1.9%) and 24 (DAG –3.6% vs. TAG –2.5%), time points prespecified for analyses, did not differ significantly between the groups. Repeated-measures ANOVA of body weight indicated significant effects for treatment ($P = 0.025$) and time ($P < 0.001$).

Changes from baseline to 12 and 24 wk of treatment in fat mass (by whole-body dual-energy X-ray absorptiometry scans) and intraabdominal fat area (by computed tomography scans) are shown in Figures 10.2 and 10.3, respectively. Fat mass and intraabdominal fat area decreased from baseline to wk 12 and 24 in both treatment groups. At wk 12, reductions from baseline fat mass were greater for subjects in the DAG group (–6.4%) compared with the TAG group (–3.5%, $P = 0.044$). However, the difference between groups in fat mass changes at wk 24 was not significant (DAG –8.3% vs. TAG –5.6%). Repeated-measures ANOVA of fat mass indicated significant treatment ($P = 0.037$) and time ($P < 0.001$) effects. In terms of intraabdominal fat area, repeated-measures ANOVA indicated no significant effects, although there was a trend for both DAG and TAG groups to show reductions from baseline ($P = 0.064$ for time effect).

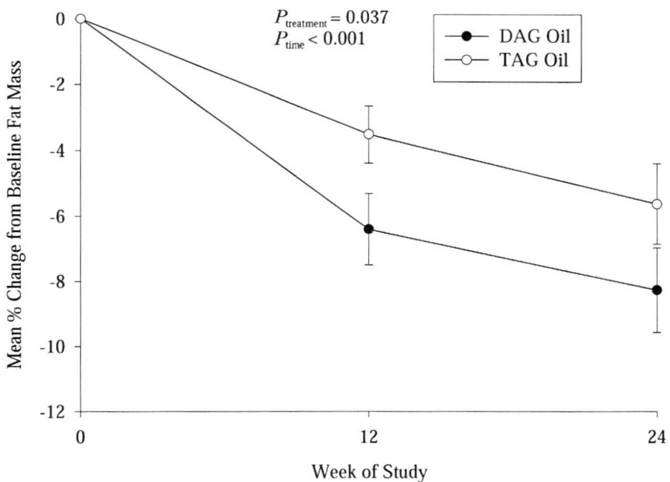

FIG. 10.2. Change in fat mass (%) from baseline to wk 12 and 24 among subjects assigned to diacylglycerol (DAG) oil or triacylglycerol (TAG) oil treatment groups. Values are means ± SEM. Fat mass values at baseline were 35.4 ± 0.9 and 34.7 ± 0.9 kg in the DAG and TAG oil groups, respectively. Source: Maki et al., 2002.

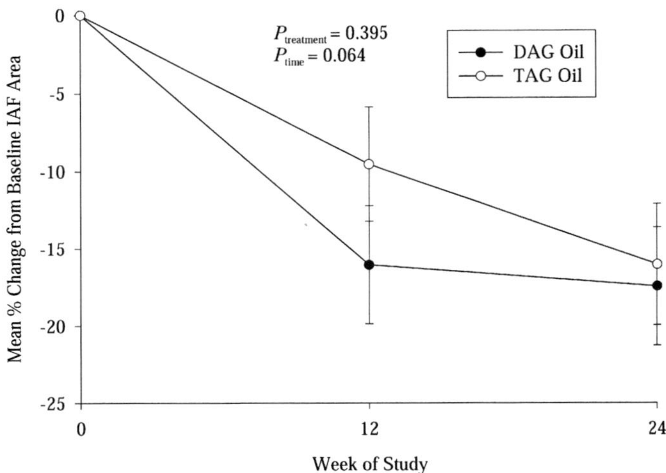

FIG. 10.3. Change in intraabdominal fat (IAF) area (%) from baseline to wk 12 and 24 among subjects assigned to the diacylglycerol (DAG) oil or triacylglycerol (TAG) oil groups. Values are means ± SEM. IAF areas at baseline were 150.4 ± 10.7 and 160.6 ± 9.9 cm^2 in the DAG and TAG oil groups, respectively. Source: Maki et al., 2002.

Total abdominal fat and subcutaneous fat area decreased from baseline to wk 24 in both the DAG and TAG oil groups, which did not differ significantly from each other. The median liver-to-spleen attenuation ratio also increased in both treatment groups from baseline to wk 24 (DAG 3.9% and TAG 1.5%), indicating no evidence of increased hepatic fat content in either group.

Results of subgroup analyses of subjects according to gender, age, and baseline BMI are shown in Figures 10.4, 10.5, and 10.6, respectively (Maki et al., 2000). These data indicate that among those with BMI ≥ 30 kg/m^2, males, or age at or above the median, reductions in body weight at wk 12 were significantly greater for DAG oil subjects than for TAG subjects ($P < 0.05$). At wk 24, only subjects with age above the median had significantly greater weight loss in the DAG group compared with the TAG group (–5.1% vs. –1.8%, $P < 0.05$). Further, at wk 12, subjects in the DAG oil group with baseline fat mass at or above median had a significantly greater reduction from baseline in fat mass compared with subjects in the TAG group. The fat mass response was also greater for subjects with baseline BMI ≥ 30 kg/m^2 and among subjects with age at or above the median. As was the case for the reduction in body weight at wk 24, only subjects with age above the median had a significantly greater loss of fat mass in the DAG group (–9.6%) compared with the TAG group (–4.2%, $P < 0.05$). Analyses also indicated significant reductions in intraabdominal fat area among the DAG subgroups with BMI ≥ 30 kg/m^2 and initial fat mass at or above the median value.

Weight loss in both treatment arms of this study was modest and approximately half of the amount anticipated, likely due in part to the problems subjects had incorporating

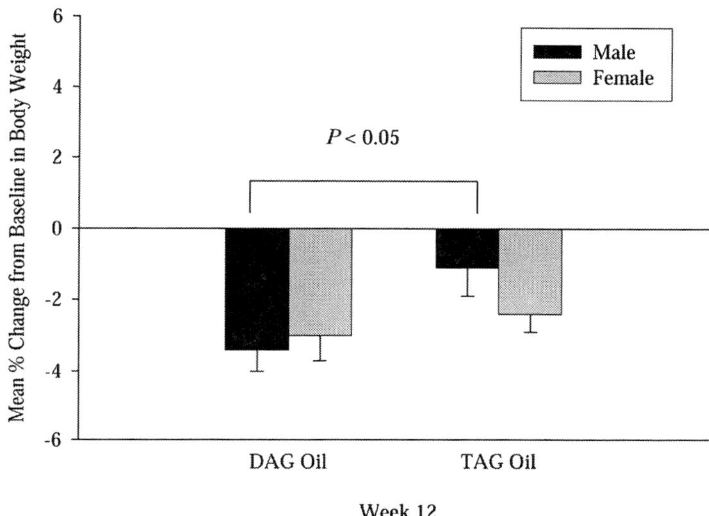

FIG. 10.4. Change in body weight (%) from baseline to wk 12 among subgroups of men and women assigned to the diacylglycerol (DAG) oil or triacylglycerol (TAG) oil groups. Values are means ± SEM. Source: Maki et al., 2000.

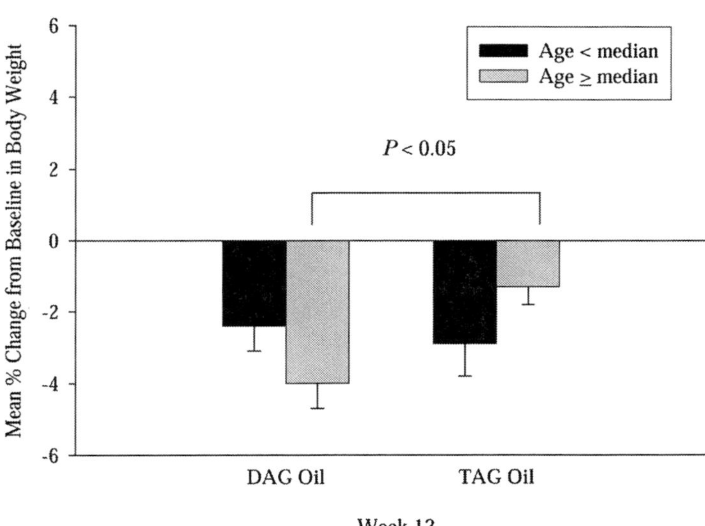

FIG. 10.5. Change in body weight (%) from baseline to wk 12 among subjects assigned to diacylglycerol (DAG) oil or triacylglycerol (TAG) oil groups divided according to age below or at or above the median. Values are means ± SEM. Source: Maki et al., 2000.

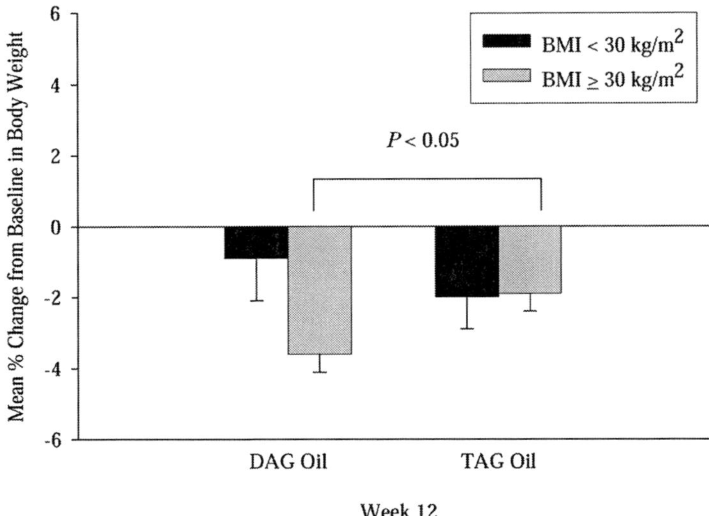

FIG. 10.6. Change in body weight (%) from baseline to wk 12 among subjects assigned to diacylglycerol (DAG) oil or triacylglycerol (TAG) oil groups divided according to baseline body mass index (BMI) < 30 kg/m^2 or BMI ≥ 30 kg/m^2. Values are means ± SEM. Source: Maki et al., 2000.

the relatively high-energy study food products into the reduced-energy diet. However, at each postrandomization time point in this study, weight loss was greater for subjects who received DAG oil compared with TAG oil. In addition, fat loss was greater in the DAG group, confirming results from previous studies in animals and humans (Nagao et al., 2000; Murase et al., 2001; Takei et al., 2001). The results of this study extended those of Nagao et al.. (2000) and Takei et al. (2001) and expanded the sample under examination to include men and women, and participants who were older and heavier than those studied previously.

More recently, Kawashima et al. (2008) reported the effects of 1-yr ad libitum consumption of DAG or TAG oil as part of a regular diet by 312 men and women with BMI ≥25 kg/m^2 and/or hypertriglyceridemia (triglycerides ≥150 mg/dL). Body weight and BMI decreased more in the DAG group, and the largest reductions occurred in participants with higher initial BMI as shown in Figure 10.7. These results are consistent with those from the prior controlled trials and extend the results to a follow-up of one year. The mechanisms responsible for the enhanced body weight and fat loss associated with DAG consumption remain to be elucidated, although several have been proposed.

Mechanisms of Action of DAG on Body Weight and Body Fat

To explain the different effects that DAG and TAG oil appear to have on body fat stores, it must be assumed that these oils are dissimilar in their effects on energy expenditure,

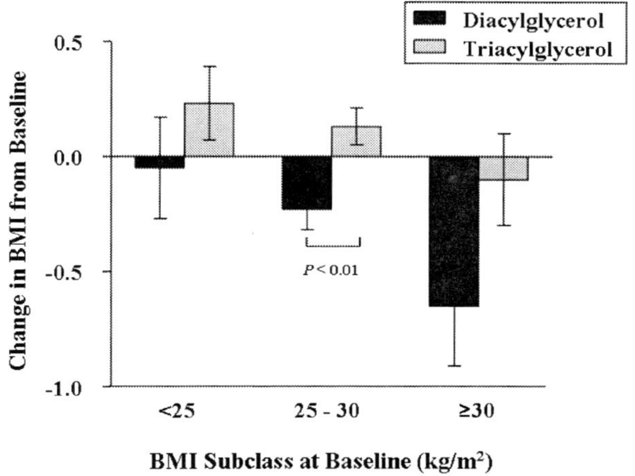

FIG. 10.7. Change in body mass index (kg/m^2) from baseline to 12 months among subjects assigned to diacylglycerol oil or triacylglycerol oil groups according to BMI subclass at baseline. Values are means ± SEM. Source: Kawashima et al. (2008).

energy intake, or both. Studies in experimental animals provide some support for increased energy usage with DAG (Watanabe et al., 1997; Murase et al., 2001). Murase et al. (2001) conducted a study in obesity- and diabetes-prone mice consuming a standard chow control diet (5% TAG), a 30% TAG diet, or a 30% DAG diet ad libitum for 5 mo. Body weight, body fat, insulin, and leptin were significantly elevated compared with control mice in mice fed the TAG diet but not the DAG diet. Interestingly, both the DAG (270 kJ/d) and TAG (286 kJ/d) diets increased total daily energy intakes compared with the control diet (242 kJ/d), and fecal fat loss did not differ between the two higher-fat treatments. These results seem to indicate that the mice consuming the DAG diet were able to maintain smaller fat stores while maintaining a greater energy intake than controls, and point to increased energy expenditure as a possible mechanism of action for DAG. Findings from Watanabe et al. (1997) of increased oxygen consumption in rats after short-term DAG administration lend strength to the proposed mechanism of increased energy expenditure. Results of a recent study by Kimura et al. (2006) in which a single dose of a DAG emulsion administered orally to rats was associated with increased oxygen consumption and fat oxidation and decreased the respiratory quotient, compared with TAG, also support this as the mechanism for the effect of DAG.

Increased energy expenditure with DAG consumption, as a mechanism for its effects on body fat, also has support at the organ level. The main digestive product of DAG is 1-monoglyceride, which is poorly reesterified into triglyceride in the small intestinal mucosa (Bierbach, 1983), so that greater amounts of free fatty acids are released into the portal circulation compared with TAG ingestion (Watanabe et al., 1997). The main digestive product of TAG is 2-monoglyceride, which is esterified to form triglyceride and incorporated into chylomicrons (Hunt & Groff, 1990). It is postulated that fat stores expand when fat oxidation is lower than fat intake (Flatt, 1978; Flatt, 1988). Enlarged fat stores result in a greater release of free fatty acids, which elevate fat oxidation, thus reestablishing an equilibrium between fat intake and oxidation. Short-term increases in TAG intake, however, do not elevate fat oxidation (Flatt, 1978; Flatt, 1988; Schutz, 1995). Because greater amounts of fatty acids from digested DAG are released into the portal circulation rather than being incorporated into chylomicrons, hepatic exposure to free fatty acids is increased postprandially during DAG intake, leading to enhanced ß-oxidation by the liver (Murata et al., 1997; Watanabe et al., 1997; Murase et al., 2001). Thus, unlike TAG consumption, DAG does promote a short-term increase in fat oxidation (Watanabe et al., 1997), which may allow a balance between fat oxidation and fat intake to occur at a higher level of fat consumption.

In a crossover study of 12 healthy women fed either a DAG-rich oil or a TAG-based oil during a 36-h respiratory chamber evaluation, fat oxidation was significantly higher with DAG than with TAG (Kamphuis et al., 2003). Further, feelings of hunger and appetite were significantly lower with DAG treatment, but 24-h energy expenditure did not differ between DAG and TAG. These results suggest that the mechanism of DAG's effect is increased fat oxidation, associated with improved appetite control and energy balance. More recently, in a crossover study with 13 healthy men, energy expenditure and respiratory quotient were measured in the first 5 hours following consumption of a DAG- or TAG-rich meal (Saito et al., 2006). Energy expenditure after the DAG meal tended to be increased more at 3 hours compared with the TAG meal. The postprandial respiratory quotient was significantly lower at 2 and 5 hours after DAG than TAG. These findings suggest that DAG increases lipid oxidation during the postprandial period.

We might expect there to be a compensatory increase in the drive to feed to replenish the body fat stores that may be depleted during chronic DAG consumption (Flatt, 1995); however, this does not appear to occur. Data from mice indicated that a smaller fat mass was associated with smaller liver glycogen stores (Acheson et al., 1984; Stubbs et al., 1993; Flatt, 1995). Increased hepatic fat oxidation, which occurs with DAG consumption (Murata et al., 1997; Murase et al., 2001), downregulates liver glycogen stores (Russek, 1981). Thus, if the hepatic glycogen "set-point" is reduced so that glycogen stores are maintained at a lower level, a smaller quantity of carbohydrate and energy may be required to achieve repletion (Flatt, 1978; Tremblay et al., 1994). It is hypothesized that the regulation of food intake is more closely linked to the maintenance of carbohydrate than to fat balance (Russek, 1981; Flatt, 1995), which may provide an explanation for the apparent lack of an increase in the drive to feed despite a loss in stored body fat.

Summary

In conclusion, several long-term clinical trials have indicated that DAG consumption results in losses of body weight and body fat (including visceral and hepatic fat) in healthy nonobese and obese men and women, and may therefore be beneficial in preventing and managing excess adiposity (Maki et al., 2000; Nagao et al., 2000; Takei et al., 2001; Maki et al., 2002). The mechanisms for these effects are not yet completely understood, but current literature suggests that DAG affects energy balance through a combined effect on energy intake and energy expenditure (Maki et al., 2002; Kamphuis et al., 2003).

Obesity is a difficult condition to treat (National Institute of Health, National Heart, Lung, and Blood Institute, 1998). Although energy restriction and increased physical activity are the first lines of therapy, these interventions have shown limited success for long-term maintenance of weight and fat loss (National Institute of Health, National Heart, Lung, and Blood Institute, 1998). The relative simplicity of incorporating DAG oil and DAG oil–containing food products into an energy-restricted diet shows promise as a means of enhancing reductions in body weight and altering body fat deposition. In reducing body fat, particularly visceral fat, lifestyle-related diseases such as diabetes mellitus, hyperlipidemia, hypertension, and atherosclerosis may be prevented.

References

Acheson, K.J.; Y. Schutz; T. Bessard; E. Ravussin; E. Jequier; J.P. Flatt. Nutritional Influences on Lipogenesis and Thermogenesis After a Carbohydrate Meal. *Am. J . Physiol.* **1984**, *246*, E62–E70.

Bierbach, H. Triacylglycerol Biosynthesis in Human Small Intestinal Mucosa. Acyl-CoA:Monoglyceride Acyltransferase. *Digestion* **1983**, *28*, 138–147.

Couillard, C.; N. Bergeron; D. Prud'homme; J. Bergeron; A. Tremblay; C. Bouchard; P. Mauriege; J.P. Despres. Postprandial Triglyceride Response in Visceral Obesity in Men. *Diabetes* **1998**, *47*, 953–960.

Flatt, J.P. The Biochemistry of Energy Expenditure. *Recent Adv. Obes. Res.* **1978**, *2*, 211–228.

Flatt, J.P. Importance of Nutrient Balance in Body Weight Regulation. *Diabetes Metab. Rev.* **1988**, *4*, 571–581.

Flatt, J.P. McCollum Award Lecture, 1995: Diet, Lifestyle, and Weight Maintenance. *Am. J. Clin. Nutr.* **1995**, *62*, 820–836.

Flegal, K.M.; M.D. Carroll; C.L. Ogden; C.L. Johnson. Prevalence and Trends in Obesity Among US Adults, 1999-2000. *J. Am. Med. Assoc.* **2002**, *288*, 1723–1727.

Hara, K.; K. Onizawa; H. Honda; K. Otsuji; T. Ide; M. Murata. Dietary Diacylglycerol-Dependent Reduction in Serum Triacylglycerol Concentration in Rats. *Ann. Nutr. Metab.* **1993**, *37*, 185–191.

Hunt, S.M.; J.L. Groff. *Advanced Nutrition and Human Metabolism.* West Publishing Company: New York, 1990.

Kamphuis, M.M.; D.J. Mela; M.S. Westerterp-Plantenga. Diacylglycerols Affect Substrate Oxidation and Appetite in Humans. *Am. J. Clin. Nutr.* **2003**, *77*, 1133–1139.

Kawashima, H.; H. Takase; K. Yasunaga; Y. Wakaki; Y. Katsuragi; K. Mori; T. Yamaguchi; T. Hase; N. Matsuo; T. Yasukawa; I. Tokimitsu; W. Koyama. One-Year Ad Libitum Consumption of Diacylglycerol Oil as Part of a Regular Diet Results in Modest Weight Loss in Comparison with Consumption of a Triacylglycerol Control Oil in Overweight Japanese Subjects. *J. Am. Diet. Assoc.* **2008**, *108*, 57-66.

Kimura, S.; H. Tsuchiyam; H. Inage; S. Meguro; N. Matsuo; I. Tokimitsu. Effects of Dietary Diacylglycerol on the Energy Metabolism. *Int. J. Vitam. Nutr. Res.* **2006**, *76*, 75-79.

Leenen, R.; K. Van Der Kooy; P. Deurenberg; J.C. Seidell; J.A. Weststrate; F.M.J. Schouten; J.G.A.J. Hautvast. Visceral Fat Accumulation in Obese Subjects: Relation to Energy Expenditure and Response to Weight Loss. *Am. J. Physiol.* **1992**, *263*, E913–E919.

Maki, K.C.; M.H. Davidson; R. Tsushima; N. Matsuo; I. Tokimitsu; D. Umporowicz; K.A. Ingram; M.R. Dicklin; M. Bell; B.D. Anderson; S.D. Frost; A. Tagala. Diacylglycerol Oil vs. Soybean Oil in a Weight Reduction Diet. North American Association for the Study of Obesity Meeting, abstract presented October 2000.

Maki, K.C.; M.H. Davidson; R. Tsushima; N. Matsuo; I. Tokimitsu; D. Umporowicz; M.R. Dicklin; G.S. Foster; K.A. Ingram; B.D. Anderson; S.D. Frost; M. Bell. Consumption of Diacylglycerol Oil as Part of a Reduced-Energy Diet Enhances Loss of Body Weight and Fat in Comparison with Consumption of a Triacylglycerol Control Oil. *Am. J. Clin. Nutr.* **2002**, *76*, 1230–1236.

Mekki, N.; M.A. Christofilis; M. Charbonnier; C. Atlan-Gepner; C. Defoort; C. Juhel; P. Borel; H. Portugal; A.M. Pauli; B. Vialettes; D. Lairon. Influence of Obesity and Body Fat Distribution on Postprandial Lipemia and Triglyceride-Rich Lipoproteins in Adult Women. *J. Clin. Endocrinol. Metab.* **1999**, *84*, 184–191.

Murase, T.; T. Mizuno; T. Omachi; K. Onizawa; Y. Komine; H. Kondo; T. Hase; I. Tokimitsu. Dietary Diacylglycerol Suppresses High Fat and High Sucrose Diet-Induced Body Fat Accumulation in C57BL/6J Mice. *J. Lipid Res.* **2001**, *42*, 372–378.

Murata, M.; K. Hara; T. Ide. Alteration by Diacylglycerols of the Transport and Fatty Acid Composition of Lymph Chylomicrons in Rats. *Biosci. Biotechnol. Biochem.* **1994**, *58*, 1416–1419.

Murata, M.; T. Ide; K. Hara. Reciprocal Responses to Dietary Diacylglycerol of Hepatic Enzymes of Fatty Acid Synthesis and Oxidation in the Rat. *Br. J. Nutr.* **1997**, 77, 107–121.

Nagao, T.; H. Watanabe; N. Goto; K. Onizawa; H. Taguchi; N. Matsuo; T. Yasukawa; R. Tsushima; H. Shimasaki; H. Itakura. Dietary Diacylglycerol Suppresses Accumulation of Body Fat Compared to Triacylglycerol in Men in a Double-Blind Controlled Trial. *J. Nutr.* **2000**, *130*, 792–797.

Naito, S.; H. Watanabe; H. Shimasaki; Y. Matuzawa; H. Ikakura. Effect of Dietary Diacylglycerols on Lipids Metabolism in Humans: I. The Suppressive Effect of Dietary Diacylglycerols on the Increase of the Serum Reconstituted Triglycerides Immediately After the Ingestion, 16th International Congress of Nutrition, Montreal, Canada. 1997, p. 341.

National Institute of Health, National Heart, Lung, and Blood Institute (1998) Clinical Guidelines on the Identification, Evaluation, and Treatment of Overweight and Obesity in Adults. Executive Summary via www.nhlbi.nih.gov (last accessed 10/7/03).

Rudkowska, I.; C.E. Roynette; I. Demonty; C.A. Vanstone; S. Jew; P.J.H. Jones. Diacylglycerol: Efficacy and Mechanism of Action of an Anti-Obesity Agent. *Obes. Res.* **2005**, *13*, 1864-1876.

Russek, M. Current Status of the Hepatostatic Theory of Food Intake Control. *Appetite* **1981**, *2*, 137–143.

Saito, S.; K. Tomonobu; T. Hase; I. Tokimitsu. Effects of Diacylglycerol on Postprandial Energy Expenditure and Respiratory Quotient in Healthy Subjects. *Nutrition*, **2006**, *22*, 30-35.

Schutz, Y. Macronutrients and Energy Balance in Obesity. *Metabolism* **1995**, *44 (Suppl. 3),* 7–11.

Stubbs, R.J.; P.R. Murgatroyd; G.R. Goldberg; A.M. Prentice. Carbohydrate Balance and the Regulation of Day-to-Day Food Intake in Humans. *Am. J. Clin. Nutr.* **1993**, *57*, 897–903.

Takei, A.; T. Toi; H. Takahashi; Y. Takeda; J. Moriwaki; H. Takase; Y. Katsuragi. Effects of Diacylglycerol-Containing Mayonnaise on Lipid Metabolism and Body Fat in Humans (English translation). *J. Nutr. Food* **2001**, *4*, 89–101.

Tremblay, A.; N. Almeras; J. Boer; E.K. Kranenbarg; J.P. Despres. Diet Composition and Postexercise Energy Balance. *Am. J. Clin. Nutr.* **1994,** *59*, 975–979.

Watanabe, H.; N. Matsuo. in *Lipids for Functional Foods and Nutraceuticals;* Gunstone, F.D., ed.; The Oily Press: Bridgwater, UK, 2003; pp. 113–148.

Watanabe, H.; K. Onizawa; H. Taguchi; M. Kobori; H. Chiba; S. Naito; N. Matsuo; T. Yasukawa; M. Hattori; H. Shimasaki. Nutritional Characterization of Diacylglycerols in Rats. *J. Jpn. Oil Chem. Soc.* **1997**, *46*, 301–307.

World Health Organization (1998) Obesity: Preventing and Managing the Global Epidemic. Report of a WHO Consultation on Obesity, Geneva, 3–5 June, 1997.

Chapter 11

Long-Term Clinical Studies of *Ad Libitum* Diacylglycerol Consumption in Subjects in a Free-Living Environment

Koichi Yasunaga, Hideto Takase, and Kenta Mori

Health Care Food Research Laboratories, Kao Corporation, Tokyo 131-8501, Japan

Introduction

Obesity is a known risk factor for coronary heart disease (Krauss & Witson, 1998) as well as part of the metabolic syndrome, leading to hypertension, diabetes, and hyperlipidemia (Ford et al., 2002). Despite the fact that obesity is a complex disorder involving interactions between genes and the environment, epidemiologic studies indicated the existence of a strong relation between the amount of dietary fat consumed and the incidence of overweight/obesity (Bray & Popkin, 1998). Numerous dietary and physical activity programs designed to promote weight loss have appeared, but obese participants rarely lose weight permanently or maintain their body mass index (BMI) in the normal range because the weight lost is often regained (Heshka et al., 2003, Jeffery et al., 1995). An explanation for this weight regain may be a lack of discipline in following a diet program. Because fat is not only an essential energy source but contributes to palatability in foods, a person is less likely to tolerate a low-fat diet for an extended period of time.

Recent studies, using both animals (Watanabe et al., 1997) and humans (Kamphuis et al., 2003), revealed that diacylglycerol (DAG) oil has unique nutritional characteristics, one of which leads to the activation of energy metabolism. The long-term consumption of DAG oil suppresses total body fat and visceral fat accumulation in humans (Nagao et al., 2000, Maki et al., 2002) in comparison with TAG oil (see Chapter 10), even at similar energy intakes (Taguchi et al., 2001). These findings regarding weight loss and body fat reduction were investigated primarily under hypoenergetic dietary conditions. DAG oil containing 80% or more diglycerides is an edible oil that is currently used as a cooking oil in the home in Japan and the United States. It is important to investigate the effect of DAG oil on body weight and fat consumed under ordinary dietary conditions for an extended period of time. In this chapter, we summarize the effect of DAG oil consumed as a part of a practical diet on the management of body weight and fat as well as blood lipid profiles in adults and children.

Studies on the Ad Libitum Consumption of DAG Oil in Adults

Double-blind, Controlled Study for 1 y in the Obese and Overweight (Kawashima et al., 2008). The aim of this study was to investigate the effects of long-term ad libitum consumption of DAG oil on body weight and body fat. This was a randomized, double-blind, controlled, parallel trial, in which 312 adults were randomly assigned to either a treatment (DAG oil, n = 155) or placebo (TAG oil, n = 157) group. Recruitment criteria in the present study were as follows: BMI ≥ 25 kg/m^2 and/or fasting serum triglyceride level ≥ 150 mg/dL and overall good health. Subjects visited the facility at 0 (baseline), 3, 6, 9, and 12 mo during the study period and received a medical check-up including anthropometric measurements and blood profiles. They were also requested to submit a consecutive 3-d dietary record to monitor dietary intake at every visit. The dietary records were collected and reviewed by a licensed nutritionist. The test oil (DAG oil) was prepared though the esterification of fatty acids derived from natural plant edible oil with glycerin in the presence of an immobilized lipase (see Chapter 23). The placebo oil (designated as TAG oil) was prepared by mixing several vegetable oils to achieve a fatty acid composition that was comparable to that of the DAG oil. The subjects were requested to replace their regular cooking oil used at home with the test oil during the study period. The test oil was used for ordinary cooking, such as deep-frying, stir-frying, and for homemade dressing and mayonnaise.

In the intention-to treat analysis, there were no differences in anthropometric measures or blood profiles between the DAG and TAG groups at the beginning of the study. Because the present study was intended to be conducted in a free-living environment, the subjects were not asked to comply with any dietary restrictions except a prohibition on drinking any alcohol 24 h before blood sampling; however, they were requested to maintain their normal level of physical activity. Their dietary intake including total energy, fat, carbohydrate, and protein was analyzed on the basis of a 3-d dietary record. There were no differences in the total energy, fat, carbohydrate, and protein intake between the groups. Also, the mean level of physical activity during the study did not differ between the groups.

Changes in body weight from baseline are shown in Figure 11.1. Body weight, BMI and skin-fold thickness decreased significantly in the DAG group compared to the TAG group. The subclass analysis indicated that the greater weight loss was found in the subjects of subgroups with higher BMI at baseline (Fig.11.2) or more replacement of cooking oil with DAG oil in their diet (Fig.11.3). There were no significant differences between the DAG and TAG groups in serum lipid and blood glucose panels. No serious adverse effects with respect to hematology, serum chemistry, and complaints of health conditions relating to study treatment were noted during the study period.

Previously, a body of clinical trials documented the effectiveness of diet intervention (Levy & Heaton, 1993, National Institute of Health, National Heart, Lung, and Blood Institute Obesity Education Initiative, 1998), fat substitutes (Bray et al., 2002), and drugs (Davidson et al., 1999, Wirth & Krause, 2001) on weight loss in overweight and/or obese subjects, showing weight reduction of ~5-10% during the treatment period. In an examination of the long-term consumption of DAG oil, Nagao et al. (Nagao et al., 2000) reported a 16-wk fat-controlled study in which subjects consumed approximately 5% of total calorie from DAG oil. Body weight decreased significantly in the DAG group compared with the TAG group with a –1.4 kg (2.1%) difference at 16 wk. Maki et al.

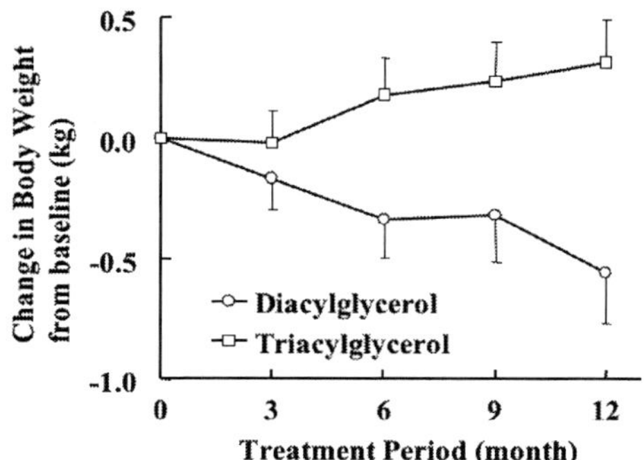

FIG. 11.1. Change in body weight from baseline to month 3, 6, 9, and 12 among participants who were randomized to the diacylglycerol (DAG) oil (n=150) or triacylglycerol (TAG) oil (n=153) groups. Repeated measures analysis of variance indicated a significant treatment effect (P=0.013) and treatment x visit interaction (P=0.001). Body weights at baseline were 72.5 ± 1.0 and 71.9 ± 0.9 kg in the DAG and TAG oil groups, respectively. Mean body weight loss from baseline at month 12 was significantly greater in the DAG oil group compared with the TAG oil group (-0.87 kg; P=0.002).

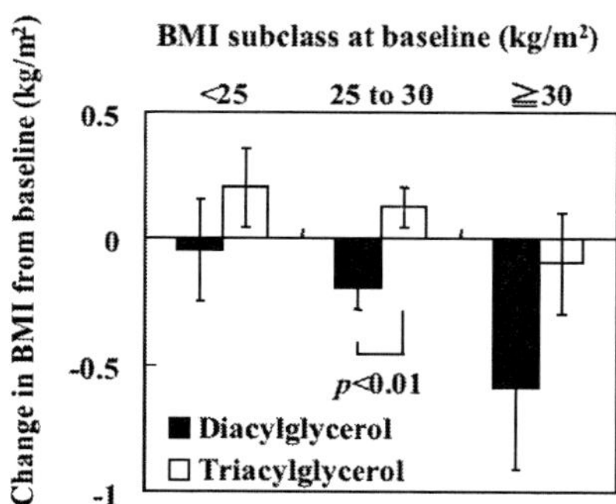

FIG. 11.2. Change in body mass index (BMI; kg/m^2) from baseline to month 12 for participants who were randomized to diacylglycerol (DAG) oil (n=150) or triacylglycerol (TAG) oil (n=153) groups. Participants were classified into three BMI subclasses by baselines, <25, 25 to 30, and ≥30. The number of participants in each subclass was 29, 104, and 17 in the DAG oil group and 28, 104, and 21 in the TAG oil group.

(Maki et al., 2002) reported on the consumption of DAG oil for 24 wk under hypoenergetic dietary conditions in which the test oil comprised 15% of the total energy intake. The percentage of body weight reduction from baseline at 24 wk was 1.1% lower in the DAG group than in the TAG group. The difference in the weight loss found in the present study between DAG oil group and TAG oil group was 0.87kg (1.2%), which was consistent with the results from the earlier diet-controlled studies. However, a longer time was

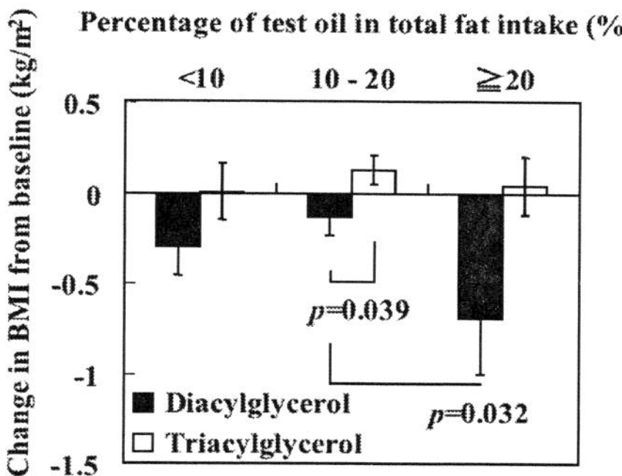

FIG. 11.3. Change in body mass index (BMI; kg/m^2) from baseline to month 12 in relation to the amount of test oil substituted. Participants were classified into three subclasses according to the amount of test oil intake (<10%, 10% to 20%, and ≥20% fat from test oil/total fat intake). The number of participants in each subclass was 31, 86, and 17 in the DAG oil or 38, 91, and 16 in the TAG oil groups, respectively.

required to achieve a similar magnitude of difference in body weight. This mild effect on body weight reduction in the present study is due in part to the relatively small amount of DAG oil intake and to the free-living environment.

Open-Labeled Studies (Post-Market Study)

The following two long-term studies were conducted as post-market studies of DAG oil.

Monadic Study for 1 y (Yasukawa & Yasunaga, 2001, Katsuragi et al., 1999). The objective of this study was to investigate the changes in anthropometric and serum lipid profiles when DAG oil was consumed as part of a regular diet without any dietary restriction for 12 mo. The study was a 12 mo, open-labeled monadic design, involving Kao employees (n = 114, 95 men, 19 women; mean age, 39.3 y). Subjects were requested to replace their regular cooking oil for home use with DAG oil supplied as Econa cooking oil (Kao Corporation, Tokyo, Japan). Blood samples were taken and anthropometric measurements were made at 0 (baseline), 3, 6, 9, and 12 mo during the study. Subjects were also requested to submit diet records for consecutive 3-d periods every 3 mo. Subjects were asked to replace their cooking oil for home use with DAG oil and to use this product primarily for breakfast and dinner. In addition to home use, subjects were requested to have lunch in the company's cafeteria as often as possible, where DAG oil (Econa cooking oil) was used for cooking.

The mean initial BMI in men and women were 24.7 and 20.7 kg/m^2, respectively. The men were categorized as overweight, whereas the women were categorized as lean according to Japanese guidelines for obesity. In addition, 26% of the subjects had serum triglyceride levels >150 mg/dL at the beginning of the study, which is considered hypertriglyceridemic by the Japan Arteriosclerosis Society. Fat energy intake was 31.3%, which was higher than the national average of 26.6%, indicating that the subjects had a tendency toward excessive intake of fat. The daily energy intake during the study period was 2005 kcal, which was 8.4% higher than the initial level. There were no significant changes from baseline in energy from protein, fat, and carbohydrate. Table 11.1 summarizes the changes in anthropometric variables. There was a modest decrease in the body weight of either men or women. The reduction from baseline was significant in the men at 12 mo. The percentage of subjects who lost >1 kg of body weight reached 22% at 3 mo, 38% at 9 mo, and 46% at 12 mo. There were also significant decreases in waist circumference in both men and women at 6, 9, and 12 mo compared with their respective baseline values. Skinfold thickness decreased significantly at 3, 6, 9, and 12 mo. The percentage of subjects with decreased skin fold thickness was 60% at 3 mo, 79% at 6 mo, 75% at 9 mo, and 80% at 12 mo.

Changes in blood variables are shown in Table 11.2. There were no significant changes from baseline in serum triglycerides, but there were significant changes in the subclass analysis. Serum triglyceride levels of subjects with normal serum triglyceride levels at baseline, i.e., <150 mg/dL, did not decrease during the study period. On the other hand, the hypertriglyceridemic subjects whose initial serum triglyceride levels were >150 mg/dL had significantly decreased serum triglyceride levels compared with baseline. This reduction in serum triglyceride levels was more evident in subjects with initial serum triglyceride levels >200 mg/dL. Total cholesterol levels did not change throughout the study period. A significant increase from baseline was observed in HDL cholesterol levels. LDL cholesterol, as determined by the Friedewald equation, was significantly decreased at 6 and 12 mo. A relationship exists between plasminogen activator inhibitor type-1 (PAI-1) and thrombosis. As shown in Table 11.2., PAI-1 declined significantly from 3 to 12 mo compared with the 3 mo level. Further subclass analysis revealed that this decrease in PAI-1 was more evident in subjects with higher PAI-1 levels.

Previously, we reported clinical data showing that postprandial serum triglyceride elevation was significantly lower after DAG oil ingestion compared with TAG oil (Taguchi et al., 2000). Although further investigation is warranted, the mechanism responsible for

TABLE 11.1
Changes in Anthropometric Variables in the 1 y Monadic Study[a]

Month	0	3	6	9	12
Body weight (kg)					
Men	72.1 ± 1.0	72.0 ± 1.0	71.9 ± 1.0	71.8 ± 1.0	71.4 ± 1.0***
Women	52.2 ± 1.7	52.0 ± 1.6	51.7 ± 1.4	51.7 ± 1.4	51.3 ± 1.4
Waist circumference (cm)					
Men	86.4 ± 0.8	86.2 ± 0.8	85.6 ± 0.8	85.6 ± 0.8	85.3 ± 0.8***
Women	66.1 ± 1.1	65.4 ± 1.2	64.6 ± 0.9	64.8 ± 0.9	64.6 ± 1.0**
Skin fold thickness (mm)					
Men	37.1 ± 1.2	35.3 ± 1.2***	33.6 ± 1.1***	33.5 ± 1.2***	32.0 ± 1.2***
Women	38.6 ± 3.0	36.6 ± 2.9**	35.0 ± 2.2**	34.3 ± 2.0**	33.8 ± 2.2***

[a]Values are means ± SEM, n = 114. Asterisks indicate different from baseline: ***$P < 0.001$; **$P < 0.01$; *$P < 0.05$.

TABLE 11.2
Changes in Blood Parameters in the 1 y Monadic Study[a,b]

Month	0	3	6	9	12
Serum triglyceride (mg/dL)					
All	127 ± 91	127 ± 9	125 ± 8	118 ± 7	120 ± 8
<150	84 ± 4	94 ± 5*	97 ± 6*	92 ± 5	91 ± 5
150–200	171 ± 4	177 ± 26	153 ± 13	166 ± 21	149 ± 13*
>200	298 ± 29	239 ± 23**	233 ± 29*	202 ± 25***	234 ± 34**
Serum cholesterol (mg/dL)					
Total	207 ± 3	208 ± 3	206 ± 3	206 ± 3	206 ± 3
HDL	54 ± 1	54 ± 1	56 ± 1***	57 ± 1***	57 ± 1***
LDL	128 ± 3	127 ± 3	125 ± 3*	124 ± 3*	125 ± 3
Plasma PAI-1 (ng/mL)	NM	36.5 ± 2.5	29.9 ± 1.6††	29.6 ± 2.2††	26.8 ± 1.8†††

[a]Values are means ± SEM, n = 114. Asterisks indicate different from baseline:****P* < 0.001; ***P* < 0.01; **P* < 0.05. Daggers indicate different from 3 mo: ††*P* < 0.001; †††*P* < 0.01.
[b]Abbreviations: PAI-1, plasminogen activator inhibitor type-1; NM, not measured.

this reduction in fasting serum triglyceride might be related to an alteration in the hydrolysis activity of LPL or the generation of endogenous triglycerides in the liver.

HDL cholesterol increased significantly after 3 mo, and HDL cholesterol is negatively correlated with serum triglycerides as well as body weight. It is particularly interesting to note that the subjects with normal serum triglyceride levels had a significantly increased HDL cholesterol level compared with baseline. Subjects who gained body weight during the study period also had increased HDL cholesterol. These findings indicate that the increase in HDL cholesterol was caused by an factor other than lowered serum triglycerides or body weight, suggesting the possibility that DAG oil consumption could directly affect HDL cholesterol metabolism.

PAI-1 decreased significantly during the study as shown in Table 11.2. PAI-1 is one of the primary inhibitors of the fibrinolytic system, and is secreted from adipose tissue and vascular endothelial cells. It is also well known that there is a positive correlation between PAI-1 vs. visceral fat mass (Oseroff et al., 1989) and serum triglycerides (Shimomura et al., 1996). In the present study, a significant decrease was observed in PAI-1 in subjects whose serum triglyceride levels were not increased. This result may indicate that DAG oil plays a role in the control of the production of PAI-1 from adipose tissues and vascular endothelial cells in the process of DAG metabolism. In conclusion, DAG oil used as a cooking oil in the daily diet helps to reduce body weight, and may improve CHD risk factors such as serum triglycerides and HDL cholesterol.

Monadic Study for 2 y (Ostuki et al., 2004). The intent of this study was to clarify the efficacy and safety of DAG oil consumption in obese and/or hypertriglyceridemia subjects in a free-living environment. This study was conducted as an open-labeled, monadic study, and all results obtained were compared with the baseline values. The subjects were recruited from participants of the human dry dock at Muroran Nikko Memorial hospital. Inclusion criteria were BMI ≥ 25 kg/m^2 and/or fasting serum triglyceride levels ≥150 mg/dL. The subjects used the test oil (Econa cooking oil, Kao Corporation, Tokyo, Japan) delivered by the hospital and were asked to replace their regular cooking oil in the home with the test oil. The subjects were instructed to visit the facility at 0 (baseline), 3, 6, 9, 12, 18, and 24 mo to receive a medical check-up. To avoid excessive energy intake during the study, subjects were instructed to consume the test food as a substitute for a meal. Total energy, fat, carbohydrate, and protein intake were calculated from a diet diary (3 consecutive days), which the subjects recorded at 3-mo intervals. There were 60 subjects (men, 51; women, 9) in the study.

The mean BMI and fasting serum triglyceride were 25.5 kg/m^2 and 143 mg/dL, respectively. No significant changes were observed in total energy and fat intake during the study period on the basis of the 3 d diet records submitted by the subjects at each clinical visit. Changes in anthropometric variables during the study can be summarized as follows: body weight tended to decrease from baseline (P = 0.083) but the difference was not significant. BMI decreased during the study period (P = 0.044) and was significantly different at 18 mo compared with baseline. Waist and hip circumferences decreased significantly during the study period and were significantly different at 18 and 24 mo (P < 0.05) and at 12, 18, and 24 mo (P < 0.001) compared with baseline, respectively. Skinfold thickness, which is the sum of the upper arm and back, changed significantly at several points during the study period (P < 0.0001) although it was not significantly different at 24 mo.

Changes in serum lipid variables are shown in Table 11.3. Serum triglycerides, total cholesterol, and LDL cholesterol did not differ from baseline. HDL cholesterol increased significantly during the study period (P < 0.0001) and was significantly different at 324 mo compared with baseline. HbA1c decreased during the study period (P < 0.0001) and was significantly different at 18 and 24 mo compared with baseline. The arteriosclerosis index was calculated according to the following equation: (total cholesterol - HDL cholesterol)/HDL cholesterol. The index decreased significantly during the study period (P = 0.010) and was significantly different at 9, 18, and 24 mo compared with baseline. No significant differences were observed in fasting blood sugar (P = 0.1190) or insulin levels (P = 0.168) during the study period.

Various risk factors for metabolic syndrome were evaluated according to the guidelines in National Cholesterol Education Program Adult Treatment Panel III (NCEP ATP-III) (National Cholesterol Education Program, 2001) with a modification for Japanese adults. The criteria were as follows: (i) Waist circumference: men, ≥ 85 cm; women, ≥ 90 cm; (ii) Blood pressure ≥ 140 mmHg at constriction and/or ≥ 90 mmHg at expansion; (iii) Serum triglyceride level ≥ 150 mg/dL; (iv) HDL cholesterol < 40 mg/dL; and (v) Fasting blood sugar ≥ 110 mg/dL.

Figure 11.4 shows changes in the risk grade for metabolic syndrome during the study period. The subjects were categorized by the number of risk factors they possessed at baseline; 49 subjects were categorized as having a risk grade ≤2 and 11 subjects were categorized at risk grade 3 or more at the start of the study. The mean risk grade in all subjects and those with Risk 2 remained unchanged during the study period. However, the mean risk grade in the group with ≥3 risk factors decreased significantly during the

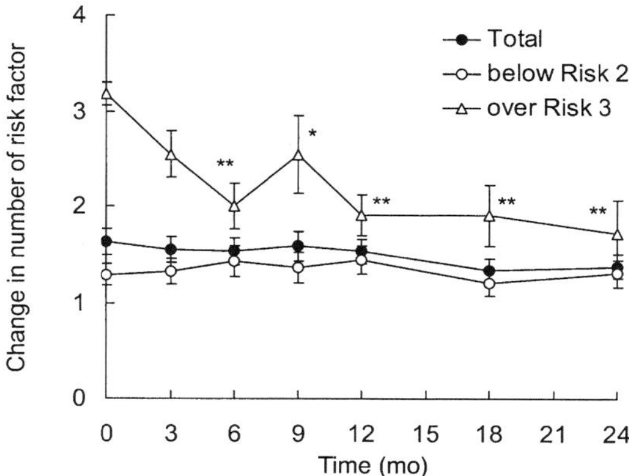

FIG. 11.4. Change in the number of risk factors (enumerative factors) in monadic study. Repeated-measures ANOVA indicated a significant effect of time ($P = 0.002$) in subjects with ≥3 risks factors at baseline. Asterisks indicate different from baseline, $*P < 0.05$; $**P < 0.01$.

study period (P = 0.002); it was significantly different at 6, 12, and 24 mo compared with baseline and was 1.5 points lower at the end of the study. There were no serious adverse effects during the study.

In conclusion, the results of the present clinical study investigating the effect of long-term ad libitum consumption of DAG in obese and/or hypertriglyceridemic subjects in a free-living environment revealed that BMI, waist and hip circumferences, and skinfold thickness decreased during the study period. The atherosclerosis index, fasting blood sugar, and HbA1c also decreased, whereas HDL cholesterol increased. The risk grade for the metabolic syndrome was significantly improved at 2 y compared with baseline. The use of DAG oil instead of ordinary cooking oil may be effective in reducing body fat and the risk of metabolic syndrome.

Study of Ad Libitum Consumption of DAG Oil in Children

DAG Efficacy in Children with Obesity and/or Hypertriglyceridemia (Matsuyama, 2001, Matsuyama et al., 2006). Increasing obesity is a serious, worldwide problem not only in adults, but also in children. Obesity is associated with many risk factors of coronary heart disease in children as well as in adults. Because obesity in childhood frequently continues into adulthood, an increase in the incidence of childhood obesity is a major contributor to adult obesity (Deckelbaum & Williams, 2001). An excessive accumulation of visceral fat is highly correlated with the occurrence of lifestyle-related disease, especially in children (Dietz & Gortmaker, 2001). One of the factors responsible for the development of childhood obesity is physical inactivity. Several clinical studies revealed that increases in body fat in children are positively correlated with an increase in total time spent watching television (Vessey et al., 1998, Anderson et al., 1998) as well as with other sedentary activities (Grund et al., 2000). On the other hand, diet is also an important factor. As mentioned earlier, dietary fat plays a critical role in the increased prevalence of obesity, leading to a positive fat balance due to an imbalance between fat oxidation and fat intake (Maffeis et al., 1997, Schrauwen & Westerterp, 2000). The key strategies in the prevention and treatment of childhood obesity are to maintain a healthy and balanced diet pattern in combination with adequate daily physical activities (William et al., 2001).

The nutritional effects of DAG in children with hyperlipidemia and/or obesity were investigated by Matsuyama et al. (Matsuyama, 2001, Matsuyama et al., 2006). The subjects participating in this clinical study were 11 volunteers who had received daily dietetic instruction at the Tokyo Metropolitan National Health Insurance Organization Fussa Hospital, Tokyo, Japan for obesity and/or hyperlipidemia. DAG oil containing 80% DAG or more (Econa cooking oil, Kao Corporation, Tokyo, Japan) was used as a test oil (see Chapter 19). The subjects consumed DAG oil, ad libitum, in the form of cooking oil for 5 mo. DAG oil was administered by substitution of the edible cooking oil used at home. Guardians of the participants were instructed to provide a similar diet for the children during the study as before its commencement. Before the beginning of the study, diet analysis, anthropometric measures (weight, height, waist and hip), a computerized tomography (CT) scan, and a blood test (fasting subjects) were conducted. After the start of the study, anthropometric measurements and a blood test (fasting subjects) were conducted at monthly intervals. Anthropometric variables were used to calculate the obesity level, BMI, and Rohrers index [body weight (kg)/ weight $(m)^3 \times 10$]. Blood samples from fasting subjects were analyzed every month for the following variables: leptin, PAI-1, glucose, glycoalbumin, insulin, creatinine, and serum lipids (phospholipids, total ketone bodies, acetoacetic acid, hydroxybutyric acid, total cholesterol, LDL-cholesterol, HDL-cholesterol, free cholesterol, serum triglycerides, total/esterified cholesterol ratio, and free fatty acids). A CT scan was also conducted to monitor visceral and subcutaneous fat at 0, 3, and 5 mo.

The mean consumption of DAG oil in the 11 volunteers was 11.3 g/d. Height increased significantly over the course of the study and a significant increase in body weight from the initial values was observed at 2, 4, and 5 mo. Waist circumference increased significantly at 3, 4, and 5 mo, as well as hip at 3 and 4 mo. The waist/hip ratio increased significantly only at 3 mo. The level of obesity, calculated on the bases of standard body weights and heights for Japanese children, generally increased over the course of the study and was significantly higher than baseline at 2, 4, and 5 mo. BMI remained relatively unchanged over the course of the study but was significantly higher than baseline at the end of the study (5 mo). However, CT scan images revealed a significant decrease in total fat area at 3 and 5 mo compared with baseline. The visceral fat area decreased from the baseline, but was not significant at 5 mo ($P = 0.065$). A significant decrease in the subcutaneous fat area was observed at 3 and 5 mo compared with baseline, as shown in Figure 11.5. No significant differences were found in Rohrer's index values.

Significant decreases were also found in serum leptin and PAI-1 levels at several time points during the study period compared with baseline (Table 11.4). Serum leptin levels are reported to be closely associated with body fat mass (Takahashi et al., 1996). Furthermore, it was reported that leptin resistance is frequently found in obese children

TABLE 11.3
Change in Blood Parameters in the 2-y Monadic Study[a]

Month	0	3	6	9	12	18	24	*P*-value[b]
Serum triglyceride (mg/dL)	143 ± 14	132 ± 11	132 ± 11	140 ± 13	135 ± 11	112 ± 9	133 ± 11	0.269
Serum Cholesterol (mg/dL)								
Total	209 ± 4	212 ± 4	215 ± 4	215 ± 4	216 ± 4	214 ± 5	217 ± 4	0.069
HDL	52 ± 2	56 ± 2**	56 ± 2**	57 ± 2**	56 ± 2**	57 ± 2**	58 ± 2**	<0.0001
LDL	128 ± 4	129 ± 4	130 ± 4	128 ± 4	128 ± 4	130 ± 4	130 ± 4	0.818
Arteriosclerosis index	3.20 ± 0.13	3.02 ± 0.14	3.08 ± 0.15	2.97 ± 0.13*	3.06 ± 0.13	2.93 ± 0.15**	2.94 ± 0.14**	0.01
Hemoglobin A1c (%)	5.1 ± 0.1	5.0 ± 0.1	5.0 ± 0.1	5.1 ± 0.1	5.0 ± 0.1	4.9 ± 0.1**	5.0 ± 0.1*	<0.0001
Insulin (mU/mL)	8.3 ± 0.5	8.2 ± 0.6	8.9 ± 0.9	8.6 ± 0.8	7.7 ± 0.6	7.5 ± 0.5	8.7 ± 0.6	0.168
Glucose (mg/dL)	103 ± 2	101 ± 2	103 ± 3	102 ± 2	104 ± 3	101 ± 2	104 ± 4	0.119

[a]Values are means ± SEM, n = 60. Asterisks indicate different from baseline:**P < 0.01; *P < 0.05 (Dunnett test).
[b]Repeated-measures ANOVA.

(Lahlou et al., 1997) and that diet therapy and exercise improve leptin levels as observed in adulthood obesity (Nakane et al., 1999). As mentioned earlier, plasma PAI-1 levels are also positively correlated with visceral fat mass. On the basis of these findings, ad libitum consumption of DAG in obese children may contribute to the suppression of body fat accumulation by improving lipid metabolism via the hormone network in adipose tissue.

Serum triglyceride levels decreased significantly at 4 mo and HDL cholesterol levels increased at 4 and 5 mo (Table 11.4). No significant differences were observed in the other serum lipids. Serum glycoalbumin, insulin, and creatinine levels did not differ from baseline. Serum glucose levels tended to decrease over the first few months and were significantly decreased at 2 mo compared with baseline. No adverse effects including effects on growth and development were found in the study, supporting the safety of DAG consumption in various populations (see Chapter 17). The positive changes in HDL cholesterol and serum triglyceride also support the potential benefits of DAG consumption.

In conclusion, the daily intake of DAG oil in place of regular cooking oil in home use had an effect on lipid metabolism in obese child patients or children with hyperlipidemia. Because of the difficulty in strict diet restriction and medication in children, DAG oil may potentially play a crucial role in the prevention and treatment of childhood obesity.

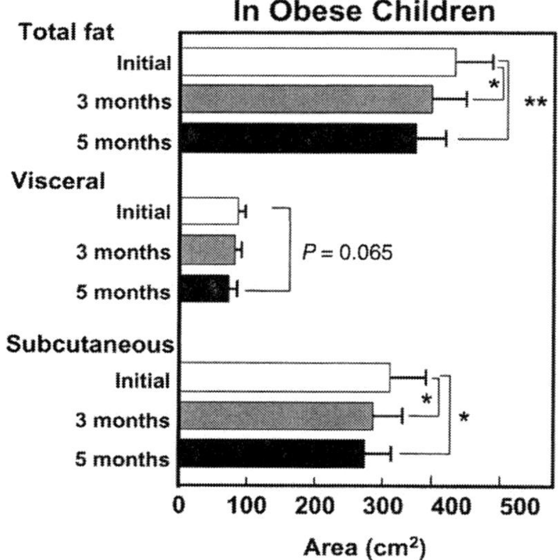

FIG. 11.5. Changes in body fat areas at the level of the umbilicus by CT scan in the study of obese children. Values are means ± SEM, n = 11. Asterisks indicate different from baseline, $^{*}P < 0.05$; $^{**}P < 0.01$.

Summary

In the present chapter, the effect of dietary DAG oil consumed practically as a cooking oil at home is summarized with respect to anthropometric and blood variables in adults and children in a free-living environment. A double-blind controlled study in obese and/or hypertriglyceridemic adults (Kawashima et al., 2008) revealed that long-term DAG oil intake significantly reduced body weight, BMI, skinfold thickness, and waist circumference compared with TAG oil intake. One of the interesting findings in this study is that the weight loss could be attributed solely to the consumption of DAG oil, because total energy and fat intake were constant during the study period. Two open-labeled studies (Yasukawa & Yasunaga, 2001, Katsuragi et al., 1999, Otsuki et al., 2004) conducted as post-market studies also indicated significant decreases in skinfold thickness, waist circumference, BMI (2 y study) and weight (1 y study) compared with baseline. Although the improvements in anthropometric variables found in the present open-labeled studies might be overestimated due to the placebo effect, these findings strongly support the effectiveness of DAG oil on weight loss and maintenance of weight loss used as a part of everyday dietary conditions. In addition to adults, several improvements in anthropometric and blood profiles were also observed in the open-labeled, uncontrolled study in obese and/or hyperlipidemia children (Matsuyama, 2001, Matsuyama et al., 2006), in which total and subcutaneous fat area, as measured by a CT scan image, significantly decreased compared with baseline, accompanied by a reduction in blood variables relating to adiposity and fat mass including leptin and PAI-1.

The potential mechanism responsible for this weight loss was reported in animal studies, showing that DAG oil stimulates energy expenditure and mRNA involved in -oxidation in the liver and small intestine (Dietz & Gortmaker, 2001). Further investigations are warranted to clarify the precise mechanism of weight loss induced by dietary DAG oil. Most of the current diet strategies for attempting weight loss as part of the treatment of obesity restrict total energy and fat intake. However, it is very difficult to continuously manage a restricted diet or to achieve and maintain body weight and fat in

TABLE 11.4
Change in Blood Variables in the Study of Obese Children[a,b]

	Baseline	1 mo	2 mos	3 mo	4 mo	5 mo
Serum triglyceride (mg/dL)	115 ± 22	117 ± 17	102 ± 18	118 ± 24	97 ± 21*	109 ± 31
Serum cholesterol (mg/dL)						
Total	173 ± 11	186 ± 10	173 ± 12	181 ± 11	175 ± 11	182 ± 14
LDL	64 ± 3	63 ± 4	66 ± 3	63 ± 4	64 ± 4	59 ± 6
HDL	48 ± 4	49 ± 4	48 ± 3	48 ± 4	52 ± 4**	54 ± 5**
Plasma PAI-1 (ng/mL)	34.6 ± 6.0	38.6 ± 5.6	30.6 ± 5.6*	34.3 ± 4.7	29.9 ± 4.4	27.4 ± 4.0
Serum leptin (ng/mL)	24.2 ± 4.0	24.5 ± 4.6	17.9 ± 2.4*	18.8 ± 2.8**	17.4 ± 2.3*	19.1 ± 2.3
Serum Insulin (mU/mL)	17 ± 2	17 ± 3	16 ± 2	18 ± 3	19 ± 4	18 ± 4

[a]Values are means ± SEM, n = 11. Asterisks indicate different from baseline:**$P < 0.01$; *$P < 0.05$.
[b]Abbreviation: PAI-1, plasminogen activator inhibitor type-1.

the normal range. In this regard, DAG oil could be one practical option for use in the prevention and treatment of obesity without any loss of palatability.

References

Anderson, R.E., Crespo, C.J., Bartlett, S.J., Cheskin, L.J., Pratt, M. Relationship of Physical Activity and Television Watching with Body Weight and level of Fatness Among Children. Results from the Third National Health and Nutrition Examination Survey. *J. Am. Med. Assoc.* **1998,** *279,* 938-942

Bray, G.A., and Popkin, B.M. Dietary Fat Intake Does Affect Obesity. *Am. J. Clin. Nutr.* **1998,** *68,* 1157-1173.

Bray, G.A., Lovejoy, J.C., Most-Windhauser, M., Smith, S.R., Volaufova, J., Denkins, Y., de Jonge, L., Rood, J., Lefevre, M., Eldridge, A.L., and Peters, J, C. A 9-mo Randomized Clinical Trial Comparing Fat-Substituted and Fat-Reduced Diets in Healthy Obese Men: The Ole Study. *Am J. Clin. Nutr.* **2002,** *76,* 928-934.

Davidson, M.H., Hauptman, J., DiGirolamo, M., Foreyt, J.P., Halsted, C.H., Heber, D., Heimburger, D.C., Lucas, C.P., Robbins, D, C., Chung, J., and Heymsfield, S.B. Weight Control and Risk Factor Reduction in Obese Subjects Treated for 2 Years with Orlistat. *J. Am. Med. Assoc.* **1999,** *281,* 235-242.

Deckelbaum, R.J., and Williams, C.L. Childhood Obesity: The Health Issue. *Obes.* **2001.** *Suppl 4,* 239S-243S.

Dietz, W.H., and Gortmaker, S.L. Preventing Obesity in Children and Adolescents. *Annu. Rev. Public Health* **2001,** *22,* 232-253.

Ford, E.S., Giles, W.H., and Dietz, W.H. Prevalence of the Metabolic Syndrome Among US Adults; Findings from the Third National Health and Nutrition Examination Survey. *J. Am. Med. Assoc.* **2002,** *287,* 356-359.

Grund, A., Dilba, B., Forberger, K., Krause, H., Siewers, M., Rieckert, H., and Muller, M.J. Relationships Between Physical Activity, Physical Fitness, Muscle Strength and Nutritional State in 5- to 11-Year-Old Children. *Eur. J. Appl. Physiol.* **2000,** *82,* 425-438.

Heshka, S., Anderson, J.W., Atkinson, R.L., Greenway, F.L., Hill, R.O., Phinney, S.D., Kolontkin, R.L., Miller-Kovach, K., and Pi-Sunyer, F.X. Weight Loss with Self-Help Compared with a Structured Commercial Program. *J. Am. Med. Assoc.* **2003,** *289,* 1792-1798.

Jeffery, R.W., Hellerstedt, W.L., French, S.A., and Baxter, J.E. A Randomized Trial of Counseling for Fat Restriction Versus Calorie Restriction in the Treatment of Obesity. *Int. J. Obes. Relat. Metab. Disord.* **1995,** *19,* 132-137.

Kamphuis, M.M., Mela, D.J., Westerterp-Plantenga, M.S. Diacylglycerols Affect Substrate Oxidation and Appetite in Humans. *Am. J. Clin. Nutr.* **2003**, *77,* 1133-1139.

Katsuragi, Y., Toi, T., and Yasukawa, T. Effects of Dietary Diacylglycerols on Obesity and Hyperlipidemia. *J. Jpn. Human Dry Dock* **1999,** *14,* 258-262.

Kawashima, H., Takase, H., Yasunaga, K., Wakaki, Y., Mori, K., Yamaguchi, T., Hase, T., Matsuo, N., Yasukawa, T., Tokimitsu, I., and Koyama, W., One-Year Ad Libitum Consumption of Diacylglycerol Oil As Part of a Regular Diet Results in Modest Weight Loss in Comparison With Consumption of a Triacylglycerol Control Oil in Overweight Japanese Subjects. *J. Am. Diet. Assoc.* **2008,** *208,* 57-66

Krauss, R.M., and Witson, M. Obesity: Impact on Cardiovascular Disease, *Circulation* **1998**; *98,* 1472-1476.

Lahlou, N., Landais, P., De Boissieu, D., and Bougneres, P.F. Circulating Leptin in Normal Children and During the Dynamic Phase of Juvenile Obesity: Relation to Body Fatness,

Energy Metabolism, Caloric Intake, and Sexual Dimorphism. *Diabetes* **1997,** *46,* 989-993.

Levy, A.S., and Heaton, A.W. Weight Control Practice in US Adults Trying to Lose Weight. *Ann. Intern. Med.* **1993,** *119,* 667-671.

Maffeis, C., Zaffanello, and M., Schutz, Y. Relationship Between Physical Inactivity and Adiposity in Prepubertal Boys. *J. Pediatr.* **1997,** *131,* 288-292.

Maki, K.C., Davidson, M.H., Tsushima, R., Tokimitsu, I., Umporowicz, D.M., Dicklin, M.R., Foster, G.S., Ingram, K.A., Anderson, B.D., Frost, S.D., and Bell, M. Consumption of Diacylglycerol Oil as Part of a Mildly Reduced-Energy Diet Enhances Loss of Body Weight and Fat Compared with a Triacylglycerol Control Oil. *Am. J. Clin. Nutr.* **2002,** *76,* 1230-1236.

Matsuyama, K. Effect of Diacylglycerol in the Obese Children of Patients. *Pediatr. Jpn.* **2001,** *43*, 928-933.

Matsuyama, K., Shoji, K., Watanabe, H., Shimizu, M., Saotome, Y., Nagao, T., Matsuo, N., Hase, T., Tokimitsu, I., and Nakaya, N. Effects of diacylglycerol on adiposity in obese children. *Journal of Pediatric Endocrinology & Metabolism.* **2006,** *19,* 795-804

Nagao, T., Watanabe, H., Gotoh, N., Onizawa, K., Taguchi, H., Matsuo, N., Yasukawa, T., Tsushima, R., Shimasaki, H., and Itakura, H. Dietary Diacylglycerol Suppresses Accumulation of Body Fat Compared to Triacylglycerol in Men in a Double-Blind Controlled Trial. *J. Nutr.* **2000,** *130,* 792-797.

Nakane, T., Asayama, K., Hayashibe, H., Uchida, N., Kodera, K., and Dobashi, K. Changes in Serum Leptin Concentration During Behavioral Therapy in Obese Children. *Endocr. J.* **1999,** *46,* 703-709.

National Institute of Health, National Heart, Lung, and Blood Institute Obesity Education Initiative. *Clinical Guidelines on the Identification, Evaluation, and Treatment of Overweight and Obesity in Adults.* **1998**; National Institutes of Health, Bethesda, MD.

NCEP, Expert Panel on Detection, Evaluation, and Treatment of High Blood Cholesterol in Adults Executive Summary of the Third Report of the National Cholesterol Education Program (NCEP) Expert Panel on Detection, Evaluation, and Treatment of High Blood Cholesterol in Adults (Adults Treatment Panel III). *J. Am. Med. Assoc.* **2001,** *285,* 2486-2497.

Oseroff, A., Krishnamurti, C., Hassett, A., Tang, D., and Alving, B. Plasminogen Activator and Plasminogen Activator Inhibitor Activities in Men with Coronary Artery Disease. *J. Lab. Clin. Med.* **1989,** *113,* 88-93.

Ostuki, K., Mori. K., Takase, H., and Onodera, Y. Two years, long-term effects of dietary diacylglycerols on the risks of the metabolic syndrome. *J. Jpn. Human Dry Dock* **2004,** *19,* 29-32.

Schrauwen, P., and Westerterp, K. The Role of High-Fat Diets and Physical Activity in the Regulation of Body Weight. *Br. J. Nutr.* **2000,** *84,* 417-427.

Shimomura, I., Funahashi, T., Takahashi, M., Maeda, K., Kotani, K., Nakamura, T., Yamashita, S., Miura, M., Fukuda, Y., Takemura, K., Tokunaga, K., and Matsuzawa Y. Enhancement Expression PAI-1 in Visceral Fat: Possible Contributor to Vascular Disease in Obesity. *Nat. Med.* **1996,** *2,* 800-803.

Taguchi, H., Watanabe, H., Onizawa, K., Nagao, T., Gotoh, N., Yasukawa, T., Tsushima, R., Shimasaki, H., and Itakura, H. Double-Blind Controlled Study on the Effects of Dietary Diacylglycerol on Postprandial Serum and Chylomicron Triacylglycerol Responses in Healthy Humans. *J. Am. Coll. Nutr.* **2000,** *19,* 789-796.

Taguchi, H., Nagao, T., Watanabe, H., Onizawa, K., Matsuo, N., Tokimitsu, I., and Itakura, H. Energy Value and Digestibility of Dietary Oil Containing Mainly 1,3-Diacylglycerol Are

Similar to Those of Triacylglycerol. *Lipids* **2001,** *36,* 379-382.

Takahashi, M., Funahashi, T., Shimomura, I., Miyaoka, K., and Matsuzawa, Y. Plasma Leptin Levels and Body Fat Distribution. *Horm. Metab. Res.* **1996,** *28,* 751-752.

Vessey, J.A., Yim-Chiplis, P.K., and MacKenzie, N.R. Effects of Television Viewing on Childrens Development. *Pediatr. Nurs.* **1998,** *24,* 483-486.

Watanabe, H., Onizawa, K., Taguchi, H., Kobori, K., Chiba, H., Naito, S., Matsuo, N., Yasukawa, T., Hattori, M., and Shimasaki, H. Nutritional Characterization of Diacylglycerols in Rats. *J. Jpn. Oil Chem. Soc.* **1997,** *46,* 301-307.

William, C.L., Gulli, M.T., Deckelbaum, R.J. Prevention and Treatment of Childhood Obesity. *Curr. Atheroscler. Rep.* **2001,** *3,* 486-497.

Wirth, A., and Krause, J. Long-Term Weight Loss with Sibutramine. *J. Am. Med. Assoc.* **2001,** *286,* 1331-1339.

Yasukawa, T., and Yasunaga, K. Nutritional Functions of Dietary Diacylglycerols. *J. Oleo. Sci.* **2001,** *50,* 427-432.

Section C

Lifestyle-Related Disease

Chapter 12

Effects of Long-Term Ingestion of Diacylglycerol Oil on Various Clinical Conditions of Diabetes

Kunio Yamamoto[a] and Hideki Asakawa[b]

[a]Department of Health Management, Koshien University, Graduate School of Nutrition, Takarazuka, Hyogo, 665-0006 Japan and [b]Asakawa Naika Clinic, Suita, Osaka, 564-0031 Japan

Introduction

Type 2 diabetes, formerly known as non-insulin-dependent diabetes, accounts for most cases of diabetes worldwide. The World Health Organization reported that the number of cases of diabetes worldwide is currently estimated to be ~150 million, and that number is predicted to double by 2025 (WHO, 2003). Type 2 diabetes has escalated recently in all age groups and is now being identified in younger age groups, including children (WHO, 2003). Overweight and obesity are associated with an increased risk of type 2 diabetes. In particular, the distribution of body fat, rather than the body mass index (BMI), may be an important risk factor (Colditz, et al., 1990; Després, et al., 2001; Chan et al., 1994; Boyko et al., 2000; Despres, 2001). Preventing type 2 diabetes is also important because of its association with hypertriglyceridemia.

Recently, the nutritional properties of a diacylglycerol-rich oil containing 80% or more DAG (DAG oil) were reported. It was shown that the postprandial increase in serum triglyceride levels is lower in humans after the ingestion of DAG oil, compared with triacylglycerol oil (TAG oil) with the same fatty acid composition (Taguchi et al., 2000). The suppression of triglyceride levels was more pronounced in the chylomicrons (Taguchi et al., 2000) and remnant lipoprotein fractions (Tada et al., 2001). A plausible explanation for these results is that DAG oil, which contains mainly 1,3-DAG, is not readily resynthesized to triglyceride by small intestinal epithelial cells (Murase et al., 2002; Kondo et al., 2003), and this suppresses the lymphatic secretion of triglycerides (Murata et al., 1994). Moreover, long-term ingestion of DAG oil was reported to decrease fasting serum glucose and serum insulin levels (Murase et al., 2002) in obese and diabetes-prone mice (C57BL/6J) fed a high-fat diet. On the basis of these nutritional characteristics of DAG oil, the consumption of DAG oil might be beneficial in the treatment of the hypertriglyceridemia that accompanies diabetes, and might relieve multiple risk factors for the development of arteriosclerosis. In this chapter, we describe the effects of DAG oil consumption on various clinical conditions of diabetes in humans.

DAG Oil Consumption Affects Fasting Serum Triglycerides in Type 2 Diabetic Patients

Yamamoto, et al. (Yamamoto et al., 2001) conducted a clinical study to determine the effect of long-term consumption of DAG oil on fasting triglyceride levels in type 2 diabetic patients. The fasting serum triglyceride levels of these patients were persistently high, despite continuous nutritional counseling at the outpatient clinic. This was a randomized, single-blind, controlled, parallel trial conducted at the Itami City Hospital, Itami, Japan to determine the influence of the long-term ingestion of DAG oil on fasting blood lipids in type 2 diabetics (n = 16) with hypertriglyceridemia. Subjects were randomly assigned to the DAG group (3 men, 5 women; age, 56.8 ± 7.3 y) or the TAG group (4 men, 4 women; age, 54.1 ± 18.8 y). The serum triglyceride levels of these patients remained high despite continuous nutritional counseling at the outpatient clinic over a period of 14.4 ± 12.7 mo. Five patients in the DAG group and six in the TAG group received only nutritional counseling. Three subjects in the DAG group each received a different medication to control blood glucose (sulfonylurea, an α-glucosidase inhibitor, and insulin). Two subjects in the TAG group each received a different medication (sulfonylurea and an α-glucosidase inhibitor). The ordinary cooking oil used in their homes was replaced by the test oils (target dose, 10 g/d). For both groups, the study period was 3 mo. Both test oils were used for cooking, such as frying food, and for homemade dressing. Blood samples were collected from fasting subjects at wks 0 and 12.

Body weight, energy intake, the percentage of energy from fat, total fat intake, total cooking oil intake, DAG oil intake, and TAG oil intake measured before and at the end of the study are shown in Table 12.1. Serum triglyceride, total cholesterol, and HDL cholesterol levels before and after the test period are shown in Table 12.2 for both groups. Baseline serum triglyceride, total serum cholesterol, and HDL cholesterol levels did not differ significantly between the DAG and TAG groups. Changes in these variables dur-

TABLE 12.1
Body Anthropometric Variables and Energy and Fat Intakes in Diabetic Patients Before and 12 wk After the Substitution of DAG Oil for Ordinary Cooking Oil[a]

	DAG group		TAG group (control)	
	Before	After	Before	After
Body weight (kg)	60.8 ± 7.6	60.7 ± 7.2	74.5 ± 15.5	73.9 ± 15.0
Body mass index (kg/m^2)	24.4 ± 1.7	24.4 ± 1.6	28.1 ± 2.9	27.9 ± 2.9
Energy intake (MJ/d)	6.93 ± 0.92	6.83 ± 0.59	7.30 ± 2.57	6.88 ± 2.36
Energy from fat (%)	27.9 ± 4.9	30.0 ± 4.0	28.3 ± 5.8	26.9 ± 6.6
Total fat intake (g/d)	52.0 ± 13.7	54.9 ± 11.4	56.9 ± 30.4	51.1 ± 30.7
Total oil intake (g/d)	27.1 ± 9.4	26.4 ± 9.3	25.1 ± 15.4	23.9 ± 19.6
TAG intake (g/d)	26.8 ± 9.3	15.7 ± 8.9*	24.9 ± 15.2	23.5 ± 19.5
DAG intake (g/d)	0.3 ± 0.1	10.6 ± 3.9**	0.3 ± 0.1	0.3 ± 0.2

[a]Values are means ± SD, n = 8/group. Asterisks indicate different from before, *$P < 0.05$; **$P < 0.001$.

TABLE 12.2
Serum Variables in Diabetic Patients Before and 12 wk After the Substitution of DAG Oil for Ordinary Cooking Oil[a]

	DAG group		TAG group (control)	
	Before	After	Before	After
Total Cholesterol (mmol/L)	5.82 ± 1.32	5.87 ± 0.80	6.00 ± 0.98	5.74 ± 0.70
Triglyceride (mmol/L)	2.51 ± 0.75	1.52 ± 0.28**†	3.22 ± 2.13	3.59 ± 1.70
HDL Cholesterol (mmol/L)	1.27 ± 0.23	1.34 ± 0.36	1.09 ± 0.20	1.22 ± 0.36
Glucose (mmol/L)	6.72 ± 0.72	6.94 ± 0.89	7.83 ± 1.78	7.77 ± 2.50
Hemoglobin A_{1c} (%)	6.41 ± 1.15	5.79 ± 0.85*	6.88 ± 0.53	6.65 ± 0.73

[a]Values are means ± SD, n = 8/group. Asterisks indicate different from before, *$P < 0.05$; **$P < 0.01$. †Different from control group, $P < 0.05$.

ing the test period were not significant in the control group. In contrast, the serum triglyceride levels in the DAG group decreased ($P < 0.01$) significantly after the 3-mo treatment, and were significantly ($P < 0.05$) lower than those of the control group. Fasting blood glucose levels before and after the test period are also shown in Table 12.2. No differences in fasting serum glucose levels were found during the study period in either group. In the DAG group, however, glycohemoglobin (HbA_{1c}) levels decreased from 6.41 ± 1.15 to 5.79 ± 0.85% (9.7% reduction, $P < 0.05$). The normal range of HbA_{1c} established by the Japan Diabetes Society is 4.3–5.8%. No differences in HbA_{1c} levels were observed in the control group. The reason for the reduction of HbA_{1c} in the DAG group without changes in fasting serum glucose levels is not known at present. A more detailed investigation is warranted to determine the effect of DAG oil on diabetic patients including postprandial responses to DAG oil.

DAG Oil Consumption Decreased LDL and Small Dense LDL-Triglyceride in Type 2 Diabetic Patients

Patients with diabetes frequently develop small dense LDL (sLDL), a subclass of LDL particles that differs in size, density, and chemical composition from normal LDL. This subclass is associated with a relative increase in serum triglyceride levels, reduced levels of HDL, and an increased risk of coronary artery disease, compared with subjects with larger LDL (Feingold et al., 1992).

To investigate in detail the effects of DAG oil consumption on lipid metabolism as well as anthropometric variables, a double-blind, placebo-controlled, clinical trial was conducted in type 2 diabetic patients (Yamamoto et al., 2006). In particular, the study objective was to determine the difference in lipoprotein levels and their sizes. Type 2 diabetic patients (n = 24) participated in the study and were required to ingest either DAG oil or TAG oil for 3 mo; 11 subjects were assigned to the DAG group (4 men, 7 women; age, 61.5 ± 6.2 y, mean ± SD) and 13 to the TAG group (7 men, 6 women; age, 54.3 ± 13.1 y). The participants were provided with bottled DAG or TAG oil with no label. The subjects consumed DAG or TAG oils by replacing their ordinary cooking oil with the test oils for 3 mo. The target dose of the test oil was 10 g/d. The actual daily test oil intake, determined at the end of the treatment period, was 15.5 ± 8.2 g/d in the DAG group and 14.6 ± 7.3 g/d in the TAG group. There were no differences between groups in daily energy intake, fat intake, or amount of test oil ingested.

During the 3 mo of treatment, changes from the initial waist circumference values were statistically significant (DAG group: –1.2 cm at 1 mo, –1.1 cm at 2 mo, –1.5 cm at 3 mo). Body mass index was not significantly different between groups. In the DAG group, serum triglyceride levels were reduced ($P < 0.05$) and serum HDL-cholesterol levels were increased ($P < 0.01$) from baselines. These changes were significantly different between treatment groups. Plasma plasminogen activator inhibitor-1 concentrations were significantly lower in the DAG oil group. Thus, DAG oil consumption decreased the fasting serum triglyceride levels and waist circumferences in diabetic patients with hypertriglyceridemia. The results were consistent with previous study results (Yamamoto et al., 2001). However, variables related to glucose metabolism, such as blood glucose and insulin levels, were not significantly different between groups in this study.

A recent study has suggested that an increased proportion of small LDL particles (<25.5 nm) is closely associated with future cardiovascular disease, even in the presence of a relatively normal LDL-cholesterol concentration, and that LDL particle size might be a factor that can be used to accurately predict the risk of ischemic heart disease (St-Pierre et al., 2001). In a study by Yamamoto et al., LDL particle size was significantly larger after 3 mo of consumption of DAG oil, compared with the TAG oil, in patients whose initial LDL particle size was not larger than 25.5 nm. DAG oil ingestion might be meaningful as it relates to a therapeutic strategy for decreasing the risk of CHD in patients with diabetes. However, its relationship to insulin resistance or other factors related to insulin resistance remains unknown.

These results suggest that the daily consumption of DAG oil may improve impaired lipid metabolism by preventing the postprandial formation of sLDL, resulting in a decrease in serum triglyceride levels and an increase in HDL cholesterol levels. Waist circumference (Janssen et al., 2002; Harris, et al., 2000) and serum PAI-1 level (Shimomura et al., 1996) were positively associated with abdominal fat mass, particularly visceral fat mass. Decreases in waist circumference and serum PAI-1 levels were considered to be the result of visceral fat reduction. Recent studies revealed that DAG oil consumption significantly suppressed body fat accumulation (Nagao et al., 2000; Maki et al., 2002) (see Chapters 10 and 11). The distribution of body fat, especially visceral fat, is important for the risk of coronary heart disease (Matsuzawa et al., 1995). From the results of these two clinical studies, it appears that DAG oil consumption may contribute to a reduction in visceral fat and a consequent improvement in variables of lipid metabolism such as serum triglyceride and HDL levels. Elucidation of the mechanism for the effects of DAG oil consumption in diabetic patients, however, requires further studies on lipid metabolism.

DAG Oil Consumption Delays the Progression of Renal Failure in Type 2 Diabetic Patients with Nephropathy

Diet therapy for diabetic nephropathy is important in preventing the progression of renal failure by controlling hyperglycemia, hypertension, hyperlipidemia, and obesity. From the results mentioned in the previous sections, we hypothesized that diet therapy with DAG oil may be useful for patients with diabetic nephropathy. We conducted a clinical study for 6 months, with an additional 3-year follow-up period, to investigate the long-term effects of DAG oil in dietary therapy in subjects with type 2 diabetes with nephropathy.

The subjects were 15 outpatients (aged 47-77 years) at the Itami City Hospital who were diagnosed with type 2 diabetes in stage IIIb (macroalbuminuric stage with renal dysfunction; n=2) and stage IV (renal failure stage; n=13) according to the classification of diabetic nephropathy (Shigeta et al., 1992). They had received nutritional counseling for 13 months on average. The duration of diabetes was 18.2±7.8 (range 5-28) years.

The study was an open labeled, parallel design. The subjects in the DAG group were requested to replace the usual cooking oil used at home with the DAG oil, whereas subjects in the control group were asked to use their usual cooking oil, containing mainly triacylglycerol, for 6 months. The target doses of test oil, energy and protein intake were 30 kcal/kg and 0.8g/kg of ideal body weight, respectively. DAG oil consumption significantly reduced fasting serum triglyceride levels. The delayed progression of renal failure with DAG consumption may have resulted from improvement in hypertriglyceridemia, a predictor of progressive renal dysfunction. In the DAG group, body weight, BMI, and serum triglyceride levels were significantly decreased from the baseline to 6 mo, and the percent changes were significantly different between the groups. The DAG group maintained their serum creatinine levels during the study period, whereas levels in the control group were significantly increased from the baseline (Table12.3). In the 3-year follow-up period after the 6-month treatment, the progression of renal failure and onset of dialysis in the DAG group were retarded as compared to the TAG group.

Thus, the long-term ingestion of DAG oil prevented the progression of renal failure in subjects with type 2 diabetes with nephropathy, delaying the onset of dialysis.

Summary

Through two clinical studies involving diabetic patients with hypertriglyceridemia, the consumption of DAG oil was clearly shown to decrease fasting serum triglyceride levels. The goal of nutritional counseling for diabetics is to maintain normal blood sugar levels and to prevent complications of diabetes. A decrease in the hypertriglyceridemia that accompanies diabetes might relieve multiple risk factors and aid in preventing arteriosclerosis. Thus, we conclude that the use of DAG oil as cooking oil in a free-living environment is useful for maintaining the quality of life for diabetic patients. These new insights may lead to a revision of the recommended standard instructions concerning fat limitation for diabetics. Diet therapy for diabetic nephropathy is also important in prevent-

TABLE 12.3

Clinical Characteristics and Laboratory Data for Subjects During the Treatment Period

	Control group			DAG group		
	Baseline	6 months	Change (%)	Baseline	6 months	Change (%)
Age (years)	66.0 ±9.2			60.3 ±7.0		
Sex (male/female)	6/1			7/1		
Duration of DM (years)	19.9 ± 7.4			16.8 ± 8.4		
Stage of nephropathy a)	1/6 (IIIb/ IV)			1/7 (IIIb/ IV)		
Weight (kg)	66.9 ± 8.1	68.1 ± 8.1	1.9 ± 3.8	67.6 ± 10.9	64.0 ± 10.4**	-5.1 ± 3.7‡
BMI (kg/m^2)	24.6 ± 2.5	25.1 ± 3.0	1.9 ± 3.8	25.1 ± 4.0	23.8 ± 4.0**	-5.1 ± 3.7‡
Systolic blood pressure (mmHg)	147 ± 26	145 ± 21	1.8 ± 24.7	141 ± 18	148 ± 21	5.5 ± 11.1
Diastolic blood pressure (mmHg)	75 ± 18	78 ±14	9.6 ± 35.7	84 ± 12	82 ± 15	-3.1 ± 7.7
Triglycerid (mmol/L)	2.05 ± 0.63	2.08 ± 0.75	6.8 ± 37.0	2.56 ± 1.21	1.72 ± 0.46*	-26.5 ± 20.5†
Glucose (mmol/L)	8.3 ± 2.9	8.3 ± 2.3	2.3 ± 16.0	7.8 ± 3.0	6.7 ± 1.2	-6.2 ± 27.0
HbA_{1c} (%)	7.4 ± 2.0	7.3 ± 1.5	1.5 ± 12.4	7.3 ± 1.8	6.3 ± 1.0	-10.4 ± 16.3
Blood urea nitrogen (mmol/L)	9.2 ± 2.6	9.7 ± 2.9	6.6 ± 19.9	11.1 ± 5.3	10.3 ± 3.0	-1.1 ± 25.5
Creatinine (μmol/L)	164.2 ± 63.3	207.1 ± 89.8*	24.4 ± 15.2	198.9 ± 99.2	189.0 ± 90.4	-3.7 ± 18.4‡

Data are expressed as mean ± SD.

a) Stage of nephropathy: IIIb, macroalbuminuric stage with renal dysfunction; IV, renal failure stage.

Significantly different from the baseline to 6 months; * $p < 0.05$, ** $p < 0.01$.

Significantly different in percent changes from the baseline to 6 months between the groups; †$p < 0.05$, ‡ $p < 0.01$.

ing the progression of renal failure by controlling hyperglycemia, hypertension, hyperlipidemia and obesity. Consumption of DAG oil might have improved these conditions and resulted in a favorable response in these patients. Further investigations will be needed to clarify the relationship between body fat loss and the prevention of renal failure by DAG oil intake.

References

Boyko, E.J., Fujimoto, W.Y., Leonetti, D.L., and Newell-Morris, L. Visceral Adiposity and Risk of Type 2 Diabetes: a Prospective Study Among Japanese Americans, *Diabetes Care*. **2000** *23*, 465-471.

Chan, J.M., Rimm, E.B., Colditz, G.A., Stampfer, M.J., and Willett, W.C. Obesity, Fat Distribution, and Weight Gain as Risk Factors for Clinical Diabetes in Men. *Diabetes Care*. **1994**, *17*, 961-969.

Colditz, G.A., Willett, W.C., Stampfer, M.J., Manson, J.E., Hennekens, C.H., Arky, R.A., and Speizer, F.E. Weight as a Risk Factor for Clinical Diabetes in Women. *Am. J. Epidemiol.* **1990**, *132*, 501-513.

Deprés, J.P. Health Consequences of Visceral Obesity. *Ann. Med.* **2001**, *33*, 534-541.

Deprés, J.-P., Lemieux, I., and Prud'homme, D. Treatment of Obesity: Need to Focus on High Risk Abdominally Obese Patients. *Br. Med. J.* **2001**, *322*, 716-720.

Feingold, K.R., Grunfeld, C., Pang, M., Doerrler, W., and Krauss, R.M. LDL Subclass Phenotypes and Triglyceride Metabolism in Non-Insulin-Dependent Diabetes. *Arterioscler. Thromb.* **1992**, *12*, 1496-1502.

Harris, T.B., Visser, M., Everhart, J., Cauley, J., Tylavsky, F., Fuerst, T., Zamboni, M., Taaffe, D., Resnick, H.E., Scherzinger, A., and Nevitt, M. Waist Circumference and Sagittal Diameter Reflect Total Body Fat Better than Visceral Fat in Older Men and Women. The Health, Aging and Body Composition Study. *Ann. N.Y. Acad. Sci.* **2000**, *904*, 462-473.

Janssen, I.B., Heymsfield, S.B., Allison, D.B., Kotler, D.P., and Ross, R. Body Mass Index and Waist Circumference Independently Contribute to the Prediction of Nonabdominal, Abdominal Subcutaneous, and Visceral Fat. *Am. J. Clin. Nutr.* **2002**, *75*, 683-688.

Kondo, H., Hase, T., Murase, T., and Tokimitsu, I. Digestion and Assimilation Features of Dietary DAG in the Rat Small Intestine. *Lipids*. **2003**, *38*, 25-30.

Maki, K.C., Davidson, M.H., Tsushima, R., Matsuo, N., Tokimitsu, I., Umporowicz, D.M., Dicklin, M.R., Foster, G.S., Ingram, K.A., Anderson, B.D., Frost, S.D., and Bell, M. Consumption of Diacylglycerol Oil as Part of a Reduced-Energy Diet Enhances Loss of Body Weight and Fat in Comparison with a Consumption of Triacylglycerol Control Oil. *Am. J. Clin. Nutr.* **2002**, *76*, 1230-1236.

Matsuzawa, Y., Nakamura. T, Shimomura, I., and Kotani, K. Visceral Fat Accumulation and Cardiovascular Disease. *Obes. Res.* **1995**, *5*, 645S-647S.

Murase, T., Aoki, M., Wakisaka, T., Hase, T., and Tokimitsu, I. Anti-Obesity Effect of Dietary Diacylglycerol in C57BL/6J Mice: Dietary Diacylglycerol Stimulates Intestinal Lipid Metabolism. *J. Lipid Res.* **2002**, *43*, 1312-1319.

Murata, M., Hara, K., Ide, T. Alteration by Diacylglycerols of the Transport and Fatty Acid Composition of Lymph Chylomicrons in Rats. *Biosci. Biotech. Biochem.* **1994**, *58*, 1416-1419.

Nagao, T., Watanabe, H., Goto, N., Onizawa, K., Taguchi, H., Matsuo, N., Yasukawa, T., Tsushima, R., Shimasaki, H., and Itakura, H. Dietary Diacylglycerol Suppresses Accumulation of Body Fat Compared to Triacylglycerol in Men in a Double-Blind Controlled Trial. *J. Nutr.* **2000**, *130*, 792-797.

Shigeta, Y., Classification of type 2 diabetic nephropathy. Report of the National Diabetes Research Group. Tokyo, Japan, Ministry of Health and Welfare. **1992**, p. 317-320 [in Japanese].

Shimomura, I., Funahashi, T., Takahashi, M., Maeda, K., Kotani, K., Nakamura, T., Yamashita, S., Miura, M., Fukuda, Y., Takemura, K., Tokunaga, K. and Matsuzawa, Y. Enhancement Expression of PAI-1 in Visceral Fat: Possible Contributor to Vascular Disease in Obesity.*Nat. Med.* **1996**, *2*, 800-803.

St-Pierre, A.C., Ruel, I.L., Cantin, B., Dagenais, G.R., Bernard, P.M., Despres, J.P., Lamarche, B. Commparison of various electrophoretic charcteristics of LDL particles and their relationship to the risk of ischemic heart disease. *Circulation*. **2001**, *104*, 2295-9.

Tada, N., Watanabe, H., Matsuo, N., Tokimitsu, I., and Okazaki, M. Dynamics of Postprandial Remnant-Like Lipoprotein Particles (RLP) in Serum After Loading of Diacylglycerols. *Clin. Chim. Acta*. **2001**, *311*, 109-117.

Taguchi, H., Watanabe, H., Onizawa, K., Nagao, T., Gotoh, N., Yasukawa, T., Tsushima, R., Shimasaki, H., and Itakura, H. Double-Blind Controlled Study on the Effects of Dietary Diacylglycerol on Postprandial Serum and Chylomicron Triacylglycerol Responses in Healthy Humans. *J. Am. Coll. Nutr*. **2000**, *19*, 789-796.

World Health Organization. Diet, Nutrition and the Prevention of Chronic Diseases, Technical Report Series no. 916, **2003**, pp. 54-134, WHO, Geneva.

Yamamoto, K., Tomonobu, K., Asakawa, H., Tokunaga, K., Hase, T., Tokimitsu, I., Yagi, N. Diet therapy with diacylglycerol oil delays the progression of renal failure in type 2 diabetic patients with nephropathy. *Diabetes Care*. **2006**, *29*, 417-9.

Yamamoto, K., Asakawa, H., Tokunaga, K., Watanabe, H., Matsuo, N., Tokimitsu, I., and Yagi, N. Long-Term Ingestion of Dietary Diacylglycerol Lowers Serum Triacylglycerol in Type II Diabetic Patients with Hypertriglyceridemia. *J. Nutr*. **2001**, *131*, 3204-3207.

Yamamoto, K., Takeshita, M., Tokimitsu, I., Watanabe, H., Mizuno, T., Asakawa, H., Tokunaga, K., Tatsumi, T., Okazaki, M., Yagi, N. Diacylglycerol oil ingestion in type 2 diabetic patients with hypertriglyceridemia. *Nutrition* **2006**, *22*, 23-9. Erratum in: *Nutrition*. 2006, 22, 343.

Chapter 13

Phytosterols in Diacylglycerol Versus Triacylglycerol Improve Serum Cholesterol and Atherosclerosis Levels

Shinichi Meguro

Health Care Food Research Laboratories, Kao Corporation, Sumida-ku, Tokyo, Japan

Introduction

Phytosterols (PS), abundant in the fat-soluble fractions of plants, are consumed at levels of 200–400 mg/d in Western diets (Jones et al., 1997). PS have become dietary ingredients of increasing importance since Peterson first reported in 1951 that the addition of soy sterols to a cholesterol-enriched diet prevented increases in plasma cholesterol levels and significantly reduced the incidence of atherosclerotic plaque in chick aortae (Peterson, 1951; Peterson et al., 1952). Clinical investigations under a wide range of study conditions indicated that the administration of PS to humans reduces plasma total cholesterol and LDL cholesterol levels (Jones et al., 1997). A number of theories have been advanced concerning the mechanism by which PS functions to reduce serum cholesterol. It is generally accepted that PS inhibit the intestinal absorption of cholesterol (Jones et al., 1997; Ling & Jones, 1995).

Although few side effects were reported in these clinical trials, relatively large amounts of PS (1.5–3.0 g/d) must be given to achieve significant cholesterol-lowering effects (Heinemann et al., 1986; Vanhanen et al., 1993). In addition, PS are barely soluble in oil or water, and are therefore usually administered in crystalline form (Sugano et al., 1976). These drawbacks have limited the wider use of PS. Recently, Pelletier et al. (Pelletier et al., 1995) reported that 700 mg/d of PS mixed with butter fat lowered serum cholesterol in normocholesterolemic subjects. In addition, Hendriks et al. (Hendriks et al., 1999) reported that 830 mg PS/d contained in a spread, mainly in esterified form, had a cholesterol-lowering effect in normocholesterolemic and mildly hypercholesterolemic subjects. These studies demonstrated the efficacy of PS at low doses, and further suggested the possibility of the wider use of PS.

On the other hand, we found that PS could be dissolved in dietary diacylglycerol (DAG) at higher levels without esterification (Meguro et al., 2001). The actual solubility of PS in DAG and triacylglycerol (TAG) was 4.2–6.0% and 0.5–1.3%, respectively (Meguro et al., 2001; unpublished data). Therefore, DAG might be a useful oil for enhancing the cholesterol-lowering effect of PS because Ostlund et al. (Ostlund et al., 1999) clearly demonstrated that solubilization of PS in phospholipids improves the bioavailability despite the administration of a very low dose. DAG is a minor component in various vegetable oils and fats (D'Alonzo et al., 1982), and is currently used in small quantities in foods as an emulsifier. Recent studies suggested that DAG consumed in larger amounts might have metabolic characteristics distinct from TAG that could have beneficial effects with regard to the prevention and management of obesity. Experimental studies in animals and humans indicate that DAG reduces postprandial triglyceridemia compared with TAG (Murata et al., 1994; Taguchi et al., 2000; Tada et al., 2001) (see Chapters 7 and 8). In controlled feeding studies, DAG reduced body weight and abdominal fat stores in humans (Nagao et al., 2000; Maki et al., 2002) (see Chapter 10). In this chapter, improvements in serum cholesterol and atherosclerosis levels by ingestion of PS in DAG vs. TAG prepared with the same fatty acid composition as DAG are discussed.

Comparison of Serum Cholesterol Levels

In an animal study, Meguro et al. (Meguro et al., 2003) investigated the effects of PS in combination with DAG (PS/DAG) compared with PS in combination with TAG (PS/TAG) on serum lipids in cholesterol-fed rabbits. Cholesterol-fed (0.3%) New Zealand white (NZW) rabbits were fed a 0.3%-PS:7%-TAG diet or a 0.3%-PS:7%-DAG diet for 14 wk. The serum total cholesterol level in the PS/DAG group was significantly lower than that in the control and PS/TAG groups, whereas serum HDL cholesterol and triacylglyceride levels did not differ between the two groups (Fig. 13.1). In a human study, Meguro et al. (Meguro et al., 2001) also investigated the effects of PS/DAG compared with PS/TAG on serum lipids in 12 normocholesterolemic or moderately hypercholesterolemic [serum total cholesterol: 4.66–6.78 mmol/L (180.3–262.4 mg/dL)] healthy men. For 2 wk before the test period (designated the control period), all subjects consumed control mayonnaise (PS-free) daily with supper and were randomly assigned to two groups. In the 2-wk test period, one group was given mayonnaise containing PS (500 mg/d) dissolved in DAG (10 g/d), and the other group was given mayonnaise containing PS (500 mg/d) dispersed in TAG (10 g/d). After a washout period during which control mayonnaise (PS-free) was consumed for 4 wk, the groups were reversed for 2 wk. PS/TAG feeding had no effect on the serum cholesterol level. In contrast, PS/DAG feeding significantly reduced the total and LDL cholesterol levels from the initial value of 5.57 mmol/L (215.6 mg/dL) to 5.31 mmol/L (205.5 mg/dL) (4.7%; $P < 0.05$) and from 3.69 mmol/L (142.8 mg/dL) to 3.39 mmol/L (131.2 mg/dL) (7.6%; $P < 0.05$), respectively (Table 13.1). Moreover, the degree of total cholesterol reduction induced by PS/DAG feeding in the test period was significantly greater than that induced by PS/TAG feeding ($P < 0.05$). In the PS/TAG feeding group, serum triglyceride levels increased from the initial value of 1.65 mmol/L (63.9 mg/dL) to 1.83 mmol/L (70.8 mg/dL), whereas in the PS/DAG feeding group, serum triglyceride levels were slightly but not significantly reduced from the initial value of 1.73 mmol/L (66.9 mg/dL) to 1.67 mmol/L (64.6 mg/dL).

It is not clear why the hypolipidemic activity of PS/DAG is greater than that of PS/TAG. Although there are many hypotheses for the serum cholesterol–lowering mechanism of PS, the most common hypothesis is that PS have a specific inhibitory effect on the micellar solubilization of cholesterol through competition with or displacement of

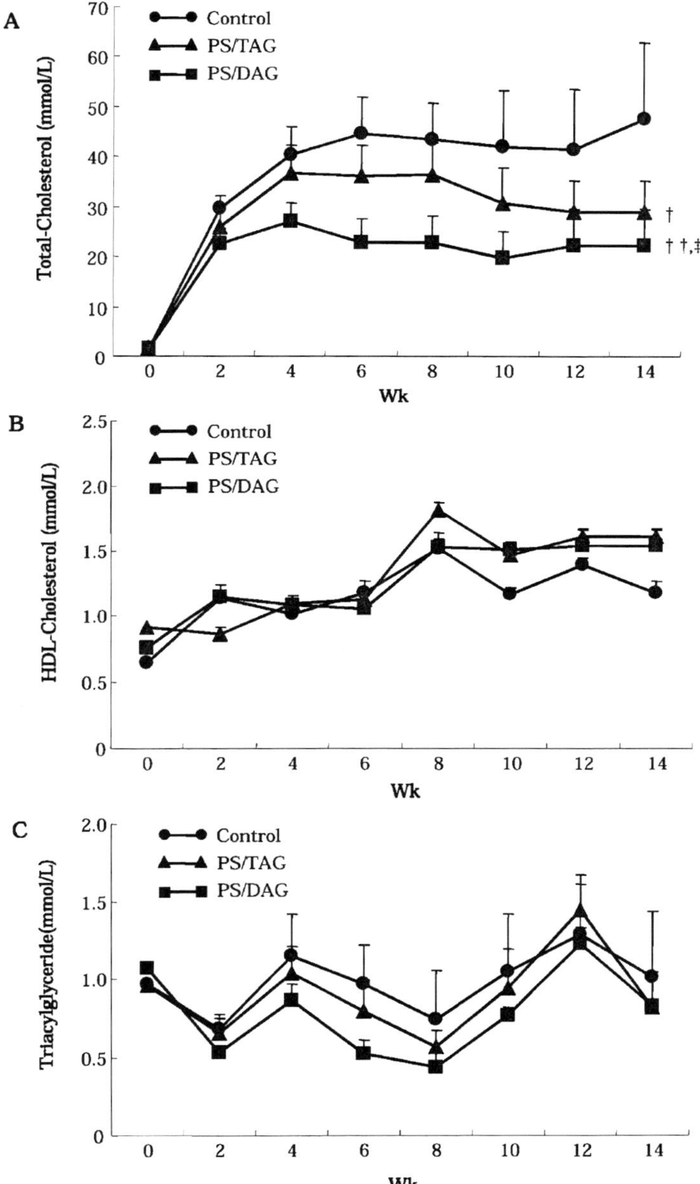

FIG. 13.1. Time course of serum total cholesterol (A), HDL cholesterol (B), and triacylglyceride (C) concentrations in cholesterol-fed NZW rabbits fed a control diet, PS/TAG diet, or PS/DAG diet for 14 wk. Values are means ± SD. Daggers indicate different from the control group $^{\dagger}P < 0.05$, $^{\dagger\dagger}P < 0.01$; ‡different from the PS/TAG group, $P < 0.05$.

TABLE 13.1
Effects of Phytosterols (PS) in Diacylglycerol (DAG) and in Triacylglycerol (TAG) on Serum Lipid Levels in Volunteers[a]

Serum lipids(mg/dL)	Oil	Initial	End	Change[b]
Total cholesterol	PS/DAG	5.57 ± 0.66	5.31 ± 0.78[†]	−0.26 ± 0.39[‡]
	PS/TAG	5.37 ± 0.64	5.40 ± 0.74	0.03 ± 0.35
LDL cholesterol	PS/DAG	3.69 ± 0.59	3.39 ± 0.66[†]	−0.30 ± 0.41
	PS/TAG	3.57 ± 0.65	3.43 ± 0.56	−0.13 ± 0.40
HDL cholesterol	PS/DAG	1.09 ± 0.24	1.14 ± 0.19	0.06 ± 0.11
	PS/TAG	1.05 ± 0.20	1.13 ± 0.23	0.08 ± 0.11
Triacylglycerol	PS/DAG	1.73 ± 0.60	1.67 ± 0.52	−0.06 ± 0.44
	PS/TAG	1.65 ± 0.71	1.83 ± 0.58	0.18 ± 0.45

[a]Values are means ± SD, n = 12. [†]Different from the initial value, $P < 0.05$; [‡]different from the PS/TAG diet group, $P < 0.05$.
[b]End value – initial value.

cholesterol in bile salt micelles (Ling & Jones, 1995; Ikeda et al., 1988). The solubility of PS in DAG is 6.0%, but only 1.3% in TAG (Meguro et al., 2003). Thus, the PS were fully dissolved in PS/DAG, but not in PS/TAG. The hypocholesterolemic activity of PS/DAG might be due to the high solubility of PS in DAG, which induces an increase in the distribution of PS in bile acid micelles after ingestion. Vanhanen et al. (Vanhanen et al., 1993) and Ostlund et al. (Ostlund et al., 1999) reported that the administration of PS in a soluble form might be important for obtaining sufficient levels in the micellar phase for the inhibition of cholesterol absorption in the intestine. Mattson et al. (Mattson et al., 1982), however, reported that ß-sitosterol is more efficient than ß-sitosterol esters in decreasing cholesterol absorption in humans. ß-Sitosterol esters dissolve more completely in dietary oils and are hydrolyzed to ß-sitosterol during lipolysis of dietary fats (Miettinen & Siurala, 1971). Also, there were no significant increases in the hypocholesterolemic activity due to differences in solubility in oil (Miettinen & Vanhanen, 1994). These results suggest that factors other than the solubility of PS in oil might be responsible for the competition of cholesterol for incorporation into bile acid micelles in the intestine. Further detailed experiments are required to clarify the mechanism of the hypocholesterolemic activity of PS/DAG.

Comparison of Serum Phytosterol Levels

Lees et al. (Lees et al., 1977) reported that the daily administration of 3 g of PS in hypercholesterolemic patients for 1 mo did not increase the plasma PS concentration. Jones et al. (Jones et al., 1999) reported that even after daily administration of 1.7 g of PS to subjects with primary hypercholesterolemia for 1 mo, the PS level did not differ between the PS-fed and control subjects. Meguro et al. (Meguro et al., 2001) reported that serum ß-sitosterol and campesterol concentrations did not change in the PS/DAG or PS/TAG feeding groups with a daily intake of 500 mg for 2 wk. Long-term effects of PS/DAG intake on serum PS levels are described in Chapter 14.

Comparison of Atherosclerosis Levels

Meguro et al. (Meguro et al., 2003) also investigated the effects of PS/DAG compared with PS/TAG on atherosclerosis levels in cholesterol-fed (0.3%) NZW rabbits given a 0.3%-PS:7%TAG diet or a 0.3%-PS:7%-DAG diet for 14 wk. The ratio of thoracic, abdominal, and total lesion areas of the PS/DAG group was smaller than that of the control group (0.3% cholesterol-fed NZW rabbits without PS in 7% TAG diet). Furthermore, the ratio of the abdominal lesion area of the PS/DAG group was smaller than that of the PS/TAG group. The ratio of the atherosclerotic lesion area between the PS/TAG and control groups did not differ (Table 13.2). The mean thickness of the intima in the thoracic, abdominal, and total aortae of the PS/DAG group was lower than that of the control group. Furthermore, the mean thickness of the intima in the thoracic and total aortae of the PS/DAG group was also significantly lower than that of the PS/TAG group. The mean thickness of the intima did not differ between the PS/TAG and control groups (Table 13.3).

Murphy et al. (Murphy et al., 1973) reported that a diet with a fourfold excess of PS over cholesterol in NZW rabbits for 190 d reduced atherosclerotic lesion formation

TABLE 13.2
The Ratio of the Atherosclerotic Lesion Area in the Aortae of Cholesterol-Fed NZW Rabbits Treated with Phytosterols (PS) in Combination with Diacylglycerol (DAG) or Triacylglycerol (TAG) for 14 wk[a,b]

	Atherosclerotic lesion area (%)		
Group	Thoracic	Abdominal	Total
Control	71.6 ± 4.5	15.8 ± 2.5	44.6 ± 3.7
PS/TAG	58.5 ± 7.8	16.9 ± 5.4	38.2 ± 6.3
PS/DAG	36.1 ± 9.4†	3.8 ± 1.1†‡	19.6 ± 5.1†

[a]Values are means ± SD and were analyzed using the Mann-Whitney U test. †Different from the control group, $P < 0.05$; ‡different from the PS/TAG diet group, $P < 0.05$.
[b]The ratio of the atherosclerotic lesion area in the aortae was measured as described in the text.

TABLE 13.3
The Mean Thickness of the Intima in the Aortae of Cholesterol-Fed NZW Rabbits Treated with Phytosterols (PS) in Combination with Diacylglycerol (DAG) or Triacylglycerol (TAG) for 14 wk[a,b]

	Mean thickness of the intima (μm)		
Group	Thoracic	Abdominal	Total
Control	258.7 ± 57.1	6.7 ± 3.1	132.7 ± 28.7
PS/TAG	175.3 ± 55.6	7.6 ± 3.7	91.4 ± 27.8
PS/DAG	60.9 ± 22.1†‡	0.6 ± 0.4†	30.7 ± 11.1†‡

[a]Values are means ± SD and were analyzed using the Mann-Whitney U test. †Different from the control group, $P < 0.05$; ‡different from the PS/TAG diet group, $P < 0.05$.
[b]The mean thickness of the intima in the aortae was measured as described in the text.

compared with the non-PS diet. On the other hand, Ntanios et al. (Ntanios et al., 1998) reported that a diet with a twofold excess of PS over cholesterol in NZW rabbits for 10 wk did not reduce atherosclerotic lesion formation compared with the non-PS diet. These reports indicate that the antiatherosclerotic activity of PS requires a diet with a more than twofold excess of PS over cholesterol in cholesterol-fed rabbits. In Meguro's study (Meguro et al., 2003), the rabbits were fed a diet containing the same amount of PS as cholesterol (0.3% each) for 14 wk. Atherosclerosis formation was inhibited in the PS/DAG group, but there was no inhibitory effect observed in the PS/TAG group. These findings suggest that DAG potentiates the inhibitory effect of PS on atherosclerosis formation.

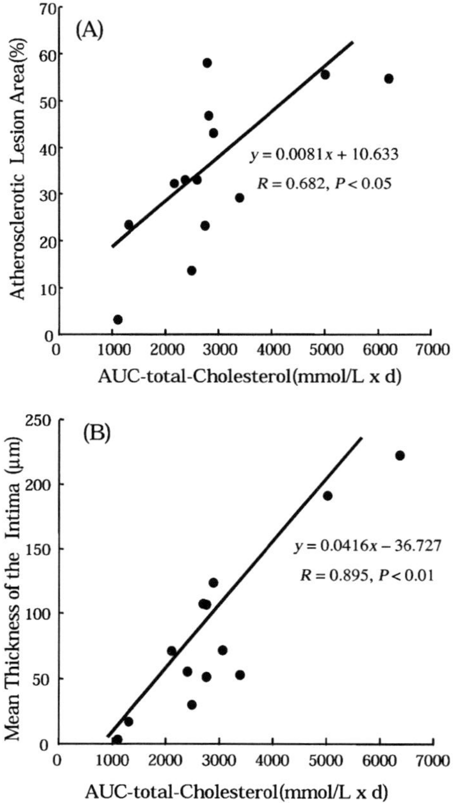

FIG. 13.2. Correlation between the ratio of the atherosclerotic lesion area (A) or the mean thickness of the intima (B) of the aortae and area under the curve (AUC)-total cholesterol. Regression lines are shown for all data points. The AUC was estimated from the area under the curve of the cholesterol concentration vs. duration of each feeding and is expressed as mmol/L × d.

Bocan et al. (Bocan et al., 1993) reported that aortic atherosclerosis formation is highly correlated with exposure of the arterial wall to cholesterol in cholesterol-fed rabbits. Exposure of the aorta to total cholesterol [area under the curve (AUC)-total cholesterol], which was decreased by ingestion of PS, was measured according to the method reported by Bocan et al. (Bocan et al., 1993). There was a positive correlation between the ratio of the atherosclerotic lesion area or the mean thickness of the intima and AUC-total cholesterol (Fig. 13.2). These findings suggest that the PS-induced reduction of the exposure of the aorta to total cholesterol inhibited atherosclerosis formation. DAG alone does not affect the serum cholesterol level (Nagao et al., 2000; Hara et al., 1993; Murata et al., 1997). Therefore, the antiatherosclerotic activity of DAG might be superior to that of TAG because the effect of PS is potentiated by DAG, which reduces the exposure of the aorta to total cholesterol.

Summary

PS in DAG have a greater ability to reduce serum cholesterol levels and inhibit the development of atherosclerosis induced by cholesterol exposure compared with PS in TAG. These results suggest that PS/DAG might be useful for the prevention of hypercholesterolemia and the initiation of atherosclerosis. It is not clear, however, why PS/DAG is more effective than PS/TAG. Further detailed experiments are required to clarify the mechanism underlying the hypocholesterolemic and antiatherosclerotic activity of PS/DAG.

References

Bocan, T.M., Mueller, S.B., Mazur, M.J., Uhlendorf, P.D., Brown, E.Q., and Kieft, K.A. The Relationship Between the Degree of Dietary-Induced Hypercholesterolemia in the Rabbit and Atherosclerotic Lesion Formation. *Atherosclerosis* **1993,** *102,* 9-22.

D'Alonzo, R.P., Kozarek, W.J., and Wade, R.L. Glyceride Composition of Processed Fats and Oils as Determined by Glass Capillary Gas Chromatography. *J. Am. Oil. Chem. Soc.* **1982,** *59,* 292-295.

Hara, K., Onizawa, K., Honda, H., Otsuji, K., Ide, T., and Murata, M. Dietary Diacylglycerol-Dependent Reduction in Serum Triacylglycerol Concentration in Rats. *Ann. Nutr. Metab.* **1993,** *37,* 185-191.

Heinemann, T., Leiss, O., and von Bergmann, K. Effects of Low-Dose Sitostanol on Serum Cholesterol in Patients with Hypercholesterolemia. *Atherosclerosis* **1986,** *61,* 219-223.

Hendriks, H.F.J., Weststrate, J.A., van Vliet, T., and Meijer, G.W. Spread Enriched with Three Different Levels of Vegetable Oil Sterols and the Degree of Cholesterol Lowering in Normocholesterolaemic and Mildly Hypercholesterolaemic Subjects. *Eur. J. Clin. Nutr.* **1999,** *53,* 319-327.

Ikeda, I., Tanaka, K., Sugano, M., Vahouny, G.V., and Gallo, L.L. Inhibition of Cholesterol Absorption in Rats by Plant Sterols. *J. Lipid Res.* **1988,** *29,* 1573-1582.

Jones, P.J.H., MacDougall, D.E., Ntanious, F., and Vanstone, C.A. Dietary Phytosterols as Cholesterol-Lowering Agents in Humans. *Can. J. Physiol. Pharmacol.* **1997,** *75,* 217-227.

Jones, P.J.H., Ntanios, F.Y., Raeini-Sarjaz, M., and Vanstone, C.A. Cholesterol-Lowering Efficacy of a Sitostanol-Containing Phytosterol Mixture with a Prudent Diet in Hyperlipidemic Men. *Am. J. Clin. Nutr.* **1999,** *69,* 1144-1150.

Lees, A.M., Mok, H.Y., Lees, R.S., McCluskey, M.A., and Grundy, S.M. Plant Sterols as Cholesterol-Lowering Agents: Clinical Trials in Patients with Hypercholesterolemia and Studies of Sterol Balance. *Atherosclerosis* **1977,** *28,* 325-338.

Ling, W.H., and Jones, P.J.H. Dietary Phytosterols: A Review of Metabolism, Benefits and Side Effects. *Life Sci.* **1995,** *57,* 195-206.

Maki, K.C., Davidson, M.H., Tsushima, R., Matsuo, N., Tokimitsu, I., Umporowicz, D.M., Dicklin, M.R., Foster, G.S., Ingram, K.A., Anderson, B.D., Frost, S.D., and Bell, M. Consumption of Diacylglycerol Oil as Part of a Mildly Reduced-Energy Diet Enhances Loss of Body Weight and Fat Compared with a Triacylglycerol Control Oil. *Am. J. Clin. Nutr.* **2002,** *76,* 1230-1236.

Mattson, F.H., Grundy, S.M., and Crouse, J.R. Optimizing the Effect of Plant Sterols on Cholesterol Absorption in Man. *Am. J. Clin. Nutr.* **1982,** *35,* 697-700.

Meguro, S., Higashi, K., Hase, T., Honda, Y., Otsuka, A., Tokimitsu, I., and Itakura, H. Solubilization of Phytosterols in Diacylglycerol Versus Triacylglycerol Improves the Serum Cholesterol Lowering Effect. *Eur. J. Clin. Nutr.* **2001,** *55,* 513-517.

Meguro, S., Hase, T., Otsuka, A., Tokimitsu, I., and Itakura, H. Effect of Phytosterols in Dietary Diacylglycerol on Atherosclerosis in Cholesterol-Fed Rabbits. *Nutrition* **2003,** *19,* 670-675.

Miettinen, T.A., and Siurala, M. Bile Salts, Sterols, Sterol Esters, Glycerides and Fatty Acids in Micellar and Oil Phase of Intestinal Contents During Fat Digestion in Man. *Z. Klin. Chem. Biochem.* **1971,** *9,* 47-52.

Miettinen, T.A., and Vanhanen, H. Dietary Sitostanol Related to Absorption, Synthesis and Serum Level of Cholesterol in Different Apolipoprotein E Phenotypes. *Atherosclerosis* **1994,** *105,* 217-226.

Murata, M., Hara, K., and Ide, T. Alteration by Diacylglycerols of the Transport and Fatty Acid Composition of Lymph Chylomicrons in Rats. *Biosci. Biotechnol. Biochem.* **1994,** *58,* 1416-1419.

Murata, M., Ide, T., and Hara, K. Reciprocal Responses to Dietary Diacylglycerol of Hepatic Enzymes of Fatty Acid Synthesis and Oxidation in the Rat. *Br. J. Nutr.* **1997,** *77,* 107-121.

Murphy, E.A., Rowsell, H.C., and Mustard, J.F. The Effects of Sitosterol on Serum Cholesterol, Platelet Economy, Thrombogenesis and Atherosclerosis in the Rabbit. *Atherosclerosis* **1973,** *17,* 257-268.

Nagao, T., Watanabe, H., Goto, N., Onizawa, K., Taguchi, H., Matsuo, N., Yasukawa, T., Tsushima, R., Shimasaki, H., and Itakura, H. Dietary Diacylglycerol Suppresses Accumulation of Body Fat Compared to Triacylglycerol in Men in a Double-Blind Controlled Trial. *J. Nutr.* **2000,** *130,* 792-797.

Ntanios, F.Y., Jones, P.J., and Frohlich, J.J. Dietary Sitostanol Reduces Plaque Formation but Not Lecithin Cholesterol Acyl Transferase Activity in Rabbits. *Atherosclerosis* **1998,** *138,* 101-110.

Ostlund, R.E., Jr., Spilburg, C.A., and Stenson, W.F. Sitostanol Administered in Lecithin Micelles Potently Reduces Cholesterol Absorption in Humans. *Am. J. Clin. Nutr.* **1999,** *70,* 826-831.

Pelletier, X., Belbraouet, S., Mirabel, D., Mordret, F., Perrin, J.L., Pages, X., and Debry, G. A Diet Moderately Enriched in Phytosterols Lowers Plasma Cholesterol Concentration in Normocholesterolemic Humans. *Ann. Nutr. Metab.* **1995,** *39,* 291-295.

Peterson, D.W. Effect of Soybean Sterols in the Diet on Plasma and Liver Cholesterol in Chicks. *Proc. Soc. Exp. Biol. Med.* **1951,** *78,* 1143-1147.

Peterson, D.W., Nichols, C.W., and Schneour, E.W. Some Relationships Among Dietary Sterols, Plasma and Liver Cholesterol Levels and Atherosclerosis in Chicks. *J. Nutr.* **1952,** *47,* 57-65.

Sugano, M., Kamo, F., Ikeda, I., and Morioka, H. Lipid-Lowering Activity of Phytostanols in Rats. *Atherosclerosis* **1976,** *24,* 301-309.

Tada, N., Watanabe, H., Matsuo, N., Tokimitsu, I., and Okazaki, M. Dynamics of Postprandial Remnant-Lipoprotein Particles in Serum After Loading of Diacylglycerols. *Clin. Chim. Acta* **2001,** *311,* 109-117.

Taguchi, H., Watanabe, H., Onizawa, K., Nagao, T., Gotoh, N., Yasukawa, T., Tsushima, R., Shimasaki, H., and Itakura, H. Double-Blind Controlled Study on the Effects of Dietary Diacylglycerol on Postprandial Serum and Chylomicron Triacylglycerol Responses in Healthy Humans. *J. Am. Coll. Nutr.* **2000,** *19,* 789-796.

Vanhanen, H.T., Blomqvist, S., Ehnholm, C., Hyvonen, M., Jauhiainen, M., Torstila, I., and Miettinen, T.A. Serum Cholesterol, Cholesterol Precursors, and Plant Sterols in Hypercholesterolemic Subjects with Different ApoE Phenotypes During Dietary Sitostanol Ester Treatment. *J. Lipid Res.* **1993,** *34,* 1535-1544.

Chapter 14

Studies of *Ad Libitum* Ingestion of Phytosterol-Enriched Diacylglycerol Oil

Masao Takeshita and Shinichiro Saito

Health Care Food Research Laboratories, Kao Corporation, Tokyo, Japan

Introduction

The excessive ingestion of dietary fats in a typical Western diet is associated with an increased risk of various metabolic disorders such as obesity, hyperlipidemia, and impaired glucose tolerance that are linked to progression of arteriosclerosis (Hennig et al., 2001). Arterosclerosis and related vascular disorders have been the leading cause of death worldwide and are key health concerns. Based on a large body of evidence, an elevated blood low-density lipoprotein (LDL) cholesterol level has been well established as an independent risk factor for arteriosclerosis (Lipid Research Clinics Program, 1984; Shepherd et al., 1995). In particular, a dietary therapy that prescribes a low cholesterol intake is remarkably effective in lowering serum LDL cholesterol levels (Ornish et al., 1998), but it requires great perseverance to maintain such a restricted diet over a long period of time. Thus, new functional foods enriched with cholesterol-lowering ingredients such as a mixture of phytosterols consisting of Δ5-phytosterols ('sterols') or 5α-reduced phytosterols ('stanols') have been developed and are now available throughout the world. The United States National Cholesterol Education Program Adult Treatment Panel III endorses phytosterol esters as an essential feature of a therapeutic lifestyle change along with diet modifications, weight reduction, intake of viscous fibers, and increased physical activity to reduce the risk of coronary heart disease (National Cholesterol Education Program, 2002).

Diacylglycerol (DAG) oil has unique metabolic characteristics that may be useful for the management of obesity, hypertriglyceridemia and postprandial triglyceridemia compared with conventional oils. Additionally, to lower blood LDL cholesterol levels in mild to moderate hypercholesterolemia, the use of diacylglycerol oil in conjunction with phytosterols is recommended as a part of dietary therapy. In this chapter, recent clinical studies with regard to incorporating phytosterols-enriched DAG oil (PS/DAG oil) into the daily diet are summarized.

Phytosterols

Phytosterols (PS) are widely distributed in vegetable foods in the form of either free sterols or fatty acid esters that function as an essential part of the cell membrane. A daily intake of natural PS varies from 167 to 437 mg (Ostlund, 2002a). The main components

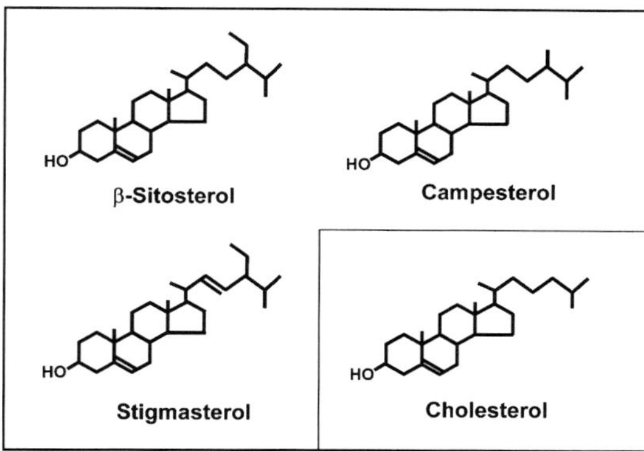

FIG. 14.1. Chemical structures of phytosterols.

of PS are β-sitosterol, campesterol, and stigmasterol, which contain a methyl or ethyl group in the molecule's side chain **(Figure 14.1)**. PS have the ability to compete with the intestinal cholesterol absorption by up to 50% as cholesterol analogs, which in turn can lower blood LDL cholesterol levels, but not triglyceride and HDL cholesterol levels (Ostlund, 2002a; Katan et al., 2003).

A meta-analysis showed that supplementation with 2 g of PS (stanols or sterols) is desirable in terms of reducing blood LDL cholesterol concentration by 10% (Katan et al., 2003). However, evidence to support this theory was concentrated close to the tail of the dose-response curve. Ostlund et al. recently showed that natural PS in commercial corn oil (Ostlund et al., 2002b) and sitostanol emulsified with lecithin (Ostlund et al., 1999) are bioactive at lower doses (150-300 mg) in single meal tests. Shin et al. also confirmed that water-dispersible micellar PS (300 mg) had a similar impact on cholesterol absorption (Shin et al., 2005). By contrast, the solubility of PS in DAG oil is ~5 times higher than in triacylglycerol (TAG) oil. Indeed, in a comparison of 500 mg/day PS dissolved in either TAG or DAG oil, the cholesterol-lowering effect of the PS was observed only when it was dissolved in DAG oil (Meguro et al., 2001). Moreover, the effective minimum dosage of either PS or PS esters dissolved in DAG oil has been shown to be 400 mg/day in Japanese men (Goto et al., 1999; Saito et al., 2006a). The Ministry of Health, Labor and Welfare in Japan approved PS/DAG oil as a Food for Specified Health Use in 1999. Thus, it is possible that DAG oil may become a useful material for incorporating PS efficiently into foods as a bioavailable formulation, although further studies are needed to elucidate the precise mechanisms underlying the cholesterol-lowering effect.

Cholesterol-Lowering Effect of PS Enriched DAG oil

Takeshita et al. (Takeshita et al., 2001) examined the issue of maintaining a cholesterol-lowering effect in Japanese adults (23–58 y, 66 men/5 women) in a study in which all

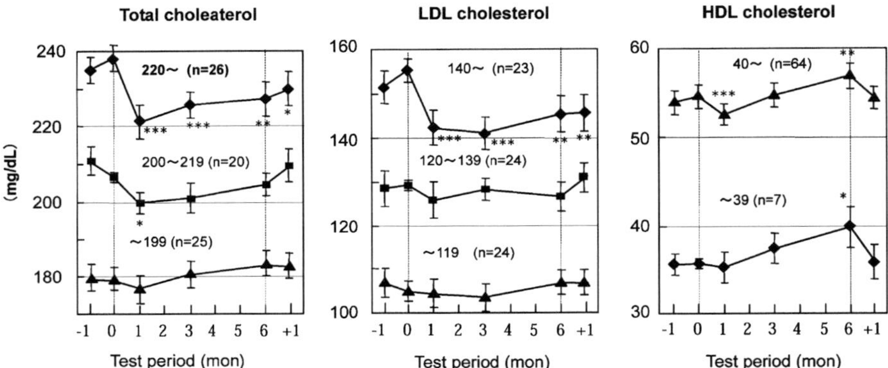

FIG. 14.2. Effects of PS/DAG oil on serum cholesterol concentrations. Values are means ± SE, n = 71. Asterisks indicate different from the baseline value (0 mo): **P* < 0.05, ***P* <0.01, ****P* < 0.001. Source: Takeshita et al., 2001.

subjects replaced their ordinary cooking oil with PS-enriched DAG oil containing 4 g PS/100 g DAG oil (PS/DAG oil), and used it as their cooking oil at home for 24 wk. In this study, the monthly intake of the test oil for all subjects was stable at 15–17 g/d during the study. No adverse events were observed throughout the study. The subjects were divided into 2–3 groups according to baseline values of each serum lipid concentration, and each group was evaluated separately (**Fig. 14.2**). Serum total cholesterol concentration in the group with high values (≥ 220 mg/dL, n = 26) were significantly reduced (7.1% (238 to 221 mg/dL, P < 0.001)) in wk 4 from the baseline value, and this level was maintained up to wk 24 (- 4.5%, P < 0.01). Furthermore, serum LDL cholesterol concentration in the group with high values (≥ 140 mg/dL, n = 23) was significantly reduced (8.4% (155 to 142 mg/dL, P < 0.001)) in wk 4 from the baseline value, and this effect was steadily maintained up to wk 24 (-6.4%, P < 0.01). In the borderline group of total cholesterol (200–219 mg/dL), serum total cholesterol concentration tended to decrease, but not in the normal group (< 200 mg/dL). In subjects whose HDL cholesterol concentration was < 40 mg/dL (n = 7), serum HDL cholesterol concentration was increased by 11% over wk 24 from the baseline value. In addition, decreases in serum triglyceride (high value group ≥ 150 mg/dL), plasminogen activator inhibitor type-1 (PAI-1), and leptin concentrations were also observed. These outcomes may have been caused by the ingestion of DAG oil (see Chapter 11).

In 16-wk randomized, double-blind controlled studies with three types of mayonnaise with control TAG oil, DAG oil, and DAG oil containing PS esters (PSE/DAG, the amount of PS intake was estimated 400 mg/day as free form molecules) conducted separately by Tomonobu et al. (Tomonobu et al., 2004) and Takeshita et al. (Takeshita et al., 2007b), cholesterol-lowering effects were observed as well as anti-obesity effects both in the free living subjects (n=184) and in subjects living under controlled condition based on the 6th revision of the recommended dietary allowances

TABLE 14.1
Summary of Serum Cholesterol-Lowering Effects due to PS/DAG Mayonnaise [a]

	TAG	DAG	PS/DAG
Total cholesterol	1.1 (–5.2 to 7.4)	–1.6 (–4.7 to 1.5)	–6.3 (–9.7 to –2.8)
LDL cholesterol	1.9 (–4.6 to 8.5)	–4.6 (–7.7 to –1.5) #	–9.3 (–12.9 to –5.7) ###
HDL cholesterol	0.8 (–5.8 to 7.4)	3.0 (–1.4 to 7.3)	–3.0 (–7.9 to 1.9)
Apolipoprotein B	1.3 (–4.7 to 7.3)	–2.8 (–6.0 to 0.3)	–7.1(–10.1 to –4.2) ##

[a] Mean percent changes (95%CI) at the end-point (average of wks 12 and 16).
Significant effect compared to TAG mayonnaise group, # $p < 0.05$, ## $p < 0.01$, ### $p < 0.001$.
TAG (control) mayonnaise group, n=18; DAG mayonnaise group, n=21; PS/DAG mayonnaise group, n=18.
Source: Takeshita et al., 2007b

for Japanese (n=57, Table. 14.1). It was suggested that PS mainly contributed to the reduction of blood cholesterol levels throughout the study period in addition to the possible effects of DAG oil on blood cholesterol levels via body weight loss over long periods. Thus, a combined daily use of DAG oil and PS may play a beneficial role for decreasing risks such as obesity and hypercholesterolemia closely linked to cardiovascular disease.

The onset of menopause dramatically increases atherogenic blood lipoprotein levels, particularly LDL, small dense LDL and lipoprotein (a) [Lp(a)], resulting in a precipitous climb in CHD rates. The cholesterol-lowering effect of PS/DAG oil (PS 4%) in postmenopausal women with mild to moderate hypercholesterolemia (n = 14, 45–62 y) was performed in a 4-wk crossover manner (Takeshita et al., 2007a). Subjects consumed test oils (PS/DAG oil or DAG oil) freely at home as a replacement for edible oils. The average intake of PS from the PS/DAG and DAG (control) oils were 563 and 50 mg/day, respectively (P = 0.001). During the control period, no effect on serum triacylglycerol and cholesterol concentrations was observed. Serum total cholesterol, LDL cholesterol and apolipoprotein B concentrations were reduced by 10.2% (P = 0.004), 12.1% (P = 0.003) and 9.7% (P = 0.001) during the PS/DAG oil period compared to the control period, respectively, as shown in Table 14.2. In addition, a slight but significant (P = 0.023) reduction in serum Lp(a) concentration was observed, particularly in subjects with higher levels for baseline values (> 20 mg/dL). No adverse events were observed in both periods. Therefore, PS/DAG oil may be a useful adjunct to first-line therapy for the management of atherogenic lipoproteins in women after menopause.

In Japan, the incidence of hyperlipidemia in children has been increasing, but drug therapy for children is limited. Matsuyama et al. examined the effect of PS/DAG oil (PS 4%) as part of a diet therapy in pediatric patients with hyperlipidemia (serum total cholesterol ≥ 200 mg/dL and/or triglyceride ≥ 150 mg/dL, n = 22) with a single-armed clinical test (Matsuyama et al., 2007). The children with hyperlipidemia ingested bread containing PS/DAG oil (total daily intake, 10g) for 6 months. Blood chemistry was examined prior to the initiation of the ingestion and at 2, 4 and 6 months after the initiation of ingestion. After 4 months of ingestion of PS/DAG oil, serum levels of LDL cholesterol and Lp(a), decreased significantly. In seven patients with familial hypercho-

TABLE 14.2
Effects of PS/DAG Oil on Serum Lipids in Postmenopausal Women with Mild to Moderate Hypercholesterolemia [a]

	Control period (DAG oil)		PS/DAG oil period		Difference [b]		P [c]
	Start	End	Start	End			
		(mmol/L)			(mmol/L)		
Total cholesterol	5.87 ± 0.85	6.01 ± 0.92	6.03 ± 0.72	5.57 ± 0.82	–0.60 ± 0.51	(–10.2 ± 8.7)	0.004
LDL cholesterol	3.55 ± 0.81	3.67 ± 0.93	3.53 ± 0.71	3.24 ± 0.91	–0.41 ± 0.40	(–12.1 ± 11..6)	0.003
HDL cholesterol	1.63 ± 0.41	1.64 ± 0.45	1.69 ± 0.41	1.61 ± 0.44	–0.09 ± 0.23	(–5.1 ± 13.6)	NS
Triacylglycerol	1.44 ± 0.67	1.32 ± 0.77	1.24 ± 0.60	1.31 ± 0.78	0.18 ± 0.63	(21.8 ± 70.9)	NS
		(mg/dL)			(mg/dL)		
ApoAI	153 ± 23	155 ± 25	157 ± 26	152 ± 22	–6.8 ± 17.7	(–4.2 ± 9.8)	NS
ApoB	114 ± 25	115 ± 26	117 ± 25	107 ± 26	–11.0 ± 7.1	(–9.7 ± 6.6)	0.001
Lp (a)	23.1 ± 18.6	23.6 ± 17.4	24.8 ± 19.2	22.5 ± 16.9	–2.8 ± 4.1	(–18.6 ± 26.8)	0.023
		(Ratio)			(Ratio)		
LDL cholesterol / HDL cholesterol	2.30 ± 0.76	2.42 ± 0.93	2.24 ± 0.80	2.18 ± 0.89	–0.19 ± 0.58	(–8.2 ± 20.7)	NS
ApoB / ApoAI	0.77 ± 0.22	0.77 ± 0.23	0.78 ± 0.25	0.72 ± 0.22	–0.05 ± 0.09	(–5.9 ± 10.8)	0.001

Abbreviations: Lp(a): lipoprotein (a), ApoAI: apolipoprotein AI, ApoB: apolipoprotein B, NS: not significant.
[a] Values are means ± standard deviation (n=14). Percentages in parentheses.
[b] Difference relative to control oil (DAG oil) is based on individual data.
[c] Difference between the changes in the two periods using the Wilcoxon matched paired signed ranks test (two-tailed).
Source: Takeshita et al., 2007a

lesterolemia, concentrations of serum total cholesterol, LDL cholesterol and Lp(a) also decreased significantly. Thus, in addition to adults, PS/DAG oil may ameliorates serum lipid metabolism in pediatric patients with hyperlipidemia as a first-line treatment in the management of blood cholesterol.

Effect of PS-Enriched DAG Oil on Serum PS Concentrations

PS are poorly absorbed from the alimentary canal and the absorbed PS are also excreted more rapidly than the cholesterol in the liver (Ikeda & Sugano, 1998). Recent human genetic studies indicate that PS are selectively pumped out of the enterocytes into the intestinal lumen through membrane proteins called ATP-binding cassette transporters (subfamily G: ABCG5 and 8), thus maintaining their low absorption (Lee et al., 2001; Berge et al., 2000). Consequently, it is likely that the PS concentrations in the human body would be extremely low. It was reported that the absorption rates of β-sitosterol, campesterol, and stigmasterol in humans are 4.2, 9.6, and 4.8%, respectively, whereas that of cholesterol is ~40% (Heinemann et al., 1993). However, it was reported that in phytosterolemia, which is an extremely rare disease, abnormally high levels of PS accumulate in the blood stream compared with normal people. The accumulation induces arteriosclerosis and causes CHD at a young age. A recent study revealed that a functional disorder in the PS excretion system in phytosterolemia leads to accumulation of PS in the blood stream due to a mutation in the gene for ABC transporters, which is closely related to the excretion of PS in the small intestine and liver (Salen et al., 2002). Therefore, a high content of PS in serum is thought to be one of the risk factors of arteriosclerosis.

In recent studies, PS concentrations in serum were investigated from the aspect of safety. The results generally concluded that serum PS concentrations in humans increase significantly when PS-enriched foods are consumed, even if the ingestion of PS is low. In subjects consuming 20 g/d of TAG margarine containing 1.82 g of PS for 3 wk, serum ß-sitosterol and campesterol concentrations increased significantly by 42 and 88%, respectively (Mussner et al., 2002).

In 6 clinical studies of PS/DAG oil for a variety of subjects including children (Matsuyama et al., 2007) and adults (Saito et al., 2006a; Takeshita et al., 2001; Tomonobu et al., 2004; Takeshita et al., 2007a; Takeshita et al., 2007b; Saito et al., 2006b), serum β-sitosterol and campesterol concentrations were increased by 20-40%. However, the ratio of serum PS to total cholesterol at the end of the PS/DAG period was within the normal range (1%). It appears likely that the increase in serum PS concentrations caused by PS/DAG oil ingestion does not differ from that caused by TAG oil. The slight increase in serum PS concentrations did not persist into the washout period, showing that the increase is transient within the range of safety (Saito et al., 2006a; Matsuyama et al., 2007; Saito et al., 2006b).

Effect of PS-Enriched DAG Oil on Cholesterol Synthesis

Serum concentrations of lathosterol and desmosterol, which are cholesterol precursors, are considered to be indicators of cholesterol synthesis in the liver (Miettinen et al., 1990). In clinical studies, it was reported that PS inhibited cholesterol absorption, which has the feedback effect of enhancing endogenous cholesterol synthesis (Mussner et al., 2002; Grundy et al., 1969; Nestel et al., 2001); however, a consensus of opinion has not been reached (Jones et al., 1999). On the other hand, in the study of PS/DAG oil for postmenopausal women (Takeshita et al., 2007a), the serum lathosterol/total cholesterol ratio (cholesterol synthesis) was slightly but significantly increased (13%) as a feedback effect of the inhibition of intestinal cholesterol absorption caused from PS ingestion, as described previously (Gylling et al., 1999). Thus, PS/DAG oil ingestion may have the potential for a slight increase in cholesterol synthesis in the liver, although further investigation is required.

Serum Fat-Soluble Vitamins

Because PS are naturally present in a balanced diet, they have demonstrated an excellent safety record through a long history of consumption as part of a normal human diet. In addition, a large number of human and animal studies have been conducted since the 1950's and have reported no serious adverse events. However, negative effects of PS ingestion on the serum fat-soluble vitamins levels have recently been examined.

It was reported that no changes in serum vitamin A, D, and K concentrations occurred as a result of PS ingestion (Plat et al., 2000). The effect of PS ingestion on serum vitamin E and carotene concentrations is currently a subject of debate. Among the trials of sterols and stanols identified in the meta-analysis on efficacy (Katan et al., 2003), 18 trials testing doses of 1.5 g/d or more reported blood concentrations of fat-soluble vitamins. Statistically significant reductions (28%) occurred, particularly in blood β-carotene levels. Part of this reduction probably is due to reduced absorption of carotenes and the rest to reduced concentrations of the lipoprotein carrier, LDL; after a correction for the decrease in total cholesterol induced by PS, a statistically significant reduction remained for β-carotene but not for α-carotene or lycopene. The reduction in vitamin E was explained similarly by a reduction in cholesterol. The decrease in β-carotene could have been prevented by adding sufficient fruits and vegetables to the diet (Noakes et al., 2002).

In an ingestion study with DAG oil alone, changes in serum fat-soluble vitamins levels such as vitamin A, E and D were not observed (Watanabe et al., 2001). In a study on the safety aspects of PS/DAG, serum retinol, α-tocopherol, and β-carotene levels were not altered by a 2-wk consumption of PS/DAG even with ingestion of a three-fold increase in the effective dose (daily intake: PS 1.2 g in DAG oil 30 g) (Saito et al., 2006b), suggesting that PS/DAG oil does not affect serum fat soluble (pro) vitamin levels (vitamin A, E and β-carotene). Furthermore, in 16-wk studies with PS/DAG mayonnaise (PS intake: 400 mg/day), no significant changes in blood vitamin A, E and carotene was shown (Tomonobu et al., 2004; Takeshita et al., 2007b). The negative effects of PS on fat-soluble (pro)vitamins are possibly dependent upon either the amount of PS or on concomitant dietary oils.

Summary

Consumption of PS-enriched DAG oil (PS 4wt%) showed total and LDL cholesterol–lowering effects in adults and in children who had begun the study with elevated blood cholesterol levels, irrespective of gender. The consumption of PS/DAG oil significantly increased serum PS concentrations; however, the subtle changes were similar to the effect observed for TAG oil ingestion. Serum lathosterol concentration, an indicator of cholesterol synthesis in the liver, was slightly increased, but the feedback effect of enhancing endogenous cholesterol synthesis is yet to be discussed in detail. Furthermore, PS/DAG oil had no significant effect on blood fat-soluble vitamin levels, presumably due to the lower effective dose of PS. In conclusion, the recent literature supports that the use of PS/DAG oil as being beneficial as a dietary therapy for reducing atherogenic LDL cholesterol concentration.

References

Berge, K.E., Tian, H., Graf, G.A., Yu, L., Grishin, N.V., Schultz, J., Kwiterovich, P., Shan, B., Barnes, R., and Hobbs, H.H. Accumulation of dietary cholesterol in sitosterolemia caused by mutations in adjacent ABC transporters. *Science.* **2000**, *290*, 1771-1775.

Goto, N, Mori, H, Katsuragi, Y, Toi, T, Yasukawa, T, and Shimasaki, H. Effects of Diacylglycerol Containing Phytosterol on Reducing Blood Cholesterol Concentration. *J. Jpn. Oil. Chem. Soc.* **1999**, *48*, 235–240.

Grundy, S.M., Ahrens, E.H., Jr., and Davignon, J. The Interaction of Cholesterol Absorption and Cholesterol Synthesis in Man. *J. Lipid. Res.* **1969**, *10*, 304–315.

Gylling, H., Puska, P., Vartiainen, E., and Miettinen, T.A. Serum sterols during stanol ester feeding in a mildly hypercholesterolemic population. *J. Lipid Res.* **1999,** *40,* 593-600.

Heinemann, T., Axtmann, G., and von Bergmann, K. Comparison of Intestinal Absorption of Cholesterol with Different Plant Sterols in Man. *Eur. J. Clin. Investig.* **1993,** *23,* 827-831.

Hennig, B., Toborek, M., and McClain, C.J. High-Energy Diets, Fatty Acids and Endothelial Cell Function: Implications for Atherosclerosis. *J. Am. Coll. Nutr.* **2001,** *20,* 97-105.

Ikeda, I., and Sugano, M. Inhibition of Cholesterol Absorption by Plant Sterols for Mass Intervention. *Curr. Opin. Lipidol.* **1998,** *9,* 527-531.

Jones, P.J., Ntanios, F.Y., Raeini-Sarjaz, M., and Vanstone, C.A. Cholesterol-Lowering Efficacy of a Sitostanol-Containing Phytosterol Mixture with a Prudent Diet in Hyperlipidemic Men. *Am. J. Clin. Nutr.* **1999,** *69,* 1144-1150.

Katan, M.B., Grundy, S.M., Jones, P., Law, M., Miettinen, T., Paoletti, R. Stresa Workshop Participants. Efficacy and safety of plant stanols and sterols in the management of blood cholesterol levels. *Mayo Clin. Proc.* **2003,** *78,* 965-978.

Lee, M.H., Lu, K, Hazard, S., Yu, H., Shulenin, S., Hidaka, H., Kojima, H., Allikmets, R., Sakuma, N., Pegoraro, R., Srivastava, A.K., Salen, G., Dean, M., and Patel, S.B. Identification of a gene, ABCG5, important in the regulation of dietary cholesterol absorption. *Nat. Genet.* **2001,** *27,* 79-83.

Lipid Research Clinics Program. The Lipid Research Clinics Coronary Primary Prevention Trial Results. I. Reduction in Incidence of Coronary Heart Disease. *J. Am. Med. Assoc.* **1984,** *251,* 351-364.

Matsuyama, T., Shoji, K., Takase, H., Kamimaki, I., Tanaka, Y., Otsuka, A., Watanabe, H., Hase, T., and Tokimitsu, I. Effects of phytosterols in diacylglycerol as part of diet therapy on hyperlipidemia in children. *Asia. Pac. J. Clin. Nutr.* **2007,** *16,* 40-48.

Meguro, S., Higashi, K., Hase, T., Honda, Y., Otsuka, A., Tokimitsu, I., and Itakura, H. Solubilization of Phytosterols in Diacylglycerol Versus Triacylglycerol Improves the Serum Cholesterol-Lowering Effect. *Eur. J. Clin. Nutr.* **2001,** *55,* 513-517.

Miettinen, T.A., Tilvis, R.S., and Kesaniemi, Y.A. Serum Plant Sterols and Cholesterol Precursors Reflect Cholesterol Absorption and Synthesis in Volunteers of a Randomly Selected Male Population. *Am. J. Epidemiol.* **1990,** *131,* 20-31.

Mussner, M.J., Parhofer, K.G., Von Bergmann, K., Schwandt, P., Broedl, U., and Otto, C. Effects of Phytosterol Ester-Enriched Margarine on Plasma Lipoproteins in Mild to Moderate Hypercholesterolemia Are Related to Basal Cholesterol and Fat Intake. *Metabolism* **2002,** *51,* 189-194.

National Cholesterol Education Program (NCEP) Expert Panel on Detection, Evaluation, and Treatment of High Blood Cholesterol in Adults (Adult Treatment Panel III). Third Report of the National Cholesterol Education Program (NCEP) Expert Panel on Detection, Evaluation, and Treatment of High Blood Cholesterol in Adults (Adult Treatment Panel III) final report. *Circulation.* **2002,** *106,* 3143-3421.

Nestel, P., Cehun, M., Pomeroy, S., Abbey, M., and Weldon, G. Cholesterol-Lowering Effects of Plant Sterol Esters and Non-Esterified Stanols in Margarine, Butter and Low-Fat Foods. *Eur. J. Clin. Nutr.* **2001,** *55,* 1084-1090.

Noakes, M., Clifton, P., Ntanios, F., Shrapnel, W., Record, I., and McInerney, J. An increase in dietary carotenoids when consuming plant sterols or stanols is effective in maintaining plasma carotenoid concentrations. *Am. J. Clin. Nutr.* **2002,** *75,* 79-86.

Ornish, D., Scherwitz, L.W., Billings, J.H., Brown, S.E., Gould, K.L., Merritt, T.A., Sparler, S., Armstrong, W.T., Ports, T.A., Kirkeeide, R.L., Hogeboom, C., and Brand, R.J. Intensive lifestyle changes for reversal of coronary heart disease. *J. Am. Med. Assoc.* **1998,** *280,* 2001-2007. Erratum: *J. Am. Med. Assoc.* **1998,** *281,* 1380.

Ostlund, R.E., Jr., Spilburg, C.A., and Stenson, W.F. Sitostanol Administered in Lecithin Micelles Potently Reduces Cholesterol Absorption in Humans. *Am. J. Clin. Nutr.* **1999,** *70,* 826-831.

Ostlund, R.E. Phytosterols in Human Nutrition. *Annu. Rev. Nutr.* **2002a,** *22,* 533-549.

Ostlund, R.E. Jr, Racette, S.B., Okeke, A., and Stenson, W.F. Phytosterols that are naturally present in commercial corn oil significantly reduce cholesterol absorption in humans. *Am. J. Clin. Nutr.* **2002b,** *75,* 1000-1004.

Plat, J., Kerckhoffs, D.A., and Mensink, R.P. Therapeutic Potential of Plant Sterols and Stanols. *Curr. Opin. Lipidol.* **2000,** *11,* 571-576.

Saito, S., Takeshita, M., Tomonobu, K., Kudo, N., Shiiba, D., Hase, T., Tokimitsu, I., and Yasukawa, T. Dose-dependent cholesterol-lowering effect of a mayonnaise-type product with a main component of diacylglycerol-containing plant sterol esters. *Nutrition.* **2006a,** *22,* 174-178.

Saito, S., Tomonobu, K., Kudo, N., Shiiba, D., Hase, T., and Tokimitsu, I. Serum retinol, alpha-tocopherol, and beta-carotene levels are not altered by excess ingestion of diacylglycerol-containing plant sterol esters. *Ann. Nutr. Metab.* **2006b,** *50,* 372-379.

Salen, G., Patel, S., and Batta, A. K. Sitosterolemia. *Cardiovasc. Drug Rev.* **2002,** *20,* 255-270.

Shepherd, J., Cobbe, S.M., Ford, I., Isles, C.G., Lorimer, A.R., MacFarlane, P.W., McKillop, J.H., and Packard, C.J. Prevention of Coronary Heart Disease with Pravastatin in Men with

Hypercholesterolemia. West of Scotland Coronary Prevention Study Group. *N. Engl. J. Med.* **1995,** *333,* 1301-1307.

Shin, M.J., Lee, J.H., Jang, Y., Lee-Kim, Y.C., Park, E., Kim, K.M., Chung, B.C., and Chung, N. Micellar phytosterols effectively reduce cholesterol absorption at low doses. *Ann. Nutr. Metab.* **2005,** *49,* 346-351.

Takeshita, M., Goto, N., Katsuragi, Y., and Yasukawa, T. Effects of Diacylglycerol Containing Plant Sterol on Risk Factors for Arteriosclerosis. *J. Jpn. Human Dry Dock* **2001,** *16,* 9-13.

Takeshita, M., Saito, S., Katsuragi, Y., Yasunaga, K., Matsuo, N., Tokimitsu, I., Yasukawa, T., and Nakamura, H. Combination of plant sterols and diacylglycerol oil lowers serum cholesterol and lipoprotein (a) concentrations in postmenopausal women with mild to moderate hypercholesterolemia. *Eur. e-J. Clin. Nutr. Metab.* **2007a,** *2,* 4-11.

Takeshita, M., Saito, S., Moriwaki, J., Takase, H., Yamada, N., Shiiba, D., Kudo, N., Nakajima, Y., and Tokimitsu, I. Effects of Dietary Diacylglycerol Oil Containing Phytosterols in Mayonnaise on Abdominal Fat and Blood Cholesterol Levels in Japanese Men. *Jpn. Pharmacol. Ther.* **2007b,** *35,* 973-987.

Tomonobu, K., Hase, T., Shiiba, D., Kudo, N., Nakajima, Y., Tokimitsu, I., and Nakamura, H. Effect of diacylglycerol oil containing phytosterol ester in mayonnaise on anthropometric parameters, serum cholesterol levels and its safety. *Progress in Medicine* **2004,** *24,* 2342-2358.

Watanabe, H., Onizawa, K., Naito, S., Taguchi, H., Goto, N., Nagao, T., Matsuo, N., Tokimitsu, I., Yasukawa, T., Tsushima, R., Shimasaki, H., and Itakura, H. Fat-Soluble Vitamin Status Is Not Affected by Diacylglycerol Ingestion. *Ann. Nutr. Metab.* **2001,** *45,* 259-264.

Chapter 15

Efficacy and Safety of the Combined Use of Diacylglycerol Oil with Other Cholesterol-Lowering Agents

Takeshi Yamashita[a], Masao Takeshita[b], Takuji Yasukawa[b], and Haruo Nakamura[a]

[a]Mitsukoshi Health and Welfare Foundation, Tokyo, Japan and [b]Health Care Food Research Labs. Kao Corporation

Introduction

The management of cholesterol is the main gateway to the prevention of coronary heart disease (CHD). Dietary treatment of mild or moderate hypercholesterolemia is important to ameliorate this disorder, and the administration of statins is also effective.

Clinical studies indicate that phytosterols (PS) and soy protein are effective in lowering LDL cholesterol levels. However, the mechanisms by which these nutrients reduce these levels are different. PS inhibit cholesterol adsorption mainly by competing for their incorporation into micelles. On the other hand, soy protein reduces LDL cholesterol primarily by stimulating LDL receptor activity, which is similar to but less effective than the action of statin.

Because the efficacy and safety of PS, soy protein, and statins have been investigated extensively, the effects of the combined use of these substances in hypercholesterolemia is not clear. As a result, we examined the use of PS-enriched diacylglycerol (PS/DAG) oil in combination with either soy protein or statin to clarify the efficacy and safety of the combined administration of these substances.

Combined Treatment of PS/DAG Oil and Soy Protein

PS, which are abundant in the fat-soluble fractions of plants, have become ingredients of increasing importance since Peterson first reported that the addition of soy sterols to a cholesterol-enriched diet prevented increases in the plasma cholesterol level and significantly reduced the incidence of atherosclerotic plaque in the chick aorta (Peterson, 1951; Peterson et al., 1952). Many studies revealed that PS administration in humans reduces plasma total and LDL cholesterol levels (Pollak, 1953). The mechanism seems to involve the fact that PS inhibit the intestinal absorption of cholesterol (Ling & Jones, 1995). Further, the enhanced solubility of PS in DAG oil compared with TAG oil improves the serum cholesterol-lowering effect (Meguro et al., 2001).

Both soy protein and estrogen have been reported to lower LDL cholesterol level (Anderson et al., 1995), to inhibit the oxidizability of LDL (Jha et al., 1985), and to enhance the activity of LDL receptors (Sirtori et al., 1984). We reported previously that soy isoflavonoids improve systemic arterial compliance in menopausal women (Nestel et al., 1997) and that a soy protein isolate favorably affects the postprandial remnant lipoprotein response, compared with casein, in the diet of normolipidemic men (Shige et al., 1998). We investigated the efficacy and safety of the combined dietary treatment of DAG oil containing PS and soy protein in mildly hypercholesterolemic patients (Katsuyama et al., 2002). Although some patients were taking regular medication, their serum cholesterol levels were stable and unchanged before the study. Seventeen subjects (8 men and 9 postmenopausal women, 57.3 ± 9.1 y old) with total cholesterol levels >220 mg/dL were analyzed for this study. After a baseline period, 10 g PS/DAG oil (Econa oil containing phytosterols, Kao Corporation), which consists of 96 g/100 g DAG oil and 4 g/100 g PS, replaced the regular cooking oils for a minimum of 4-8 wk (first PS/DAG period). After the first PS/DAG period, subjects were required to consume 10 g of PS/DAG oil and 10 g of soy protein (Soy Milk, Fuji Oil Co. Ltd.) daily for 4 wk (PS/DAG + SOY period). After this, 10 g of PS/DAG alone was given to the subjects again for 4 wk (second PS/DAG period). The subjects were instructed not to alter their dietary habits during the test and were supervised closely by a dietician. Body weight and blood pressure were measured and blood samples were collected from fasting subjects after each period.

Body weight and blood pressure results are summarized in Table 15.1. Body weight decreased by 2.6 kg during the first PS/DAG period and PS/DAG + SOY period, and then returned to the baseline level during the second PS/DAG period; however, these changes were not significant. Serum total cholesterol and LDL cholesterol decreased by 1.0 and 5.0%, respectively, during the first PS/ DAG period compared with baseline level, and decreased significantly by 2.6 and 6.6% after the PS/DAG + SOY period, respectively, and tended to increase after ceasing soy protein (Table 15.2). Serum triglyceride tended to increase after ingestion of PS/DAG and/or soy protein; however, these changes were not significant. Serum HDL cholesterol, serum glucose, high-sensitivity C-reactive protein (hs-CRP), alanine amino transferase (ALT), blood urea nitrogen (BUN), and peripheral blood cells did not change significantly throughout the test period. The serum aspartate amino transferase (AST) level decreased after ingestion of PS/DAG and/or soy protein (Table 15.3). Serum creatine phosphokinase (CPK) tended to increase during the PS/DAG + SOY period. It might be expected that DAG would contribute to inhibiting the synthesis of very low density lipoprotein (VLDL), reflected by a reduction in visceral fat and resulting in a reduction of the transfer to LDL (Nagao

TABLE 15.1
Changes in Body Weight and Blood Pressure[a]

	Baseline	1st PS/DAG	PS/DAG + SOY	2nd PS/DAG
Body weight (kg)	63.5 ± 11.0	60.9 ± 9.8	60.9 ± 9.6	63.2 ± 11.2
Blood pressure (mmHg)				
Systolic	129.5 ± 9.3	127.9 ± 8.1	130.2 ± 8.2	130.4 ± 7.0
Diastolic	80.4 ± 5.2	81.5 ± 5.7	83.1 ± 5.3	81.2 ± 2.2

[a]Values are means ± SD, n = 17.

TABLE 15.2
Changes in Serum Lipids and Blood Glucose Levels

	Baseline	1st PS/DAG	PS/DAG + SOY	2nd PS/DAG
	(mg/dL)			
Total cholesterol	241.0 ± 19.5	238.6 ± 26.8	230.4 ± 19.3**	234.2 ± 20.2
Triglyceride	132.2 ± 61.1	143.8 ± 76.0	137.1 ± 70.1	154.1 ± 81.7
LDL cholesterol	151.5 ± 15.9	144.8 ± 24.9†	141.3 ± 15.7***	143.3 ± 21.8
HDL cholesterol	62.2 ± 13.7	64.2 ± 13.7	61.5 ± 13.2	60.2 ± 15.1
Blood glucose	102.1 ± 16.2	102.5 ± 23.7	100.9 ± 16.3	99.3 ± 16.9

Values are means ± SD, n = 17. Asterisks indicate different from baseline, †$P < 0.1$; **$P < 0.01$; ***$P < 0.001$.

TABLE 15.3
Changes in High-Sensitivity CRP, Liver, and Kidney Function

	Baseline	1st PS/DAG	PS/DAG + SOY	2nd PS/DAG
hs-CRP (mg/L)	0.39 ± 0.23	0.47 ± 0.30	0.40 ± 0.30	0.39 ± 0.28
AST (IU/L)	21.9 ± 9.8	20.1 ± 7.2	19.1 ± 9.4**	18.1 ± 6.6*
ALT (IU/L)	21.3 ± 11.0	22.1 ± 9.7	20.5 ± 10.1	20.9 ± 10.4
BUN (mg/dL)	15.8 ± 3.5	15.0 ± 3.2	16.0 ± 2.4	15.4 ± 2.6
CPK (IU/L)	108.8 ± 38.6	119.1 ± 54.6	130.9 ± 67.9†	113.7 ± 42.6*

Values are means ± SD, n = 17. Asterisks indicate different from baseline, †$P < 0.1$; *$P < 0.05$; **$P < 0.01$.

et al., 2000). In this study, DAG oil containing PS was used as the test oil; 10 g of this oil contain 400 mg of PS, which includes 150 mg β-sitosterol. A meta-analysis of the effects of PS indicated that an intake of 1000-2000 mg/d decreased LDL cholesterol by 7-9% (Ostlund, 2002). In the present study, a 400-mg intake of PS decreased LDL cholesterol levels by 5%. We hypothesize that this reduction rate is due to the additive effect of PS with DAG. It was suggested that PS reduce serum cholesterol by competitively blocking cholesterol absorption from the intestinal lumen, displacing cholesterol from bile salt micelles, increasing bile salt excretion, and hindering the rate of esterification of cholesterol in the intestinal mucosa. Beta-Sitosterol has a stronger affinity than cholesterol for incorporation into bile micelles (Ostlund, 2002). Further investigation is warranted to elucidate the mechanism and biological consequences of the combined intake of PS and DAG.

In the present study, the addition of soy protein significantly decreased the LDL cholesterol level. The effect of dietary soybeans or soy protein on plasma lipids is controversial. Several well-controlled comparisons of soy vs. animal protein (generally casein) did not clarify the difference in the cholesterol-lowering effects because of the diversity in serum lipid profiles, the amount of soy protein intake, and the active components of soy protein in previous clinical studies (Anderson et al.,1995). Anderson and co-workers (Anderson et al.,1995) performed a meta-analysis and concluded that 25 g soy protein/d reduced serum cholesterol concentration by a mean of 0.23 mmol/L and 50 g soy protein/d reduced serum cholesterol concentration by 0.45 mmol/L; HDL cholesterol increased modestly. Kestin and co-workers (Kestin et al., 1989) reported that the consumption of a soybean-enriched vegetarian diet led to lower serum LDL cholesterol levels than when meat protein was substituted for soy protein. Baum et al. (Baum et al., 1998) reported that long-term intake of soy protein increases the levels of mononuclear cell LDL receptor messenger RNA. Because of its enhanced catabolism, LDL cholesterol seems to be decreased by the intake of soy protein. Soybean phytoestrogen (soy isoflavonoids) intake inhibits LDL oxidation (Tikkanen et al., 1998) and improves arterial compliance (Nestel et al., 1997). Furthermore, we reported previously that the intake of 10 g soy protein isolated decreases remnant cholesterol and triglyceride levels in serum. However, in the present study, the combined intake of PS/DAG and soy protein did not affect the serum triglyceride level, although soy protein decreased the serum cholesterol level.

We conclude that the cholesterol-lowering action observed in this study was more effective than the extrapolated action based on a meta-analysis of PS because of a combination effect of the several different cholesterol-lowering activities of beta-sitosterol, soy protein, and DAG oil (Ostlund, 2002).

Combined Treatment of PS/DAG Oil and Pravastatin

Statins reduce cholesterol synthesis primarily in the liver and have contributed greatly to the lowering of elevated serum cholesterol concentration that is associated with the development of atherosclerosis, and to the prevention of the mortality of CHD. However, an analysis by the Scandinavian Simvastatin Survival Study reported that statins tend to increase the absorption of cholesterol in the small intestine by a feedback effect that arises from the inhibition of cholesterol synthesis (Miettinen et al., 2000). Furthermore, hypercholesterolemic patients with high levels of cholesterol absorption did not receive the beneficial effect of statin in decreasing the recurrence of CHD (Miettinen et al., 1998). Therefore, cholesterol absorption inhibitors may contribute to reducing LDL cholesterol concentration in patients with high levels of cholesterol adsorption but who are low responders to statin.

We assessed the efficacy and safety of the combined treatment of pravastatin and PS/DAG oil in mildly hypercholesterolemic patients receiving pravastatin treatment (10 mg/d) whose serum cholesterol concentration were stable before the start of the study (Takeshita et al., In press). The study was a randomized, double-blind, controlled, parallel study. The subjects (n = 61), who took 10 mg pravastatin/d and maintained a stable serum cholesterol level, were randomly assigned to consume one of the three oils: triacylglycerol oil (TAG oil) as control, diacylglycerol oil (DAG oil), and PS/DAG oil with pravastatin (10 mg/d) for 12 wk. The subjects were required to use the test oils to replace their ordinary cooking oil. We examined the serum lipid levels in 44 subjects (aged 59 9 y, 37 women and 7 men) whose baseline serum total cholesterol concentration was not <200 mg/mL, in spite of the pravastatin treatment. The mean ingestion of test oil in each group was 12-13 g/d. No adverse events were observed during the study. Serum total and LDL cholesterol levels in the PS/DAG group (n = 14) decreased 4.6% (241 to 230 mg/dL, $P < 0.01$) and 5.2% (156 to 148 mg/dL, $P < 0.05$) , respectively, compared with the baseline value (Table 15.4). The percentage change in serum cholesterol concentrations in the PS/DAG group was similar to that in a previous clinical study in which subjects consumed 10 g PS/DAG oil/d (Meguro et al., 2001. Goto et al., 1999). Therefore, the cholesterol-lowering activity of PS/DAG oil for subjects taking pravastatin appeared to result from an additive effect. On the other hand, no changes in serum total and LDL cholesterol concentration in the DAG oil and TAG oil groups were found. The serum triglyceride concentration remained constant in all groups in this study because the baseline values were at normal levels (see Chapter 11).

The safety parameters with respect to liver function and blood-clotting (fibrinogen) in the DAG and PS/DAG groups decreased during this study. Although PS/DAG oil increased serum campesterol and lathosterol concentrations by 5 and 12% from their respective baseline values, there were no significant differences. The relation between serum PS concentration and PS/DAG intake level is discussed in Chapter 14. The findings of this study suggest that PS/DAG oil had no effect on the homeostasis of cholesterol metabolism, especially cholesterol synthesis in the liver because cholesterol synthesis of the subjects is naturally low and also inhibited by pravastatin.

We conclude that PS/DAG oil might be effective in lowering cholesterol, making possible a decrease in the statin dose for mildly hypercholesterolemic patients with high cholesterol absorption and a low response to statin.

Summary

We conclude that the combined treatment of PS/ DAG oil, consisting of 4 g/100 g PS and 96 g/100 g DAG oil, and soy protein is safe and effective in reducing blood cholesterol concentrations. The addition of PS/DAG oil to pravastatin treatment was also safe and further reduced blood cholesterol concentrations, thus having an additive effect. The long-term simultaneous use of these substances might be useful for the primary prevention of atherosclerosis as an adjunct to alimentary therapy.

TABLE 15.4
Effect of PS/DAG oil on Serum Lipid Parameters[a]

		TAG group	DAG group	PS/DAG group
n		15	15	14
		(mg/dL)		
Triglyceride	Baseline	93 ± 22	105 ± 36	115 ± 41
	12wk	117 ± 49*	111 ± 51	110 ± 41
Total cholesterol	Baseline	246 ± 36	242 ± 33	241 ± 25
	12wk	248 ± 33	242 ± 37	230 ± 20**
LDL cholesterol	Baseline	152 ± 37	152 ± 32	156 ± 21
	12wk	150 ± 33	153 ± 37	148 ± 19*
HDL cholesterol	Baseline	75 ± 17	70 ± 15	62 ± 17
	12wk	75 ± 21	67 ± 17	60 ± 14

[a]Values are means ± SD. Asterisks indicate different from baseline, $*P < 0.05$; $**P < 0.01$. Subjects were those with total cholesterol concentration ≥ 200mg/dL (n = 44).

References

Anderson, J.W., Johnstone, B.M., and Cook-Newell, M.E. Meta-Analysis of the Effects of Soy Protein Intake on Serum Lipids. *N. Engl. J. Med.* **1995**, *333*, 276-282.

Baum, J.A., Teng, H., Erdman, J.W., Jr., Weigel, R.M., Klein, B.P., Persky, V.W., Freels, S., Surya, P., Bakhit, R.M., Ramos, E., Shay, N.F., and Potter, S.M. Long-Term Intake of Soy Protein Improves Blood Lipid Profiles and Increases Mononuclear Cell Low-Density-Lipoprotein Receptor Messenger Rna in Hypercholesterolemic, Postmenopausal Women. *Am. J. Clin. Nutr.* **1998**, *68*, 545-551.

Goto, N., Mori, H., Katsuragi, Y., Toi, T., Yasukawa, T., and Shimasaki, H. Effects of Diacylglycerol Containing Phytosterol on Reducing Blood Cholesterol Level. *J. Jpn. Oil. Chem. Soc.* **1999,** *48*, 235-240.

Jha, H.C., von Recklinghausen, G., and Zilliken, F. Inhibition of In Vitro Microsomal Lipid Peroxidation by Isoflavonoids. *Biochem. Pharmacol.* **1985,** *34*, 1367-1369.

Katsuyama, H., Honma, Mori, K., Terada, N., Namioka, M., Mochizuki, Y., Tomita, M., Hamana, M., Funatsu, K., Kondou, S., Yamashita, T., Miyajima, E., and Nakamura, H. Efficacy and Safety of Combined Clinical Treatment of Diet with Diacylglycerol Containing Plant Sterols and Soy Protein. *Prog. Med.* **2002,** *22*, 2782-2785.

Kestin, M., Rouse, I.L., Correll, R.A., and Nestel, P.J. Cardiovascular Disease Risk Factors in Free-Living Men: Comparison of Two Prudent Diets, One Based on Lactoovovegetarianism and the Other Allowing Lean Meat. *Am. J. Clin. Nutr.* **1989,** *50*, 280-287.

Ling, W.H., and Jones, P.J.H. Dietary Phytosterols: A Review of Metabolism, Benefits and Side Effects. *Life Sci.* **1995,** *57*, 195-206.

Meguro, S., Higashi, K., Hase, T., Honda, Y., Otsuka, A., Tokimitsu, I., and Itakura, H. Solubilization of Phytosterols in Diacylglycerol Versus Triacylglycerol Improves the Serum Cholesterol-Lowering Effect. *Eur. J. Clin. Nutr.* **2001,** *55*, 513-517.

Miettinen, T.A., Gylling, H., Strandberg, T., and Sarna, S. Baseline Serum Cholestanol as Predictor of Recurrent Coronary Events in Subgroup of Scandinavian Simvastatin Survival Study. Finnish 4S Investigators. *Br. Med. J.* **1998,** *316*, 1127-1130.

Miettinen, T.A., Strandberg, T.E., and Gylling, H. Noncholesterol Sterols and Cholesterol Lowering by Long-Term Simvastatin Treatment in Coronary Patients: Relation to Basal Serum Cholestanol. *Arterioscler. Thromb. Vasc. Biol.* **2000,** *20*, 1340-1346.

Nagao, T., Watanabe, H., Goto, N., Onizawa, K., Taguchi, H., Matsuo, N., Yasukawa, T. Tsushima, R., Shimasaki, H., and Itakura, H. Dietary Diacylglycerol Suppresses Accumulation of Body Fat Compared to Triacylglycerol in Men a Double-Blind Controlled Trial. *J. Nutr.* **2000,** *130*, 792-797.

Nestel, P.J., Yamashita, T., Sasahara, T., Pomeroy. S.E., Dart, A.M., Komesaroff, P., Owen, A., and Abbey, M. Soy Isoflavones Improve Systemic Arterial Compliance but Not Plasma Lipids in Menopausal and Perimenopausal Women. *Arterioscler. Thromb. Vasc. Biol.* **1997,** *17*, 3392-3398.

Ostlund, R.E., Jr. Phytosterols in Human Nutrition. *Annu. Rev. Nutr.* **2002,** *22*, 533-549.

Peterson, D.W. Effect of Soybean Sterols in the Diet on Plasma and Liver Cholesterol in Chicks. *Proc. Soc. Exp. Biol. Med.* **1951,** *78*, 1143-1147.

Peterson, D.W., Nichols, C.W., and Schneour, E.W. Some Relationships Among Dietary Sterols, Plasma and Liver Cholesterol Levels and Atherosclerosis in Chicks. *J. Nutr.* **1952,** *47*, 57-65.

Pollak, O.J. Reduction of Blood Cholesterol in Man. *Circulation* **1953,** *7*, 202-706.

Shige, H., Ishikawa, T., Higashi, K., Yamashita, T., Tomiyasu, K., Yoshida, H., Hosoai, H., Ito, T., Nakajima, K., Ayaori, M., Yonemura, A., Suzukawa, M., and Nakamura, H. Effects of Soy Protein Isolate (SPI) and Casein on the Postprandial Lipemia in Normolipidemic Men. *J. Nutr. Sci. Vitaminol.* **1998,** *44*, 113-127.

Sirtori, C.R., Galli, G., Lovati, M.R., Carrara, P., Bosisio, E., and Kienle, M.G. Effects of Dietary Proteins on the Regulation of Liver Lipoprotein Receptors in Rats. *J. Nutr.* **1984,** *114*, 1493-1500.

Takeshita, M., Katsuragi, Y., Kusuhara, M., Higashi, K., Miyajima, E., Mizuno, K., Mori, K., Obata, T., Ohmori, R., Ohsuzu, F., Onodera, Y., Sano, J., Sawada, S., Tabata, S., Tokimitsu, I., Tomonobu, K., Yamashita, T., Yasukawa, T., Yonemura, A., Nakamura, H. Phytosterols dissolved in diacylglycerol oil reinforce the cholesterol-lowering effect of low-dose pravastatin treatment. *Nutr. Metab. Cardiovasc. Dis.* (In Press)

Tikkanen, M.J., Wahala, K., Ojala, S., Vihma, V., and Adlercreutz, H. Effect of Soybean Phytoestrogen Intake on Low Density Lipoprotein Oxidation Resistance. *Proc. Natl. Acad. Sci. USA* **1998,** *95*, 3106-3110.

Part IV

Regulatory Status and Safety

Chapter 16

Regulatory Status of Diacylglycerol Oil in North America, the European Union, Latin America, Australia/ New Zealand, and Japan

Mark W. Empie

Archer Daniels Midland Company, Decatur, IL 62521

Introduction

During the past 10 years, there has been a rapid expansion of the number of regulatory hurdles enacted, governing the ability to place new products on the market as foods and food ingredients. Previously, foods were grouped into categories of familiar foods commonly consumed and new food additives. Familiar foods were placed on the market after safety reviews; however, these reviews did not have as a basis the knowledge of nutritional biochemistry, physiology, and toxicology that we have today. In the United States, natural food components were evaluated as Generally Recognized As Safe (GRAS), and did not necessarily require premarket approval by the Food and Drug Administration (FDA), whereas chemical food additives required premarketing approval, often a lengthy and expensive process. With the development of globalized food markets and the use of modern biotechnology to produce foods, increasing regulatory controls have been introduced to evaluate the safety of new food materials not previously present in a nation's food supply at the proposed use levels. The regulatory systems put in place have followed models enacted by the United States, Canada, The European Union, and Japan.

The components of diacylglycerol oil (DAG oil) have been part of the food supply for centuries, as natural diglycerides in common vegetable oils, and as components of monodiglyceride emulsifier systems. Human consumption of diglycerides from these sources is estimated to be ~3 g/d in the U.S. population. Although these are commonly consumed materials, DAG oil as a cooking oil by itself and as a major ingredient in prepared foods represents a different level of exposure to consumers. Consequently, this higher exposure requires the product to be subjected to a formal safety evaluation according to modern standards.

In addition to approvals for marketing, regulatory schemes are in effect for health benefit labeling and advertising claims, and for nutritional labeling. The various countries use different systems; because health claims for food products are a controversial subject, interpretations of these regulations are continually changing. Generally, the systems use either a structure-function (system benefit) claim or a true health claim in which disease prevention or alleviation may be stated.

United States Regulatory Status for Diacylglycerol Oil

There are two options for regulatory approval of DAG oil in the United States, GRAS status and Food Additive status. Given that DAG oil is present in the common food supply in measurable quantities and has a long safe history of use, it is a candidate for GRAS status. GRAS status may be achieved by a formal process of self-affirmation following set rules, or it may be affirmed directly by the FDA. The process of FDA affirmation requires a formal dossier to be submitted following the rules set forth in the Code of Federal Regulations (CFR) under Title 21, Section 170.30 (CFR, 2001). This process historically has taken as long as 5 years for the FDA to complete before a formal affirmation and codification in the CFR are accomplished. With the GRAS self-affirmation process, the assembly of a GRAS dossier, followed by review and evaluation of the safety information contained therein by scientific experts, in effect completes the self-affirmation requirement before submission to the FDA. Hence, with this self-affirmation, the product is able to be legally placed on the market. In fact, some common products today are on the market as self-affirmed GRAS products even while their corresponding GRAS dossiers are still under FDA evaluation. The FDA recognized this disparity, and in 1997, the Agency proposed a GRAS Notification procedure (FR, 1997) whereby an entity may "notify" the FDA of their self-affirmation of a substance as GRAS through submission of a summary dossier. This notification would contain the basic elements of the dossiers submitted under the formal FDA affirmation process, but the FDA has no requirement to issue a formal approval. Instead, the Agency issues a letter acknowledging the ability to self-affirm the substance as GRAS and that the agency upon review of the information has no questions at this time about the basis and information submitted in support of GRAS status. As of January 1, 2008, FDA has recieved 236 notifications. The majority of notifiers have received FDA "no objection" letters, some notifications have been rejected as not having a basis for GRAS, and others have had more information requested. The basis for all these determinations always rests upon the consumer safety of the product in the intended use categories. Few products have been granted unrestricted general use.

DAG oil has had two FDA notifications, once in 1999 for two use categories (FDA, 2000), and again in 2002 for expansion to 10 food use categories (FDA, 2003b).

1999 Notification. The initial self-affirmation as GRAS claimed two categories: home cooking oil use and use in margarine spreads. The dossier was submitted by the Kao Corporation of Japan and included sections on product history of use; product composition; manufacturing process; genotoxicity, acute toxicity, and subchronic toxicity safety studies; and animal and human metabolic studies of the oil.

Diacylglycerides have been part of the human diet for many years. They may be present in vegetable oils from 1 to 15% of the composition and they are found as emulsifiers in bakery products and other foods. Consequently, they have a safe history of use in the U.S. population. The specification for DAG oil is a product that is composed of at least 80% diglycerides; of these, 70% are in the form of 1,3-diglycerides, and the remainder of the oil is comprised predominantly of vegetable oil triglycerides. All of these components are part of the U.S. diet, albeit in different proportions, and do not represent the introduction of a "new" dietary component. It was also noted in this dossier that studies demonstrate that the various diglyceride forms are fully absorbed and metabolized using the body's common lipid metabolic pathways. This was important to show that the oil is metabolized in the same way as a conventional vegetable oil, and that there were no consequences of gastrointestinal upset. Furthermore, by being completely absorbed, there is no disturbance in the uptake of fat-soluble vitamins.

The manufacturing process makes use of an immobilized lipase enzyme as shown in Chapter 23. This enzyme is derived from Rhizomucor miehei and is expressed in a recombinant Aspergillus oryzae host. This enzyme and expression system have been used separately for many years in the food industry, and this particular recombinant enzyme has been notified to the FDA as GRAS (FDA, 2003a). Being immobilized, the enzyme is used as a processing aid in order to prevent addition of enzyme or immobilization resin residues into the final commercial product. The production process for the DAG oil itself employs purified vegetable oil fatty acids, which are recombined with glycerol to make the 1,3-diglyceride. The resulting oil is purified using conventional vegetable oil refining techniques, and consequently does not make use of new manufacturing techniques that must be evaluated for safety.

A 2-y, one-generation rodent feeding study was conducted, and the results established a No Observed Adverse Effect Level (NOAEL) of 2.3 g/(kg body weight)/d (Soni et al., 2001). Consumption of the foods in the two selected categories, home cooking oil and margarine spreads, was calculated using the 1989–1992 USDA Continuing Survey of Food Intakes by Individuals (CSFII) database. The calculated consumption of DAG oil use in the two categories selected, for the mean and 90th percentile of the all consumer groups, yielded values of 0.23 and 0.42 g/(kg body weight)/d, respectively. These values are well within the safety margin of the NOAEL. Additional feeding studies in humans, both as a single oral dose (44 g/60 kg body weight) (Watanabe et al., 1997a) and as a 9mo free-living use study (Katsuragi et al.,1999), further supported safe use at these levels. Chapter 17 will describe these toxicology studies in further detail.

2002 Notification. The limitation of the initial GRAS determination to two categories was predicated upon the daily consumption values and the toxicology studies on hand at the time. Subsequent studies provided support for an expansion of GRAS categories, and 10 food groups were selected: home cooking oil and margarine spreads from the existing GRAS status; bakery products; dressings for salad; pizza; mayonnaise; breakfast/snack/power bars; soup/gravies; meal replacements; and frozen dinner/entrees. Consumption calculations for these categories using the CSFII 1994-96, 98 database raised the 90th percentile values to 0.49 g/(kg body weight)/d. The consumption values were only moderately higher than the original two-category calculation because home cooking oil and margarine use represent high consumption foods and the substitution of one product category for another in the diet would result in a person not consuming all 10 categories in 1 d. On the basis of additional new safety data (see Chapter 17) and the fact that the oil, at that time, had been consumed by the general population in Japan for

>4 y without notation of serious adverse effects, the higher consumption value [0.50 g/(kg body weight)/d of DAG oil] was determined to be within the safety limits of the preponderance of the evidence. On the basis of the evidence provided, FDA issued a no objection letter for the expansion of these categories (FDA, 2003b).

In the US, future expansion to additional categories or to unlimited Good Manufacturing Practices use will require additional studies. These studies will have to demonstrate safe usage at higher consumption levels. This is difficult to model using animal high- dose studies because the oil is a macronutrient; high replacement values to establish a safety factor are not practical because other nutrients will become limiting. Consequently, detailed mechanistic studies and studies in susceptible populations are required, most notably for growth and development in children.

Labeling. The proposed primary use of diacylglycerol oil is as a direct one-to-one replacement of vegetable oil. According to the FDA, vegetable oil is considered to be composed predominantly of triglycerides, although this definition is somewhat different from the FAO/WHO CODEX definition of being composed of glycerides (CODEX, 1981). Consequently, the FDA ruled that the common and usual name to be used on ingredient statements for DAG oil is diacylglycerol oil.

The nutritional labeling statements are conventionally calculated. The energy content of DAG oil is 120 kcal/serving, equivalent to vegetable oil, and this is borne out by studies using both bomb calorimetry and direct animal measurements (Taguchi, 2001). The fatty acid profile of the oil reflects the input acids fed to the enzyme reactor during the manufacturing process. Typically, DAG oil contains (per serving) 0.5 g saturated fat, 8 g polyunsaturated fat, and 5 g monounsaturated fat. Of the fatty acids in the oil itself, <2% are in the trans form, which results in a content that on a per serving basis is below the FDA threshold for labeling (FR, 2003b). Vitamin E is added to the product for stability enhancement. The concentration per serving is sufficient to contribute to an individual's recommended daily intake.

Claims. Claims for foodstuffs relating to health comprise a controversial and continually evolving area. Claims are regulated both by the FDA as attendant to labeling, and by the Federal Trade Commission under the advertising and consumer protection statutes. Recently, both agencies increased enforcement actions against companies making false and misleading health statements associated with food materials. The underlying principle guiding both organizations' enforcement decisions is whether there is a significant body of evidence accepted by the scientific community to support a given claim.

The extensive evidence gathered to date on diacylglycerol oil, much of which is presented in this book, supports several specific structure-function claims. From a mechanistic standpoint, the published studies in animals (Murata et al., 1994. Murata et al., 1997. Watanabe et al., 1997b) and humans (Kamphuis et al., 2003) demonstrate that DAG fatty acids are preferentially shunted directly to the liver for oxidation. This forms the basis for the direct label and advertising claim: "less likely to be stored as body fat". Sufficient evidence exists to be able to make a second label claim: "one is able to maintain a healthy weight when using DAG oil as part of a sensible diet." Several direct studies have been conducted as shown in Chapters 10 and 11. In free-living individuals using DAG oil for 1 y, the test group lost 0.5–0.9 kg compared with a fatty acid matched vegetable oil (see Chapter 11). In a second, randomized, double-blind, placebo-controlled clinical trial, with individuals consuming a Step I diet, the DAG oil subject group lost 1.36–1.81 kg more that the matched triglyceride oil group over a 6-mo period (Maki et al., 2002). The combination of a proven mechanism and several long-term clinical studies provides solid substantiation for making the listed claims.

Additionally, there is evidence to support a third claim: "triglycerides in the blood stream are lowered following a meal compared to conventional oil" As evidence continues to build, a future true health claim is possible. The FDA recently proposed a scientific study and claim rating scheme to grade the quality of studies supporting specific health claims (FR, 2003a). The objective of future studies will be to raise the scientific evidence to the "A" or highest grading level, thereby providing the strongest claim.

European Union

In 1997, the European Union enacted a Novel Foods regulation (EC, 1997), which stated that any food not heretofore present on the European market and not substantially equivalent to an existing product, or produced by a substantially different process, was to be subjected to a safety review and premarket approval at the member state and Community levels. To date, there have been 78 Novel Food submissions, and of these, less than 30 of the 78 submissions have received authorizations for marketing. The 1997 Novel Foods regulation is being updated, and has been amended in October 2003 to include GMO animal feed products (EC, 2001). A new Novel Food regulation is expected to be enacted in 2008.

During 2001, ADM Kao L.L.C. submitted a Novel Foods application and dossier to the competent authorities of The Netherlands, requesting approval to place diacylglycerol oil on the European market. This dossier was reviewed by scientific experts of The Netherlands, and an initial positive assessment was forwarded to the EU Commission working group on Novel Foods (DG SANCO) (HCN, 2002) for distribution and review by the other member states. The application requested approval for use of DAG oil in 8 food categories

A critical element in the assessment of a Novel Foods dossier is the calculation of consumer exposure. Categorical food consumption data are not available for all countries in the EU. General disappearance data are available through the Eurostat database, and The Netherlands has a database for its population, maintained by the TNO research institute. However, these databases provide approximations for consumption and may be less accurate for countries with different dietary patterns. Overall, the calculated consumption numbers for DAG oil from the Netherlands database and the disappearance data from the Eurostat database were not significantly different from those calculated in the U.S. GRAS application, which permitted an assessment of intake in relation to the toxicology study data. The competent authorities of The Netherlands concluded that the diacylglyceride oil was safe for consumption at an intake level of 140 g/d, a value higher than that proposed for the United States and well above the expected consumption levels in the various member states. On October 23, 2006, the European Commission issued an authorization permitting the placement of diacylglycerol oil on the market (Commission Decision 2006/720/EC). Food categories which were approved are: home cooking oil, fat spreads/margarines, mayonnaise, salad dressings, health drinks as meal replacements, yoghurt, and baked goods.

Labeling and Claims. The EU currently has specific legislation on labeling, but claims are not yet permitted on packaging. This issue is under discussion and proposals have been made to enact legislation permitting claims (EC, 2003). As of December, 2007, diacylglycerol oil claims were still under review, and specific claims texts were not available. However, claims have been requested for maintaining or lowering body weight, maintaining healthy triglyceride levels and reduction of body fat storage.

Canada

Canada has enacted a Novel Foods regulation similar to that of the European Union. Foods in Canada require a positive listing because there is no GRAS category in Canadian law. A full safety dossier is required, following the requirements set forth in the Novel Foods law, before marketing authorization may be achieved (HC, Website). A full dossier for DAG oil was submitted to the Canadian Food Inspection Agency on November 23, 2002, and included specific food consumption estimates calculated from databases available for Saskatchewan and Québec provincial populations. Categories for use were the same as those approved in the United States. This dossier passed the major chemistry, manufacturing, nutrition, and toxicology reviews, and on September 10, 2004, diacylglycerol oil was approved as a Novel Food in Canada. In addition to the human health safety aspects, the Canadian government also conducted an environmental assessment to determine the effect of exposure outside of the consumption evaluation. Because the components of DAG oil are common in other foods and vegetable oils, an argument has been forwarded and accepted that the components of DAG oil are already listed on the Canadian Domestic Substances List (DSL) and, therefore, a full environmental assessment is not required.

Labeling and Claims. Canada has specific laws governing the composition of package labels, and these are similar to those in the United States. However, Canada has not yet enacted laws permitting label and advertising claims for health benefits of a Novel Food. There are proposals tabled but not yet enacted.

Latin America

Brazil. Brazil operates a positive authorization system for new or novel foods. The authorization process also requires a dossier submission, and the documentation needed is similar to that required by other countries. The submission is made in the Portuguese language. A document has been submitted, and proposed approval categories were not specified, but left as general. Consumption intake data were not readily available, but were suggested to be similar to that in the U.S. estimates. Unlike the other countries, Brazil allows label claims. The approval dossier for DAG oil was submitted in December, 2003, together with a number of label and advertising claims, and approval for use as an ingredient in foods without claims was obtained in 2004. Approval for use as a home bottled oil with claims is pending.

Remainder of Latin America. The Brazilian dossier is a very key element in the approval process for other Latin American countries. There are no formal review or approval procedures legislated in other Latin American countries, only local product registration. Other countries may follow applications made to Brazil, and authorization there could support product registration in the other nations. Authorization for this type of product in Mexico, together with corresponding claims, is under discussion, but this has not yet been legislated.

Australia and New Zealand

Australia and New Zealand have formed a joint food safety authority, Food Safety Australia New Zealand (FSANZ) to review novel foods and processes. The predecessor of this agency enacted a Novel Foods regulation to require premarket approval for new food entities entering either country (ANZFA, Website). The requirements are similar to those in the E.U. and Canada. However, specific population intake data are calculated by the agency and specific health claims are not permitted. An approval dossier was submitted to FSANZ in June, 2003, and was approved in October 14, 2004, by formal gazette amendment number 74 to the Australia New Zealand Food Standards Code.

Japan

DAG oil has been on the market in Japan as an approved human food since February 1999, and has become a leading bottled oil sold on the Japanese market. The original approval was obtained by the Kao Corporation after consultation and submission of technical information to the Ministry of Health in May 1998. Japan maintains a system of positive approvals for health claims, operated by the Ministry of Health under the Foods for Special Health Use (FOSHU) review system. Kao Corporation obtained FOSHU status for their Econa brand DAG oil, and DAG oil is permitted to display two claims on their labels and advertising. The officially approved claim is that "A diacylglycerol is the main ingredient of this oil, it suppresses the increase of serum triglycerides after meals compared with other edible oils and it is less likely to become body fat." In addition, the oil bottle label is allowed to display the FOSHU symbol, signaling to the customer that the Ministry of Health has granted a "healthy" status to this product. The Econa oil product has achieved wide acceptance not solely for its quality and ingredient characteristics, but also its recognized health benefits.

Since the time DAG oil was approved in 1998, two DAG-related products have been approved as FOSHU. One is DAG oil containing phytosterols, with the claim of lowering serum cholesterol level in addition to the diacylglycerol claim described earlier. This product claim was approved in June 1999. The efficacy of this oil is described in Chapters 13–15. The other product is a mayonnaise-like substance containing DAG oil that was approved in September 2003, and that has the claim: "DAG mayonnaise type is less likely to become body fat" (see Chapter 10).

Conclusions

Regulations worldwide to control new food product innovations are constantly increasing. Many of the new Novel Food regulations have been enacted to control and evaluate the safety of new products derived from modern recombinant DNA technology. However, more conventionally produced new products have become caught in the same approval requirements, necessitating substantial safety and toxicological studies, and lengthy review times of authorization dossiers. Moreover, due to the lack of a globally harmonized review system, redundancy in the construction, submission and review of these dossiers occurs, wasting significant corporate and national scientific and governmental resources.

Fortunately, the dossier content is redundant; therefore, submissions may be adapted to specific national requirements by adding the sections specifically required. The major sections that require adaptation are the regional food consumption estimates, the specific food categories for authorization, and the label statements or claims. The main sections on product consumption history, manufacturing process outline, toxicology, safety and health effect assessments are similar among dossiers.

As of December 2007, DAG oil has been authorized for marketing in Japan, the United States, the European Union, Canada, Australia/New Zealand, and Brazil. With the exception of a few specific countries, other major regions either do not have formal novel food approval requirements or require only a simple product registration. The challenge for future regulatory action will be to achieve label and advertising health claims as each country enacts such legislation.

References

Australia New Zealand Food Authority, Guidelines for Amending the Australian Food Standards Code; Standard A19-Novel Foods. (Internet address: http://anzfa.gov.au.)

Code of Federal Regulations 2001, Title 21 170.30, U.S. Government Printing Office, Washington.

CFSAN/FDA. Letter dated December 4, 2000 to Sandra Dennis, Morgan, Lewis & Bockius, LLP, from Alan Rulis, Food and Drug Administration, Re: GRAS Notice No. GRN 000056, (Internet address:http://www.cfsan.fda.gov/~rdb/opa-g056.html.)

CFSAN/FDA. Letter dated January 27, 2003a to Jack Harris, Enzyme Technical Association, from Alan Rulis, Food and Drug Administration, Re: GRAS Notice No. GRN 000113, (Internet address: http://www.cfsan.fda.gov/~rdb/opa-g113.html.)

CFSAN/FDA. Letter dated February 24, 2003b to Mark Empie, Archer Daniels Midland Company, from Alan Rulis, Food and Drug Administration, Re: GRAS Notice No. GRN 000115, (Internet address:http://www.cfsan.fda.gov/~rdb/opa-g115.html.)

Codex Standard for Edible Fats and Oils Not Covered by Individual Standards, 19-1981, Rev. 2-1999 (Internet address: http://ftp.fao.org/codex/standard/en/CXS_019e.pdf.)

EC, Commission , Proposal for a Regulation of the European Parliament and of the Council 2003/0165 on nutrition and health claims made on foods. (Internet address:http://www.europa.eu.int/comm/food/fs/fl/fl07_en.pdf)

EC, Commission, Proposal for a Regulation on Traceability and Labelling of GMOs and Products Produced from GMOs. COM 2001 -1821 final, 25 July 2001, (Internet address: http://europa.eu.int/comm/food/fs/biotech/biotech09_en.pdf).

EC, Regulation No. 258/97 of The European Parliament and of the Council of 27 January 1997 Concerning Novel Foods and Novel Food Ingredients. (Internet address: http://www.bio-safety.be/GB/Dir.Eur.GB/FF/258_97/258_97.html)

Federal Register, Vol. 62, No. 74, April 17, 1997, Proposed Rules, p. 18938.

Federal Register, Vol. 68, No.41,. Reg. 41, 2003a, July 11, Rules and Regulations, p.387

Federal Register, Vol. 68, No. 133, July 11, 2003b, Rules and Regulations, p. 41434.

Health Canada, Section B28.002 of the Novel Foods Regulations of the Food and Drug Regulations, (Internet address: http://www.hc-sc.gc.ca/fn-an/legislation/acts-lois/fdr-rad/division-titre28_e.html)

Health Council of the Netherlands, Committee on the Safety Assessment of Novel Foods, 2002 Diacylglycerol Oil, publication no. 2002/07VNV. Health Council of the Netherlands, The Hague.

Kamphuis, M.M, Mela, D.J. and Westerterp-Plantenga, M.S. Diacylglycerols Affect Substrate Oxidation and Appetite in Humans. *Am. J. Clin. Nutr.* **2003**, *77*, 1133-1139.

Katsuragi, Y., Toi, T., and Yasukawa, T. Effects of Dietary Diacylglycerols on Obesity and Hyperlipidemia. *J. Jpn. Soc. of Human Dry Dock* **1999**, *14*, 258-262.

Maki, K.C., Davidson, M.H., Tsushima, R., Matsuo, N., Tokimitsu, I., Umporowicz, D.M., Dicklin, M.R., Foster, G.S., Ingram, K.A., Anderson, B.D., Frost, S.D. and Bell, M. Consumption of Diacylglycerol Oil as Part of a Reduced-Energy Diet Enhances Loss of Body Weight and Fat in Comparison with Consumption of Triacylglycerol Control Oil. *Am. J. Clin. Nutr.* **2002**, *76*, 1230-1236.

Murata, M., Hara, K., and Ide, T. Alteration by Diacylglycerols of the Transport and Fatty Acid Composition of Lymph Chylomicrons in Rats. *Biosci. Biotechnol. Biochem.* **1994**, *58*, 1416-1419.

Murata, M., Ide, T., and Hara, K. Reciprocal Responses to Dietary Diacylglycerol of Hepatic Enzymes of Fatty Acid Synthesis and Oxidation in the Rat. *Br. J. Nutr.* **1997**, *77*, 107-121.

Soni, M.G., Kimura, H., and Burdock, G.A. Chronic Study of Diacylglycerol Oil in Rats. *Food Chem. Toxicol.* **2001**, *39*, 319-329.

Taguchi, H., Nagao, T., Watanabe, H., Onizawa, K., Matsuo, N., Tokimitsu, I., and Itakura, H. Energy Value and Digestibility of Dietary Oil Containing Mainly 1,3-Diacylglycerol are Similar to those of Triacylglycerol. *Lipids* **2001**, *36*, 379-382.

Watanabe, H., Onizawa, K., Taguchi, H., Fujimori, N., Naito, S., Gotah, N., Yasukawa, T., Hattori, M., and Shimasaki, H. Effect of Diacylglycerol on Lipid Metabolism in Human. *J. Jpn. Oil Chem. Soc.* **1997a**, *46*, 309-317.

Watanabe, H., Onizawa, K., Taguchi, H., Kobori, M., Chiba, H., Naito, S., Matsuo, N., Yasukawa, T., Hattori, M., and Shimasaki, H. Nutritional Characterization of Diacylglycerols in Rats. *J. Jpn. Oil Chem. Soc.* **1997b**, *46*, 301-307.

Chapter 17

Safety Aspects of Diacylglycerol Oil

Joseph F. Borzelleca[a], Walter Glinsmann[b], and Elke Kennepohl[c]

[a]Virginia Commonwealth University, Richmond, VA 23229–7319;
[b]Glinsmann, Incorporated, Arlington, VA 22207–4231; and
[c]Write-Tox Consulting, Spruce Grove, AB T7Y 1G1, Canada

Introduction

Diacylglycerol (DAG) oil is an edible oil composed primarily of 1,3- and 1,2-diacylglycerol (ratio of 7:3) with characteristics similar to plant-derived fats and oils such as rapeseed oil; however, is more hydrophilic and soluble than triacylglycerol (TAG) oil. The fatty acid composition of DAG oil comprises mainly the unsaturated fatty acids, such as oleic acid (20–65%), linoleic acid (15–65%), and linolenic acid (≤15%), with small amounts of saturated fatty acids [palmitic and/or stearic acid (≤10%)] (see Chapters 18 and 19). DAG oil is produced commercially consistent with current Good Manufactuing Practice (=GMP) through esterification of fatty acids derived from natural edible plant oils in the presence of a lipase preparation; it contains >80% DAG, <20% TAG, <5% monoacylglycerol (MAG), and small amount of food additives such as antioxidants and emulsifiers as shown in Chapter 19.

Humans naturally consume diacylglycerols as constituents of edible oils such as soybean, palm, corn, and rapeseed oils; diacylglycerols are also produced endogenously (Douglas et al., 1953; Borgstrom et al., 1957; Hamosh & Scow, 1973; Small, 1991) as a product of triglyceride metabolism. Independent reviews of the safety of mono- and diglycerides, conducted by the Joint WHO/FAO Expert Committee on Food Additives (JECFA, 1974) and the Federation of American Societies for Experimental Biology (FASEB,1975), concluded that these partial glycerides presented no safety concerns at the intended levels of use. Since that time, several additional toxicological and clinical studies were conducted to further assess the safety of DAG oil and/or its components. Several acute, repeated dosing, and chronic studies in rats, in vitro and in vivo mutagenicity studies in bacterial and mammalian systems, and a series of human studies are discussed below and support the safety of DAG oil ingestion through its intended use in foods.

Metabolism of DAG Oil

Because DAG oil metabolism was discussed in detail in Part II of this book, only a brief review is provided below. In rats, the absorption of 20% dietary DAG oil was calculated to be ~96% and was identical to that of TAG oil with a similar fatty acid composition fed in similar concentrations over 15 d (Taguchi et al., 2001). In addition, the calculated energy values of DAG and TAG oils were comparable [38.9 and 39.6 kJ/g (9.3 and 9.5 kcal/d) for DAG oil and TAG oil, respectively]. The metabolism of DAG is similar to that of triglycerides and mono- and diglycerides, which are partially or completely hydrolyzed by lipases in the gastrointestinal tract (Mattson & Volpenhein, 1964; Grundy, 1996). The resulting partial glycerides and/or fatty acids are absorbed for re-esterification to triglycerides, and/or are oxidized as a fuel (e.g., glucose) to CO_2 through ß-oxidation to acetyl-CoA (Small, 1991; Mattson & Volpenhein, 1964; Grundy, 1996; Watanabe et al., 1997). 2-Monoglycerides are rapidly esterified to triglycerides, whereas the rate of esterification of 1-monoglycerides is much slower; therefore, it is more likely that they undergo hydrolysis to glycerol and free fatty acids. Because the primary metabolic product of DAG oil is 1-monoglyceride, it is expected that after absorption and hydrolysis, DAG oil metabolism favors ß-oxidation of the fatty acids rather than triglyceride re-synthesis (Watanabe et al., 1997; Murata et al., 1997).

Toxicological Studies

The safety database of DAG oil consists of a number of standard toxicological studies including acute, repeated dosing, and chronic studies in rodents and dogs. In addition, in vitro and in vivo mutagenicity/genotoxicity studies were conducted in bacterial and mammalian systems. Unless otherwise stated, the toxicological studies were conducted following OECD, Redbook II, and ICH guidelines under GLP (Good Laboratory Practice). Any findings reported in these studies generally were associated with increased dietary lipid concentrations compared with lower fat content control diets, rather than inherent toxicity of DAG oil. As summarized in Table 17.1, the results of these studies indicated that DAG oil was well tolerated and produced no adverse effects.

Acute Studies

The oral toxicity of DAG oil was determined in 3 studies using rats (Ishida, 1996a, 1996b; Ishii, 2003). In the first 2 studies, 15,000 mg/kg body weight of DAG oil or DAG oil without additives was administered by gavage to 5 male and 5 female Crl:CD®(SD)IGS rats (Ishida, 1996a, 1996b). After a 14-d observation period, the rats were euthanized and necropsied. No toxic effects were observed in either study and the 50% lethal dose (LD_{50}) of DAG oil or DAG oil without additives was in excess of 15,000 mg/kg body weight. In the third study, groups of 5 male and 5 female CD(SD)IGS rats were gavaged with 5,000 mg heated DAG oil/kg body weight (i.e., oil obtained after frying potatoes), observed for 14 d, euthanized and necropsied (Ishii, 2003). No toxic effects were reported and the LD_{50} was considered to be >5,000 mg/kg body weight.

Repeat-Dose Studies

4-wk Dietary Rat Study. The test oil used in this study was DAG oil without additives. In a preliminary 4-wk dietary study (Serabian, 1991), groups of 10 male and 10 female

TABLE 17.1
Summary of Toxicity Studies on DAG oil Conducted in Laboratory Animals

Test Substance	Species	n	Study Duration	Route	Dose (mg/kg bw/d)	Results	Reference
DAG oil	Rat	5/sex	acute	gavage	15000	LD_{50}>15000 mg/kg bw	Ishida, 1996a
DAG oil (no additives)	Rat	5/sex	acute	gavage	15000	LD_{50}>15000 mg/kg bw	Ishida, 1996b
DAG oil (Heated)	Rat	5/sex	acute	gavage	5000	LD_{50}>5000 mg/kg bw	Ishii, 2003
DAG oil (no additives)	Rat	10/sex/group	4 wk	diet	140, 700, 3500	No adverse effects	Serabian, 1991
DAG oil	Dog	4/sex/group	1 y	diet	326, 1227, 2541 (M) 348, 1487, 2300 (F)	NOAEL=2541 (M) and 2300 (F) mg/kg bw/d	Chengelis et al., 2006a
DAG oil	Rat	60/sex/group	105 wk	diet	885, 1770 (M) 1175, 2350 (F)	NOAEL=1770 (M) and 2350 (F) mg/kg bw/d	Soni et al., 2001
DAG oil	Rat	50/sex/group	24 mo	diet	356.2, 984.8, 1982.4 (M) 477.5, 1325.2, 2645.1 (F)	NOAEL=1946 (M) and 2507 (F) mg/kg bw/d	Chengelis et al., 2006b
DAG oil	Mouse	50/sex/group	24 mo	diet	1792, 3773, 7412 (M) 2509, 5286, 9796 (F)	NOAEL=7412 (M) and 9796 (F) mg/kg bw/d	Chengelis et al., 2006c

M=males; F=females

Crl:CD® rats were fed diets containing 10% fat (corn oil was used to correct the fat content to 10% in the diet) as follows: low-dose group received 0.2% test oil; mid-dose group received 1.0% test oil; high-dose group received 5.0% test oil; control group A received 4.3% from feed (other groups were fed fat-free feed); and control group B received 5.0% rapeseed oil. All rats survived to the end of the study and no overt clinical signs were reported. Mean feed consumption was significantly decreased in male rats from the low- and high-dose test oil groups during wk 1–4 compared with control group A, but not control group B. Because the decrease was not dose dependent and no statistically significant differences were reported in females, this finding was not considered to be related to the test oil. In addition, mean body weights did not differ among treated and control groups of either sex. Ophthalmoscopic examination and the evaluation of clinical pathology data (hematology, clinical chemistry, and urinalysis) and gross pathology findings revealed no treatment-related toxicological effects. Blood urea nitrogen (BUN) was significantly decreased in females from control group B and the test oil groups compared with females in control group A; however, the differences were small and the values were within the normal range (~15 to 21 mg/dL) for BUN concentrations in rats (Harkness & Wagner, 1989) and therefore considered incidental. Compared with control group A, the mean relative kidney weights in females fed the test oil and mean relative liver weights in males fed the test oil were significantly decreased. These values were comparable to those of control group B. In addition, the statistical significance of these findings was likely due to slightly higher than normal relative kidney and liver weights in female and male control group A rats, respectively. No treatment-related histomorphological alterations were reported in the rats examined. The absence of consistent treatment-related, dose-dependent, adverse histopathological effects supports a NOAEL (no-observed-adverse-effect level) of 5.0% in the diet, the highest concentration fed, which is ~3500 mg DAG oil/kg body weight/d.

One-year Dietary Dog Study. Groups of beagle dogs (4 male and 4 female/group) were fed diets containing 9.5% dietary fat (control group 1), 9.5% TAG oil (control group 2), 1.5% DAG oil/8.0% TAG oil (low-dose group), 5.5% DAG oil/4.0% TAG oil (mid-dose group), or 9.5% DAG oil (high-dose group) daily for 52 wk (Chengelis et al., 2006a). The average DAG oil consumption over 52 wk for the low-, mid- and high-dose groups was 326, 1227, and 2541 mg/kg body weight/d for males and 348, 1487, and 2300 mg/kg body weight/d for females, respectively. All dogs survived to the end of the study. No effects on clinical condition, body weights, body weight gains, and feed consumption related to DAG oil consumption were reported. No DAG-related changes in hematology were reported at any time point.

Serum chemistry results showed no toxicologically significant effects. All DAG oil-treated dogs and those in control group 2 tended to have higher mean alkaline phosphatase (AP) levels than dogs in control group 1, frequently reaching statistical significance in males at various time points throughout the study. AP levels decreased, as is normal, throughout the course of the study in all dogs, but tended to decrease more slowly in DAG oil-treated and control group 2 dogs. At wk 13, mean glucose concentration values did not differ among the DAG oil-treated dogs, but often were significantly increased compared with one or both control groups. These increases were sporadic and no dose response was evident. Without accompanying microscopic changes, the changes in mean AP and glucose levels were not considered to be toxicologically relevant. During the study, cholesterol, triglycerides, and total protein concentrations were consistently higher in dogs from control group 2 and the DAG oil-treated groups than dogs from control group 1, but this difference was significant only sporadically throughout the study. No dose-related trends were evident, and these findings appear to be related to the differences in the diet fed to the dogs rather than a direct effect of DAG oil itself. No treatment-related effects were found from urinalysis, ophthalmoscopic examinations, or electrocardiographic examinations. Macroscopic examination revealed enlarged uteri and cervices in 2 high-dose females (1 also had ovarian enlargement), but these were considered to be individual variations of the estrous cycle and not attributable to DAG oil ingestion. The only statistically significant difference in organ weights was increased mean brain weight relative to final body weight in low- and high-dose female dogs compared with dogs in control group 2. There were no treatment-related histopathological findings reported. Cytoplasmic vacuolation of the adrenal cortex that was not dose responsive was reported in dogs from control group 2 and DAG oil-treated groups, but not control group 1. Cytoplasmic vacuolation in the renal corticomedullary junction and inflammation in the lungs were observed at varying time points in all dogs, excluding control group 1 males. Only one mid-dose female had vacuolation in the kidney. Hepatocellular vacuolization also was reported in all dogs except for control group 1 females. There were no consistent, dose-dependent, statistically significant treatment-related adverse effects in the dogs fed DAG oil at dietary concentrations up to 9.5%, the highest concentration fed. The NOAEL was ~2541 and 2300 mg/kg body weight/d for males and females, respectively.

Two-year Dietary Rat Studies. Two recent, long-term dietary rat studies were conducted to assess the safety of DAG oil consumption. In the first study (Soni et al., 2001), groups of Sprague-Dawley [Crj:CD(SD) SPF] rats (60/sex/group) were fed test diets containing a total of 7% fat (5.3% fat from test oils + 1.7% fat from basal diet) for up to 105 wk. The test groups were as follows: control group 1 was fed 5.3% edible oil composed of rapeseed, corn, high linoleic safflower, and high oleic safflower oils (equivalent fatty acid composition to DAG oil); control group 2 was fed 5.3% edible oil composed of rapeseed and soybean oils; the low-dose group was fed 2.65% DAG oil + 2.65% edible oil (as described for control group 1); and the high-dose group was fed 5.3% DAG oil. Blood samples were taken and urinalysis conducted at wk 30 and 77 on 10 rats of each sex/group, after which rats were killed and necropsied. All surviving rats were euthanized at week 105 and necropsied. There were no statistically significant differences in cumulative survival rate in either sex among the test groups and no treatment-related clinical signs of toxicity were reported. No treatment-related significant changes in body weights of rats that were consistent or dose related were reported at any time point. No consistent, treatment-related, dose-dependent effects were reported in any of the variables evaluated. Any statistically significant findings from hematological analyses (e.g., activated partial thromboplastin time, prothrombin time, mean corpuscular hemoglobin concentration, eosinophilic hemogram values, and platelet counts) and clinical chemistry (e.g., alanine aminotransferase, insulin, aspartate aminotransferase, alkaline phosphatase, HDL cholesterol, sodium ion, and lactate dehydrogenase concentrations) were considered incidental and not treatment related because they were limited to one sex, one collection period, or lacked a dose response. In low- and high-dose males at 77 wk, 40% of the rats had marked or moderate occult blood in the urine compared with no or slight hematuria in the controls (20% of males in control group 2). The hematuria may be indicative of urinary tract infections common to aging rats. Of the low- and high-dose males, 60% had slight-to-moderate nephropathy, but no accompanying lesions in the bladder or decreases in hematocrit or hemoglobin values to indicate anemia were reported. Moreover, these findings were not reported in female rats. At 30 wk, statistically significant changes in absolute and/or relative organ weights (thyroid, heart, liver, and kidneys) compared with controls were reported in low-dose male rats; these were not seen in high-dose males or in any DAG-treated males at wk 77. Similarly, the statistically significant increase in absolute and relative pituitary weight of high-dose females was not seen at wk 77 or 105. Any statistically significant differences in organ weights in high-dose males and treated females compared with controls at wk 77 (brain, spleen, and heart) or at wk 105 (thyroid) were limited to one sex, one time interval, or showed no dose response and therefore were considered to be not toxicologically significant. Histopathological examination of rats euthanized or found dead (including control rats) at wk 30, 77, and 105 revealed only nonneoplastic findings that were considered common observations in Sprague-Dawley rats of this age. Slight-to-moderate necrosis of the liver and slight-to-marked extramedullary hematopoiesis of the spleen were reported among all decedent female rats with a slightly higher incidence (statistical significance not specified) in females fed high-dose DAG oil. At wk 77, no statistically significant differences in tumor incidence were reported among all groups. Pituitary tumors were reported to occur at a low incidence in all groups, and a low incidence of mammary tumors was reported in all female groups including controls. An increased incidence of benign and malignant epithelial mammary gland tumors was reported in high-dose female rats compared with values from control group 2; this was statistically significant only when these tumor numbers (benign and malignant) were combined. There was no clear dose response and the incidence values were within historical control ranges. An independent pathology review (Hall, 2000) concluded that the reported increase was not toxicologically significant. The NOAEL in this study was the highest concentration tested, 5.3% in the diet, which is ~1770 and 2350 mg DAG oil/kg body weight/d for males and females, respectively.

In another, more recent study (Chengelis et al., 2006b), 7 groups of Sprague-Dawley [Crl:CD®(SD)IGS BR] rats (50/sex/group) were fed diets containing various combinations of dietary fat, including DAG oil for a period of 24 mo. Groups 1–5 were fed fixed amounts of the diet (20 and 15 g/rat/d for males and females, respectively); groups 6 and 7 consumed the diet ad libitum. The diets were composed as follows: group 1 only was fed a standard rodent diet with a total fat content of 4.5%; group 2 was fed 5.5% TAG oil; group 3 was fed 4.5% TAG oil/1% DAG oil; group 4 was fed 2.75% TAG oil/2.75% DAG oil; group 5 was fed 5.5% DAG oil; group 6 was fed 5.5% TAG oil (ad libitum); and group 7 was fed 5.5% DAG oil (ad libitum). Survival rates were lower in rats fed the diets with 5.5% fat content (Groups 2–7) compared with rats fed the lower fat diet of 4.5% (Group 1). The average DAG oil consumption over 2 years for group 3, 4, 5, and 7 was 356.2, 984.8, 1982.4, and 1946.3 mg/kg body weight/d for males and 477.5, 1326.2, 2645.1, and 2507.2 mg/kg body weight/d for females, respectively. Due to the decreased survival rates in rats that consumed the diets ad libitum (Groups 6 and 7) and female rats fed the restricted 5.5% DAG oil diet (group 5), the rats were euthanized earlier than scheduled (for groups 6 and 7 at wk 91 and 87 for males and females, respectively; and for group 5 at wk 99). All other surviving rats were euthanized during wk 104 or 105. The number of rats with palpable masses was slightly lower in males (statistical significance not stated) fed the 4.5% restricted diet (Group 1) compared with all other groups.

Body weights were significantly increased in rats fed the higher fat diets of 5.5% compared with the 4.5% standard diet control; however, body weights of DAG oil-treated rats were comparable to those of the corresponding TAG oil controls. Rats consuming the diets ad libitum had higher body weights than those fed restricted diets. No treatment-related effects were noted for feed consumption. There were no DAG oil-related changes in hematological or serum chemistry parameters, urinalysis, or ophthalmoscopic examinations. Some statistically significant differences in serum chemistry variables (e.g., slightly higher levels of serum cholesterol, triglycerides, free fatty acids, and ß-hydroxybutyrate) were reported in groups fed the higher-fat diets (groups 2–7) compared with values from the standard 4.5% control rats (group 1), which concurred with the expected serum chemistry findings in rats fed increased levels of dietary lipids. Overall, the values were lowest in rats fed the low-fat diet (group 1), intermediate in the rats fed the restricted diets (groups 2–5) and highest in the rats that consumed the diets ad libitum (groups 6 and 7).

Five rats/sex/group were scanned with a densitometer to assess lean and non-lean body mass. Rats fed the restricted high-fat diets (groups 2–5) tended to have a higher amount of total soft tissue (fat + lean tissue) than the low-fat controls (group 1). Reported macroscopic changes (e.g., rough surface of the kidneys, swelling and/or red and white areas of the liver, enlarged spleen and pituitary gland, uterine cysts, and galactocele) were related to dietary fat content and did not result from DAG oil ingestion because the incidences did not differ between DAG oil-treated rats and the corresponding TAG oil controls. Corresponding to increased body weights due to the higher dietary fat in groups 2–7, increased absolute and relative organ weights (mean liver, kidney, spleen, and heart) were reported, and differences were often statistically significant in groups 2–5 (restricted diets) compared with the low-fat control (group 1). Mean relative brain, testes and/or epididymides weights were decreased in rats fed the 5.5% fat restricted diets (groups 2–5) compared with those fed the 4.5% fat restricted diet (group 1).

Histopathological examination revealed non-neoplastic findings in the heart (myocardial degeneration), kidneys (tubular cell vacuolation and chronic progressive nephropathy), liver (hepatocellular vacuolation, congestion, dilated sinuses, necrosis and extramedullary hematopoiesis, and spongiosis hepatis), pituitary (pituitary gland adenomas with secondary effects in the brain (compression) and mammary glands (galactoceles)), spleen and bone marrow (extramedullary hematopoiesis and myeloid hyperplasia), and testes (degeneration of testicular seminiferous tubules) as a result of increased dietary fat (5.5 vs. 4.5%) rather than specific effects of DAG oil consumption. Neoplastic findings were limited to an increased incidence (statistical significance not specified) and overall dimensions of benign pituitary gland adenomas in rats fed 5.5% fat diets compared with those fed the standard 4.5% fat diet (group 1). Unlike the previous rat study, the individual and total incidence of benign and malignant mammary gland tumors in females, although common, was not statistically different among DAG oil-treated rats compared with any of the controls. The findings reported in this study were expected as a result of increased dietary lipids (5.5 vs. 4.5%) and often were noted to occur more often or with greater severity in groups consuming the 5.5% diets ad libitum. There was no evidence of systemic toxicity or carcinogenicity. The NOAEL in this study was the highest concentration tested, 5.5%, which is ~1982 and 2645 mg DAG oil/kg body weight/d for the restricted diet males and females, respectively, and 1946 and 2507 mg/kg body weight/d for the males and females consuming the diet ad libitum, respectively.

Modification of Tumor Development Rat Study. Groups of male F344 rats (20/group) were fed various edible oils including DAG oil following administration of known carcinogens in order to determine the modifying potential of DAG oil on tumor development (Ichihara et al., 2008). It was not indicated in the publication whether or not the study followed OECD, Redbook II, and ICH guidelines under GLP. For 4 weeks prior to oil supplementation of the diet, the rats were treated with 5 carcinogens (*N*-nitrosodiethylamine, *N*-methyl-*N*-nitrosourea, 1,2-dimethylhydrazine dihydrochloride, *N*-*n*-butyl-*N*-butan-4-ol-nitrosamine, and dihydroxy-di-*N*-propylnitrosamine) in order to initiate tumors at different target sites. After initiation, the rats were fed special diets for 24 weeks as follows: basal diet (group 1; control); 5.5% TAG oil (group 2); 1.375% DAG oil (group 3), 2.75% DAG oil (group 4); 5.5% DAG oil (group 5); 5.5% high linolenic acid TAG oil (group 6); 5.5% high oleic acid TAG oil (group 7); and medium-chain TAG oil (group 8). The average DAG oil consumption over 24 wk for group 3, 4, and 5 was 730, 1460, and 2940 mg/kg body weight/d, respectively. Ingestion of DAG oil showed no difference from TAG oil controls with respect to survival, body weights, clinical condition, feed consumption, and organ weights. Quantitative evaluation of hepatic glutathione-*S*-transferase placental form positive foci showed DAG oil to have no effect. Rats in the low-dose DAG oil group (group 3), but not the mid- or high-dose DAG oil groups (groups 4 and 5), showed a statistically significant increase in the incidence of colon adenomas when compared to the TAG oil control group (group 2). Incidences and severity of other neoplastic lesions were comparable across all groups. The authors concluded that "DAG oil may not exert modifying potential on tumor development, even in the colon because of the lack of dose-dependence."

Two-year Dietary Mouse Study. Five groups of Crl:CD®-1(ICR) BR mice (50/sex/group) were fed diets containing various combinations of dietary fat, including DAG oil for a period of 24 mo (Chengelis et al., 2006c). The diets were composed as follows: group 1 only was fed a standard rodent diet with a total fat content of 4.5%; group 2 was fed 6.0% TAG oil; group 3 was fed 4.5% TAG oil/1.5% DAG oil; group 4 was fed 3% TAG oil/3% DAG oil; and group 5 was fed 6% DAG oil as part of a semi-purified rodent diet. The average DAG oil consumption over 2 years for group 3, 4, and 5 was 1792, 3773, and 7412 mg/kg body weight/d for males and 2509, 5286, and 9796 mg/kg body weight/d for females, respectively. At least 20 animals/sex/group survived to the end of week 87 and there were no DAG oil-related differences in survival compared to control mice receiving TAG oil. The causes of early death of the mice were similar among all groups (treated and control) and included amyloidosis, malignant lymphoma, mouse urologic syndrome, cystic ovaries/endometrial hyperplasia, hepatocellular carcinoma, bronchio-alveolar carcinoma, and histiocytic sarcoma. Clinical signs were similar among all groups (control and treated). Ingestion of DAG oil did not affect ophthalmic evaluations, body weights, body weight gain, or feed consumption when compared with TAG oil controls. Animals receiving the higher fat diets (6.0%) tended to show (statistical significance not stated) increased body weight, body weight gain and feed consumption than those receiving the lower fat diet (4.5%).

There was no difference between treated and control groups in the number of animals with palpable masses. The only statistically significant findings in hematology at wk 51 were an increased mean corpuscular hemoglobin concentration in group 4 females and increased mean absolute neutrophils counts in group 4 and 5 females. These findings were considered to be within normal limits and not related to DAG oil ingestion. Macroscopic findings were similar between DAG oil-treated animals and TAG oil-treated controls. Ingestion of higher fat diets (6.0% versus 4.5%) tended to present a greater prevalence of gross lesions including enlarged seminal vesicles, distended urinary bladder (males), trichobezoar (in stomach; both sexes), and hepatic masses and nodules (both sexes). Organ weights were similar between DAG oil groups and TAG oil control with the exception of a statistically significant increase in the mean absolute and relative (to final body and to brain weight) thyroid weight of females in group 3. No corroborating histological changes were noted and the finding showed no dose response.

There were no DAG oil-related changes in the incidence of non-neoplastic and neoplastic lesions. Male mice fed the higher fat diets (6.0% versus 4.5%) showed increases in the incidence of renal tubular degeneration, mineralization, tubular cell vacuolation, dilatation of the urinary bladder, and enlargement of seminal vesicles and coagulating gland. Hepatocellular adenomas and/or carcinomas were also more prevalent in these animals. Histiocytic sarcomas were reported in most of the female groups but were considered common malignant tumors and showed no statistically significant increase over TAG oil and standard diet controls. There was no evidence of systemic toxicity or carcinogenicity. The NOAEL in this study was the highest concentration tested, 6.0%, which is 7412 and 9796 mg/kg body weight/d for the restricted diet males and females, respectively.

Mutagenicity/Genotoxicity Studies

In addition to the toxicological studies in animals, DAG oil or its components were tested for possible mutagenicity using in vitro bacterial systems, cultured mammalian cells, and in vivo mammalian systems. In the in vitro reverse mutation Ames assay employing a pre-incubation method, unheated and heated DAG oil at concentrations of up to 5000 μg/plate showed no mutagenic activity when tested in *Salmonella typhimurium* strains TA1535, TA1537, TA98, and TA100 or *Escherichia coli* WP2 uvrA with or without metabolic activation (Kasamatsu et al., 2005).

Unheated or heated DAG oil did not induce chromosomal aberrations in cultured Chinese hamster lung fibroblasts (CHL/IU) at concentrations of up to 5000 μg/mL when incubated for short-term periods of 6–18 h or with continuous treatment for 24 or 48 h (Kasamatsu et al., 2005).

In a bone marrow micronucleus study using male ICR (Crj:CD-1) mice gavaged twice over a 24-h interval with 500, 1000, or 2000 mg DAG oil/kg body weight, no significant increases in micronucleated polychromatic erythrocytes were induced (Kasamatsu et al., 2005).

The results of the mutagenicity/genotoxicity assays on DAG oil or its components consistently produced negative results indicating that DAG oil is not mutagenic.

Clinical Studies

In the clinical investigations conducted to examine human tolerance and the nutritional effects of ingested DAG oil, no adverse effects were reported at daily doses as high as 45 g/person/d (~450 mg DAG oil/kg body weight/d for a 100 kg person) for up to 1 year, and DAG oil was well tolerated by the participating subjects (see Table 17.2).

Single-Dose Study

DAG and TAG oils were compared for possible effects on blood and plasma profiles in a randomized, crossover, double-blind study (Taguchi et al., 2000). Groups of healthy male volunteers (n = 10–17), in their thirties, were given a test emulsion containing 10, 20, or 44 g of test lipid (DAG or TAG oil) (~170, 330, or 730 mg/kg body weight, respectively). After a 7-d interval, a crossover occurred (i.e., volunteers receiving DAG oil, now received TAG oil at the same dose). Blood samples were taken beforehand to test lipid ingestion (baseline) and at 2-h intervals for 6 (10 g lipid group) or 8 h (44 g lipid group), or at 4 and 6 h (20 g lipid group) and analyzed. A significant reduction in serum triglyceride concentration was reported after DAG oil ingestion compared with ingestion of an equal amount of TAG oil. The volunteers tolerated the ingested test lipids

TABLE 17.2
Summary of Repeated Dose Clinical Studies Conducted Using DAG Oil

Study Design	Subjects	Exposure Period	Treatment	Estimated Intake (mg/kg bw/d)	Result	Reference
Double-blind crossover	healthy men (8/group)	1 mo	44 g DAG oil/d	520-780	Well tolerated, no adverse effects	Watanabe et al., 1997b
Double-blind parallel	healthy men (12-15/group)	3 mo	20 g DAG or TAG oil/d	300	Well tolerated, no adverse effects	Watanabe et al., 2001
Double-blind parallel	healthy men (19/group)	16 wk	10 g DAG or TAG oil/d	130-150	Well tolerated, no adverse effects	Nagao et al., 2000
Double-blind parallel	healthy men and women (18-21/sex/group)	12 wk	DAG or TAG oil incorporated into diet	up to 500	Well tolerated, no adverse effects	Yasunaga et al., 2004
Double-blind parallel	healthy women (28/group)	8 wk	20 g DAG or TAG oil/d	up to 450	Well tolerated, no adverse effects	Hasegawa, 2000
Overweight study population	95 men 19 women	1 y	normal edible oils (10.6 g/d) substituted with DAG oil	~177[a]	Well tolerated, no adverse effects	Katsuagi & Yasukawa, 1994
Randomized dietary trial	women (98-100/group)	1 y	normal edible oils substituted with DAG oil (10-12.5 g/d)	~167-208[a]	Well tolerated, no adverse effects	Yasunaga et al., 2002
Hyperlipidemic adults on dialysis	7 men 3 women	3 mo	normal cooking oil substituted with DAG oil (~9 g/d)	~150[a]	Well tolerated, no adverse effects	Teramoto et al., 2000
Uremic adults on dialysis	7 men 3 women	3 mo	normal cooking oil substituted with DAG oil (~9.8 g/d)	~163[a]	Well tolerated, no adverse effects	Teramoto et al., 2004
Obese and/or hyper-lipidemic children (aged 7-17 y)	4 boys 9 girls	5 mo	normal cooking oil substituted with DAG oil (~11 g/d)	150-270	Well tolerated, no adverse effects	Matsuyama et al., 2000
Obese children (aged 7-17 y)	4 boys 7 girls	5 mo	normal cooking oil substituted with DAG oil (~24 g/d)	~320	Well tolerated, no adverse effects	Matsuyama et al., 2006

Double-blind parallel	overweight/obese men and women (62-65/group)	24 wk	normal cooking oil substituted with DAG oil (~16-45 g/d)	160-450	Well tolerated, no adverse effects	Maki et al., 2002
Single-blind parallel	Type II diabetics with hypertriglyceridemia (3-5/sex/group)	12 wk	normal cooking oil substituted with DAG oil (~10.6 g/d) or continued use of TAG oil (~23.5 g/d)	130-160	Well tolerated, no adverse effects	Yamamoto et al., 2001
Diet therapy	Type II diabetics with nephropathy 7 men 1 women	6 mo	normal cooking oil substituted with DAG oil (~10 g/d)	~150	Well tolerated, no adverse effects	Yamamoto et al., 2006

[a]assuming an individual weighs 60 kg

well and without adverse effects at doses up to 730 mg/kg body weight, the highest dose tested.

Repeat-Dose Studies

Healthy Volunteers. Clinical studies were conducted to examine potential effects of repeated ingestion of DAG oil for up to 12 mo. Three of these studies were double-blind (1 crossover and 2 parallel) designs in which healthy male volunteers ingested up to 44 g of DAG oil as part of their diet (Watanabe et al., 1997; Watanabe et al., 2001; Nagao et al., 2000). In all three studies, an acclimation period using a controlled test diet preceded the study.

In the first of these studies (Watanabe et al., 1997b), 8 male volunteers/group (aged 30–47 y; mean, 37.8 ± 5 y) received 44 g of DAG or TAG oil (~520–780 mg/kg body weight/d) in their breakfast for a period of 1 mo for each test lipid in a crossover design. The diets were controlled such that total daily fat intake was 1330 mg/kg body weight; >50% of the fat intake was from the test lipid. Ingestion of DAG or TAG oil did not produce adverse symptoms in the volunteers and no statistically significant differences between DAG oil- and TAG oil-treated groups were reported with respect to hematology and clinical chemistry results.

In the second study, 27 male volunteers (aged 27–47 y) were divided into 2 groups (parallel design) and received 20 g of either DAG (n = 15; mean age, 33.8 ± 4.5 y) or TAG (n = 12; mean age 33.3 ± 5.7 y) oil daily (~300 mg/kg body weight/d) in an emulsified beverage for 3 mo (Watanabe et al., 2001). Volunteers underwent clinical pathology and hematological analyses before the onset of the study and monthly thereafter to the end of the study. No significant differences between the test groups were noted. In addition, the potential effect on the absorption of fat-soluble vitamins by ingestion of DAG or TAG oil was determined from blood samples taken monthly from fasting subjects. There was a statistically significant decrease in serum vitamin E concentrations in both groups at wk 4 and 8, but these levels returned to baseline values by wk 12. Increases in serum 25-hydroxyvitamin D concentrations also occurred in the DAG-treated (nonsignificant) and TAG-treated (statistically significant) volunteers during the first 2 months; however, these levels returned to baseline values by the end of the study. In addition, 1-α,25-dihydroxyvitamin D serum concentrations were significantly increased over initial values in DAG-treated volunteers (at wk 8) and TAG-treated volunteers (at wk 8 and 12), whereas serum concentrations of the nonactive form of vitamin D (24,25-dihydroxyvitamin D) were significantly decreased in both lipid test groups throughout the study. The ingestion of 300 mg DAG oil/kg body weight/d over a period of 3 mo was well tolerated and produced no adverse effects in the men.

In the third study, 38 healthy male volunteers (aged 27–49 y) received 10 g DAG (n = 19; mean age, 35.0 ± 1.0 y) or TAG oil (n = 19; mean age, 31.0 ± 1.0 y) daily as part of their diet for 16 wk (Nagao et al., 2000). Actual total lipid intake during the study was 43 g/person/d. By the end of the study period, decreased body weight, BMI, and waist circumference occurred in both groups and the difference was statistically significant in the DAG-treated volunteers. Similarly, total fat, visceral fat area, and subcutaneous fat area of the abdominal traverse images of computed tomography (CT) were significantly decreased in DAG-treated volunteers compared with those in the TAG oil group. Also in the DAG oil group, hepatic fat content (determined from CT imaging) was decreased significantly, whereas no changes were noted in the TAG oil group. No statistically significant changes were reported in serum lipid concentrations (triglycerides, total cholesterol, free fatty acid) and related metabolites (glucose, insulin, total ketone body). Daily ingestion of 10 g of DAG oil (~130–150 mg/kg body weight/d) for 16 wk was well tolerated with no reported adverse effects.

Male (n = 42; aged 25–50 y) and female (n = 39; aged 23–47 y) volunteers were divided by sex and randomly assigned to groups receiving DAG (n = 21 men; mean age, 34.7 ± 6.1 y; n = 18 women; mean age, 31.1 ± 6.8 y) or TAG oil (n = 21 men; mean age, 34.6 ± 6.9 y; n = 21 women; mean age, 32.0 ± 6.1 y) at a target dose of ~500 mg/kg body weight/d) as part of a controlled diet over a period of 12 wk in a double-blind, parallel design study (Yasunaga et al., 2004). Some statistically significant changes in hematological and serum chemistry parameters were reported in both the DAG and TAG oil groups (e.g., decreased glycohemoglobin, and increased mean corpuscular hemoglobin and mean corpuscular hemoglobin concentrations). However, these values were within the normal range for the population and were not considered biologically relevant. DAG oil ingested at doses up to 500 mg/kg body weight/d for 12 wk was well tolerated by the male and female volunteers and no adverse effects were reported.

Two groups of 28 women (aged 18–35 y; mean age, 20.0 ± 2.3 y) received either 20 g of DAG (n = 28; mean age, 20.5 ± 3.2 y) or TAG oil (n = 28; mean age, 19.6 ± 0.8 y) (~260–450 mg/kg body weight/d) as part of a controlled diet for a period of 8 wk (Hasegawa, 2000). Mean total lipid intake was 58 g/person/d. Various anthropometric measures and blood samples (serum triglyceride, total cholesterol, HDL cholesterol, free fatty acids, total ketone bodies, acetoacetic acid, 3-hydroxybutyrate, fasting blood sugar, insulin, glutamic oxalacetic transaminase (GOT), glutamic pyruvic transaminase (GPT) and γ-glutamyl transpeptidase (γ-GTP)) were taken before the dosing regime and at 4 and 8 wk. Body weights did not change throughout the study. Serum triglyceride concentrations were increased from initial values throughout the study in both groups and were statistically significantly different from baseline at wk 8. There was no difference between the groups. Initial cholesterol levels were significantly different between the DAG and TAG oil groups, but did not differ during the course of the study. Decreases in total cholesterol from initial values occurred in both DAG (statistically significant) and TAG (non-significant) oil groups during the study. A nonsignificant decrease in HDL cholesterol also was noted in both groups by the end of the study. Both groups had a significant increase in free fatty acid levels from initial values at wk 4 and 8, which were significantly different between the 2 groups at wk 8. During the study, total ketone, acetoacetic acid, and 3-hydroxybutyrate concentrations tended to be higher than initial values but were not different between the 2 groups. Significant increases in insulin levels from initial values were reported in both groups at wk 4 and 8, but did not differ between the 2 groups. No changes in fasting blood sugar, and serum GOT, GPT, and γ-GTP were reported. No adverse effects or lack of tolerance was reported in women consuming DAG oil at doses up to 450 mg/kg body weight/d for 8 wk.

In a 1-y study, 95 male volunteers and 19 female volunteers (age range not reported; mean age, 39.3 ± 8.7 y) consumed DAG oil ad libitum as part of their diet through substitution of typical dietary edible oils with DAG oil (Katsuragi et al., 1994). Based on the National Nutrition Survey in Japan in 1999, the average intake of cooking oil was 10,600 mg/person/d. The men had a mean BMI of 24.7 and 26% of the subjects had serum triglyceride levels > 150 mg/dL. Study variables were examined before the start of the study (initial values) and every 3 mo thereafter. Compared with initial values, total energy intake increased by 8.4% during the study, fat intake increased slightly at 3, 9, and 12 mo, and body weight decreased during the study; the difference was statistically significant only for male volunteers. The efficacy, such as changes in serum lipid profiles and anthropometrics parameters, were discussed in detail in Chapter 11. Slight, incidental changes in some clinical chemistry parameters (i.e., phospholipids, esterified cholesterol, free cholesterol, total bile acid, GOT, γ-GTP, red blood cell counts, hemoglobin, and hematocrit) were reported to reach statistical significance at various time points during the study, but the values remained within the normal range for each variable and the changes were not time responsive. GPT and amylase levels were significantly increased in the middle of the study, but returned to initial values by study end. Substitution of typical dietary edible oils (estimated intake of 10,600 mg/person/d) with DAG oil for 1 y was well tolerated with no reported adverse effects.

Housewives (n = ~198; aged 20–50 y) were randomly divided into 2 groups and were given DAG (n = 98, mean age, 40.2 ± 8.9 y) oil or TAG oil (n = 100, mean age, 39.8 ± 9.3 y) in unmarked containers as a substitute for their normal cooking oil over a period of 1 y (Yasunaga et al., 2002). Specific dietary consumption of the test oils was not conducted, but the authors estimated that ~10,000–12,500 mg test oil/person/d was consumed. The parameters evaluated were limited to effects on skin and gastrointestinal symptoms. No significant differences in the frequency of acne, rough dry skin, constipation, stomach ache, stomatitis, or headache were reported. Substitution of normal cooking oil with DAG oil for 1 y had no adverse effects.

Physiologically Compromised Volunteers. Clinical studies were conducted to examine the possible effects of DAG oil ingestion in adult and juvenile volunteers with hyperlipidemia (Teramoto et al., 2000; Matsuyama et al., 2000; Matsuyama et al., 2006), uremia (Teramoto et al., 2004), overweight subjects (Maki et al., 2006), and subjects with diabetes (Yamamoto et al., 2001;Yamamoto et al., 2006).

Hyperlipidemic adults (7 men and 3 women; mean age, 55.5 ± 10.8 y) undergoing dialysis treatment were given DAG oil as a replacement for their normal cooking oil for 3 mo (Teramoto et al., 2000). The mean DAG oil intake was ~9000 mg/person/d. Before dialysis treatment, blood samples were taken from fasting volunteers at the following time points: before DAG oil dietary regime (baseline), monthly during DAG oil; dietary regime (test period), and monthly for 3 mo after cessation of the DAG oil diet (washout period). Serum triglyceride levels showed an initial increase then returned to baseline values by the end of the washout period. Serum HDL tended to increase, whereas serum VLDL tended to decrease (both non-significantly) during the study period and returned to baseline values by the end of the washout period. The only statistically significant finding was an increased concentration of serum free fatty acids during the study period, which remained above baseline values during the washout period. No changes in total or HDL cholesterol were reported during the study, whereas lipoprotein (a) levels tended to be slightly decreased. The elevation in free fatty acids was not considered an adverse reaction because of its magnitude and continuing elevation above baseline related to DAG oil exposure following DAG oil exposure.

Uremic adults (7 men and 3 women; mean age, 53.8 y) undergoing dialysis treatment were given DAG oil as a replacement for their normal cooking oil for 3 mo (Teramoto et al., 2004). Based on 3-d dietary records, DAG oil intake was 9800 mg/person/d throughout the study. Fasting blood samples were taken from each subject before hemodialysis just prior to study commencement (baseline; 0 mo) and at 3 and 5 mo following study commencement. Each subject also underwent CT scanning of the abdomen at 0, 3, and 6 mo. Total calories and fat intake increased significantly from baseline values taken prior to DAG oil replacement. Serum TAG, total cholesterol, and insulin levels during the study did not change significantly from baseline values; however, serum lipoprotein (a) decreased statistically significantly at 3 mo and did not increase at 5 mo (2 mo after DAG oil replacement cessation). Other statistically significant changes were decreased serum VLDL composition and increased serum HDL composition at 3 mo compared to baseline values. The CT scans showed a statistically significant decrease in visceral fat and total fat, but not subcutaneous fat, at 3 mo compared to baseline values. At 6 mo, fat values were similar to baseline values. The results indicated that 9800 mg DAG oil/person/d (~163 mg/kg body weight/d for a 60 kg individual) was well tolerated, reduced the amount of abdominal fat, and improved serum lipid profiles.

Obese and/or hyperlipidemic children (4 males and 9 females, aged 7–17 y; mean age, 12.4 ± 3.2 y) consumed DAG oil ad libitum in their diet as a substitute for regular cooking oil for 5 mo (Matsuyama et al., 2000). Blood samples from fasting subjects and anthropometric measurements were taken before DAG oil treatment (baseline) and monthly throughout the treatment period. During the 5-mo treatment period, the average consumption of DAG oil was 11.3 g/person/d. Six of the children (aged 7–11 y) consumed an average of 270 mg DAG oil/kg body weight/d. The adolescents (aged 13–17 y) consumed an average of 150 mg DAG oil/kg body weight/d. During the study, height increased significantly with a statistically significant increase in body weight at 2, 4, and 5 mo compared with baseline values. Body height and weight were comparable to Japanese standard growth curves for all but 2 of the children. The height of one 13-y-old girl did not change throughout the study; however, her height at the study onset was equivalent to that of 17-y-old girl; therefore her lower growth curve was not considered abnormal. An 11-y-old girl also had a lower growth curve than the Japanese standard; however, she grew 1.5 cm during the study and therefore her lower rate of change in height was not considered to be biologically significant. Obesity levels generally increased over baseline values during the study and were statistically significant at 2, 4, and 5 months. BMI values were relatively constant throughout the study, but by the end of the 5 mo, they were significantly higher than baseline values. Serum triglyceride, glucose, plasminogen activator inhibitor-1, and leptin levels decreased significantly, whereas HDL cholesterol increased significantly from baseline values, at various time points throughout the study, as discussed in Chapter 11. No other statistically or biologically significant differences in blood variables were reported. DAG oil substitution in the diet for 5 mo at doses of ~150–270 mg/kg body weight/d produced no adverse effects and was well tolerated by the children.

In another study with obese children, 4 males and 7 females (aged 7–17 y; mean age, 12.9 y) consumed DAG oil ad libitum in their diet as a substitute for regular cooking oil for 5 mo (Matsuyama et al., 2006). The mean daily DAG oil intake was 24±3 g (approximately 320 mg/kg body weight/d for a child with an average weight of 75 kg). Fasting blood samples were taken and anthropometric measurements were recorded immediately prior to study commencement (baseline) and monthly thereafter to the end of the study. CT scans of the abdomen were conducted immediately prior to study commencement and in the 5th mo of the study. Height increased statistically significantly at 2, 3, and 5 mo and waist measurements were statistically significantly increased at 3, 4, and 5 mo. The Rohrer index (calculated from body weight and height) did not change significantly during the study. The CT scans showed a statistically significant decrease in total and subcutaneous fat at 5 mo. Total, subcutaneous and visceral fat were statistically significantly increased 4 mo after DAG oil ingestion cessation. Leptin levels significantly decreased throughout the study determined by the Friedman test; however, using Bonferoni's multiple comparison tests, significant differences were not observed between baseline and monthly points during the ingestion period. Serum triglyceride, free fatty acids, total cholesterol, and LDL-cholesterol were not affected by DAG oil ingestion; however, HDL-cholesterol showed a statistically significant increase at 5 mo when compared with baseline values. Serum insulin, plasma glucose, total ketone bodies, AST, ALT, and γ-GTP also were not affected by DAG oil ingestion. DAG oil substitution in the diet for 5 mo at doses of ~320 mg/kg body weight/d produced no adverse effects and was well tolerated by the children.

A randomized, double-blind, parallel intervention study was conducted among overweight or obese male and female volunteers (n = 131) (Maki et al., 2002). Test subjects were divided by sex and then into 2 groups receiving DAG (n = 65; mean age, 45.9 ± 11.4 y; mean initial body weight, 98.0 ± 1.6 kg) or TAG oil (n = 62; mean age, 48.1 ± 11.2 y; mean initial body weight, 97.6 ± 1.8 kg) as part of their diet (substitute for normal cooking oil) for a period of 24 wk to assess the percentage of change in body weight, fat mass, intra-abdominal fat area (IAF), and serum lipids through anthropometric measurements, CT scans, dual-energy X-ray absorptiometry, and serum sampling. Based on the oil content of one serving of a food product and the number of servings recommended per day, the total daily intake of the DAG and TAG oils was estimated to range from 16 to 45 g/person/d). Volunteers in the DAG oil group showed a statistically significant decrease in body weight and fat mass compared with those in the TAG oil group. At wk 24, body weight decreased 3.6 and 2.5% in the DAG and TAG oil groups, respectively. Similarly, fat mass decreased 8.3 and 5.6% in the DAG and TAG oil groups, respectively. No significant differences in IAF or in serum concentrations of total, LDL, HDL, and non-HDL cholesterol were reported between DAG and TAG oil groups. Neither DAG nor TAG oil subjects had an increased hepatic fat content based on an increased median liver-spleen attenuation ratio in the two groups. Although some mild-to-moderate adverse events, particularly upper respiratory infections, headaches, and gastrointestinal symptoms, were reported during the study, substitution of typical dietary fats with DAG oil or TAG oil did not affect the reporting of symptoms between the 2 groups. DAG oil as part of the diet for 24 wk providing doses of 16 to 45 g/person/d (~160–450 mg/kg body weight/d for a 100 kg individual) was well tolerated.

The effect of the daily consumption of DAG oil by persons with Type II diabetes and hypertriglyceridemia was investigated in a randomized, single-blind, controlled, parallel study (Yamamoto et al., 2001). Test volunteers (n = 8; 3 males and 5 females; mean age, 56.8 ± 7.3 y) incorporated DAG oil into their daily diet by substituting it for ordinary cooking oil to reach a target dose of 10 g/person/d; control subjects (n = 8; 4 men and 4 women; mean age, 54.1 ± 18.8 y) continued using regular TAG cooking oil for 12 wk. At wk 12, the mean DAG and TAG oil intake was 10.6 and 23.5 g/person/d, respectively. Body weight, total fat intake, and total oil consumption did not differ between the DAG and control groups. Compared with baseline values (measured before the treatment period), no significant changes in body weight, BMI, energy intake, percentage of energy from fat, total fat intake, total cooking oil intake, plasma glucose levels, or serum total and HDL cholesterol levels were reported in either group. Although serum glycohemoglobin A1C levels were significantly decreased from baseline values within the DAG oil group, but not within the control group at wk 12, there was no significant difference between the DAG oil and control groups. Serum triglyceride levels in the DAG oil group also were significantly decreased from baseline values and from control group values at wk 12. Ingestion of DAG oil incorporated into the diet at 10 g/person/d (~130-160 mg/kg body weight/d) for 12 wk was well tolerated and did not produce any adverse effects.

A 6-mo diet therapy study was conducted using subjects with Type II diabetes and nephropathy (Yamamoto et al., 2006). Subjects were divided into 2 groups: control (6 males and 1 female; mean age, 66.0±9.2) and DAG oil group (7 males and 1 female; mean age, 60.3±7.0). Control subjects used their usual cooking oil which contained mostly TAG oil and the DAG oil group replaced their usual cooking oil with DAG oil for 6 mo. The target intake was 10 g oil/person/d or approximately 150 mg/kg body weight/d for a 67 kg individual. Subjects were followed for up to 3 years after the end of the study. Fasting blood samples were taken at study commencement (baseline) and at 6 mo. At study initiation, there were no significant differences between the control and DAG oil groups in the duration of diabetes, the stages of nephropathy, or other measurements. In the DAG oil group, body weight, body mass index, and serum triglyceride were statistically significantly decreased from baseline at 6 mo. The only statistically significant change reported in the control group was a statistically significant increase from baseline in plasma creatinine at 6 mo. In the 3-y follow-up, 8 subjects from the DAG oil group continued to use DAG oil and 2 subjects from the control group switched to DAG oil.

Out of these 10 subjects, 8 maintained their renal function without requiring dialysis and 2 required dialysis after 2.5 and 2.8 y. Of the 5 subjects continuing to use their own cooking oil, 3 required dialysis after 1.5, 2.2, and 2.5 y. The use of DAG oil in the diet at 10 g/person/d (~150 mg/kg body weight/d) for 6 mo did not produce any adverse effects.

The ingestion of medium-chain triglycerides (C8, C10) produced an increase in ketone bodies as a result of conversion of some of the acetyl-CoA derived from the oxidation of the medium-chain fatty acids into acetoacetate and D-ß-hydroxybutyrate (ketone bodies) in the liver which is perfused with these fatty acids after the digestion and metabolism of the parent triglyceride (Greenberger et al., 1968; Pi-Sunyer et al., 1969; Bougneres et al., 1989). Longer-chain triglycerides generally show an increase that is 5–8 times less than that of the medium-chain triglycerides, comparable to the ketone body concentration characteristic of a fasted state (Owen & Cahill, 1973). It is possible that this increase could pose a health risk to persons with diabetes; however, all of the toxicological and clinical studies on DAG oil, even in diabetic patients, consistently showed that it produced no adverse health effects under its intended conditions of use. In the single-dose clinical study, no significant alterations in ketone levels were reported, whereas in one of the repeat-dose studies, DAG and TAG oil were associated with a slight increase in ketone concentration that may have been influenced by overnight fasting before blood sampling. The data provide evidence that ingestion of DAG oil should not have any adverse effects in diabetics.

Summary

The primary component of DAG oil, diacylglycerol, occurs naturally in the human diet and appears to have unique nutritional properties related to serum lipids and body fat metabolism compared with consumption of conventional TAG oil. It is rapidly digested to 1-monoglyceride and fatty acids and tends to follow a metabolic route that forms a source of energy (i.e., glucose) rather than a pathway that re-esterifies it to triglycerides.

The results of acute, repeated dosing and long-term toxicological studies in rats and dogs are consistent in supporting the lack of toxicity of ingested DAG oil. Any reported effects are uniformly related to what would be expected with an increase in dietary lipids rather than a direct toxic effect of DAG oil. No evidence of mutagenicity or genotoxicity was found in in vitro and in vivo assays. Additionally, in the 2 long-term rat studies, DAG oil showed no evidence of carcinogenicity at doses up to 2500 mg/kg body weight/d for 2 years in rats and up to 2500 mg/kg body weight/d for 1 year in dogs

Multiple clinical studies with healthy adult volunteers showed DAG oil to be well tolerated at doses as high as 45 g DAG oil/kg body weight/d or ~450 mg DAG oil/person/d for a 100 kg person and without reported adverse effects, with outcomes similar to those seen with TAG oil controls. In studies of children and persons with diabetes or uremia no signs of adverse effects after incorporation of DAG oil into their diet were reported. A critical evaluation of all available information on DAG oil supports the conclusion that DAG oil is safe for use in food when used in a manner similar to all other edible oils.

References

Borgstrom, B., Dahlqvist, A., Lundg, G., and Sjovall, J. Studies on Intestinal Digestion and Absorption in the Human. *J. Clin. Investig.* **1957**, *36*, 1521-1536.

Bougneres, P.F., Castano, L., Rocchiccioli, F., Gia, H.P., Leluyer, B., and Ferre, P. Medium-Chain Fatty Acids Increase Glucose Production in Normal and Low Birth Weight Newborns. *Am. J. Physiol.* **1989**, *256*, E692--E697.

Chengelis, C.P., Kirkpatrick, J.B., Marit, G.B., Morita, O., Tamaki, Y., and Suzuki, H. A Chronic Dietary Toxicity Study of DAG (Diacylglycerol) in Beagle Dogs. *Food. Chem. Toxicol.* **2006a**, *44*, 81-97.

Chengelis, C.P., Kirkpatrick, J.B., Bruner, R.H., Freshwater, L., Morita, O., Tamaki, Y., and Suzuki, H. A 24-Month Dietary Carcinogenicity Study of DAG (Diacylglycerol) in Rats. *Food. Chem. Toxicol.* **2006b**, *44*, 98-121.

Chengelis, C.P., Kirkpatrick, J.B., Bruner, R.H., Freshwater, L., Morita, O., Tamaki, Y., and Suzuki, H. A 24-Month Dietary Carcinogenicity Study of DAG in Mice. *Food. Chem. Toxicol.* **2006c**, *44*, 122-137.

Douglas, G.J., Jr., Reinauer, A.J., Brooks, W.C., and Pratt, J.H. The Effect of Digestion and Absorption of Excluding the Pancreatic Juice from the Intestine. *Gastroenterology* **1953**, *23*, 452-459.

FASEB Evaluation of the Health Aspect of Glycerin and Glycerides as Food Ingredients. Federation of American Societies for Experimental Biology, U.S. Department of Commerce. National Technical Information Service. **1975**; PB-254536.

Greenberger, N.J., Tzagourniz, M., and Graves, T.M. Stimulation of Insulin Secretion in Man by Medium Chain Triglycerides. *Metabolism* **1968**, *17*, 796-801.

Grundy, S.M. in Present Knowledge in Nutrition. **1996**; 7th edn. (Ziegler, E.E., and Filer, L.J., eds.) pp. 44-57, ILSI Press, Washington.

Hall, W.C. Independent Pathology Peer Review of Diacylglycerol (DAG). *[Unpublished - Report of Pathology Associates International]* **2000**.

Hamosh, M., and Scow, R.O. Lingual Lipase and Its Role in Digestion of Dietary Lipid. *J. Clin. Investig.* **1973**, *52*, 88-95.

Harkness, J.E., and Wagner, J.E. The Biology and Medicine of Rabbits and Rodents. **1989**; 3rd edn., pp. 47-54, Lea & Febiger, Philadelphia.

Hasegawa, K. Long term DAG intake study with young females. *[Unpublished—Office Report, Kao Corporation; Japan]* **2000**.

Ichihara, T., Yoshino, H., Doi, Y., Nabae, K., Imai, N., Hagiwara, A., Tamano, S., Morita, O., Tamaki, Y., and Suzuki, H. No Enhancing Effects of Diacylglycerol Oil on Tumor Development in a Medium-Term Multi-Organ Carcinogenesis Bioassay using Male F344 Rats. *Food. Chem. Toxicol.* **2008**, *46*, 157-167.

Ishida, S. Single Administration Toxicity Study of Kao Diglyceride in Rats. Final Report. Bozo Research Center, Inc.; Tokyo. *[Unpublished—Internal Report, Kao Corporation; Japan]* **1996a**.

Ishida, S. Single Administration Toxicity Study of Diglyceride Healthy Oil in Rats. Final Report. Bozo Research Center, Inc.; Tokyo. *[Unpublished—Internal Report, Kao Corporation; Japan]* **1996b.**

Ishii, T. Acute Oral Toxicity Study of DG-5 in Rats. Final Report. Drug Safety Testing Center Co. Ltd.; Saitama. *[Unpublished—Internal Report, Kao Corporation; Japan]* **2003**.

JECFA Toxicological Evaluation of Some Food Additives Including Anticaking Agents, Antimicrobials, Antioxidants, Emulsifiers and Thickening Agents, 17th Report of the Joint FAO/WHO Expert Committee on Food Additives, FAO Nutrition Meetings Report Series,

1974, No. 53. WHO Technical Report Series, No. 539. WHO Food Additives Series. **1974**; No. 5, pp. 238-240. WHO, Geneva, Switzerland.

Kasamatsu, T., Ogura, R., Ikeda, N., Morita, O., Saigo, K., Watabe, H., Saito, Y., and Suzuki, H. Genotoxicity studies on dietary diacylglycerol (DAG) oil. *Food. Chem. Toxicol.* **2005**, *43*, 253-260.

Katsuragi, Y., Toi, T., and Yasukawa, T. Effects of Dietary Diacylglycerol on Obesity and Hyperlipidemia. *J. Jpn. Soc. Human Dry Dock* **1994**, *14*, 258-262.

Maki, K.C., Davidson, M.H., Tsushima, R., Matsuo, N., Tokimitsu, I., Umporowicz, D.M., Dicklin, M.R., Foster, G.S., Ingram, K.A., Anderson, B.D., Frost, S.D., and Bell, M. Consumption of Diacylglycerol Oil as Part of a Reduced-Energy Diet Enhances Loss of Body Weight and Fat in Comparison with Consumption of a Triacylglycerol Control Oil. *Am. J. Clin. Nutr.* **2002**, *76*, 1230-1236.

Matsuyama, K., Nakaya, N., Watanabe, H., Nagao, T., Matsuo, N., and Tokimitsu, I. Effects on Serum Leptin Concentrations and Abdominal Fat Level of Child Obesity Outpatients or Child Hyperlipemia Patients by Intake of Diacylglycerol. *The 54th Japanese Society of Nutrition & Food Science, Matsuyama, Japan*, **2000**; p. 52.

Matsuyama, T., Shoji, K., Watanabe, H., Shimizu, M., Saotome, Y., Nagao, T., Matsuo, N., Hase, T., Tokimitsu, I., and Nakaya, N. Effects of Diacylglycerol Oil on Adiposity in Obese Children: Initial Communication. *J. Pediatr. Endocrinol. Metab.* **2006**, *19*, 795-804.

Mattson, F.H., and Volpenhein, R.A. The Digestion and Absorption of Triglycerides. *J. Biol. Chem.* **1964**, *239*, 2772-2777.

Murata, M., Ide, T., and Hara, K. Reciprocal Responses to Dietary Diacylglycerol of Hepatic Enzymes of Fatty Acid Synthesis and Oxidation in the Rat. *Br. J. Nutr.* **1997**, *77*, 107-121.

Nagao, T., Watanabe, H., Goto, K., Onizawa, K., Taguchi, H., Matsuo, N., Yasukawa, T., Tsushima, R., Shimazaki, H., and Itakura, H. Dietary Diacylglycerol Suppresses Accumulation of Body Fat Compared to Triacylglycerol in Men in a Double-Blind Controlled Trial. *J. Nutr.* **2000**, *130*, 792-797.

Owen, O.E., and Cahill, G.F., Jr. Metabolic Effect of Exogenous Glucocorticoids in Fasted Man. *J. Clin. Investig.* **1973**, *52*, 2596-2605.

Pi-Sunyer, F.X., Hashim, S.A., and Van Itallie, T.B. Insulin and Ketone Responses to Ingestion of Medium and Long-Chain Triglycerides in Man. *Diabetes* **1969**, *18*, 96-100.

Serabian, M.A. 4-Week Subacute Oral Toxicity Study in Rats: Final Report. Hazleton Washington, Inc.; Rockville, MD. HWA Study No. 2408-125. [Unpublished—Internal Report, Kao Corporation; Japan] **1991**.

Small, D.M. The Effects of Glycerides Structure on Absorption and Metabolism. *Annu. Rev. Nutr.* **1991**, *11*, 413-434.

Soni, M.G., Kimura, H., and Burdock, G.A. Chronic Study of Diacylglycerol in Rats. *Food. Chem. Toxicol.* **2001**, *39*, 317-329.

Taguchi, H., Watanabe, H., Onizawa, K., Nagao, T., Gotoh, N., Yasukawa, T., Tsushima, R., Shimasaki, H., and Itakura, H. Double-Blind Controlled Study on the Effects of Dietary Diacylglycerol on Postprandial Serum and Chylomicron Triacylglycerol Responses in Healthy Humans. *J. Am. Coll. Nutr.* **2000**, *19*, 789-796.

Taguchi, H., Hagao, T., Watanabe, H., Onizawa, K., Matsuo, N., Tokimitsu, I., and Itakura, H. Energy Value and Digestibility of Dietary Oil Containing Mainly 1,3-Diacylglycerol Are Similar to Those of Triacylglycerol. *Lipids* **2001**, *36*, 379-382.

Teramato, T., Nagao, T., Watanabe, H., Ito, K., Omata, Y., Furukawa, T., Shimoda, K., and Hoshino, M. Effect of Diacylglycerol on the Hyperlipidemia of Hemodialysis Patients. *J. Jpn. Soc. Clin. Nutr.* **2000**, *21*, 35-38.

Teramoto, T., Watanabe, H., Ito, K., Omata, Y., Furukawa, T., Shimoda, K., Hoshino, M., Nagao, T., and Naito, S. Significant Effects of Diacylglycerol on Body Fat and Lipid Metabolism in Patients on Hemodialysis. *Clin. Nutr.* **2004**, *23*, 1122-1126.

Watanabe, H., Onizawa, K., Naito, S., Taguchi, H., Goto, N., Matsuo, N., Tokimitsu, I., Yasukawa, T., Tsushima, T., Shimasaki, H., and Itakura, H. Fat-Soluble Vitamin Status Is Not Affected by Diacylglycerol Consumption. *Ann. Nutr. Metab.* **2001**, *45*, 259-264.

Watanabe, H., Onizawa, T., Taguchi, H., Kobori, M., Chiba, H., Naito, S., Matsuo, N., Yasukawa, T., Hattori, M., and Shimasaki, H. Nutritional Characterization of Diacylglycerols in Rats. *J. Jpn. Soc. J. Jpn. Oil Chem. Soc.* **1997a**, *46*, 301-307.

Watanabe, H., Goto, N., Fujimori, N., Toi, T., Kobori, M., Onizawa, T., Taguchi, H., Naito, S., Hattori, M., Shimasaki, H., Miyazawa, Y., Yashiro, N., Matsuzawa, Y., Itakura, H. Effect of DAG on human long term intake. *[Unpublished—Internal Report, Kao Corporation; Japan]* **1997b**.

Yamamoto, K., Asakawa, H., Tokunaga, K., Watanabe, H., Matsuo, N., Tokimitsu, I., and Yagi, N. Long-Term Ingestion of Dietary Diacylglycerol Lowers Serum Triacylglycerol in Type II Diabetic Patients with Hypertriglyceridemia. *J. Nutr.* **2001**, *131*, 3204-3207.

Yamamoto, K., Tomonobu, K., Asakawa, H., Tokunaga, K., Hase, T., Tokimitsu, I.,and Yagi, N. Diet Therapy with Diacylglycerol Oil Delays the Progression of Renal Failure in Type 2 Diabetic Patients with Nephropathy. *Diabetes Care* **2006**, *29*, 417-419.

Yasunaga, K., Glinsmann, W.H., Seo, Y., Katsuragi, Y., Kobayashi, S., Flickinger, B., Kennepohl, E., Yasukawa, T., Borzelleca, J.F. Safety Aspects Regarding the Consumption of High-Dose Dietary Diacylglycerol Oil in Men and Women in a Double-Blind Controlled Trial in Comparison with Consumption of a Triacylglycerol Control Oil. *Food. Chem. Toxicol.* **2004**, *42*, 1419-1429.

Yasunaga, K., Tomonobu, K., Katsuragi, Y. Assessment of safety and physical complaints based on the long-term practical using evaluation by outside housewives. *[Unpublished—Internal Report, Kao Corporation; Japan]* **2002**.

Part V

Physicochemical Properties and Application Technologies

Chapter 18

Physicochemical Properties of Diacylglycerol

Yoshinobu Nakajima[a], Junichi Fukasawa[a], and Atsuko Shimada[b]

[a]Health Care Food Research Laboratories, Kao Corporation, Sumida-ku, Tokyo, Japan
[b]Graduate School of Human Life Science, Showa Women's University, Setagaya-ku, Tokyo, Japan

Introduction

Diacylglycerol (DAG), a well known component of food emulsifier, mono- and diglycerides (Benson, 1967; National Archives and Records Administration, 1998), and a byproduct of cooking oil, is produced during processing and storage (Young, 1987). The crystal structure and equilibration phenomena of purified DAG have been extensively investigated (Lutton, 1972; Hernqvist, 1988; Sedarevich, 1967; Mank et al., 1976; Dorset, 1987) . However, little is known concerning its basic physicochemical properties, an important aspect of the application of DAG-enriched oil as a cooking oil. When fatty materials are used as food products, a knowledge of the physical properties of the materials is important, especially with regard to interfacial tension, solubility, viscosity, melting behavior, and other physicochemical characteristics of the material because these affect the texture and flavor, suitability for processing, and storage stability of food products containing fatty materials.

In this chapter, the general physicochemical properties of DAG and liquid DAG (DAG oil), which can be used as a cooking oil, are described as well as the physicochemical properties of the commercial product referred to as "DAG cooking oil." DAG cooking oil contains a small amount of additives such as antioxidants and glycerin fatty acid esters. These additives affect the O/W interfacial tension, the initial acid value, and the oxidative stability of DAG oil, but virtually do not affect the other physicochemical properties of DAG oil (data not shown). The typical glyceride distribution and fatty acid composition of the main DAG oil described in this text and the most common salad oil (TAG oil) in Japan are shown in Table 18.1, and their basic physicochemical properties are shown in Table 18.2 (Masui et al., 2001).

Structure

DAG consists of a glycerol molecule with two fatty acids bound to it via ester bonds; two types of isomers, 1(3),2-DAG and 1,3-DAG, are possible. Because 1(3),2-DAG has an asymmetric carbon atom at position 2, it can exist as *sn*-1,2-DAG and *sn*-2,3DAG enantiomers (Fig. 18.1). The DAG skeleton that occurs naturally in phospholipids is *sn*-1,2-DAG. Monoacylglycerol acyltransferase (MGAT), which acts on 2-monoglyceride (MAG) during the TAG resynthesis process after the digestion and absorption of fats and oils, is more likely to produce *sn*-1,2-DAG than *sn*-2,3-DAG (Lehner et al., 1993).

TABLE 18.1
Typical Glyceride Distribution and Fatty Acid Composition[a]

	DAG oil		TAG oil
Glyceride distribution		(%)	
Triacylglycerol	13.7		94.4
Diacylglycerol	85.6		1.4
(1,2-DAG)	(28.3)		(0.6)
(1,3-DAG)	(57.3)		(0.8)
Monoacylglycerol	0.7		0.2
Fatty acid composition		(%)	
C16	3.0		6.5
C18	1.3		2.8
C18:1	38.9		45.4
C18:2	47.3		33.5
C18:3	8.0		9.3
C20	0.1		0.6
C20:1	0.2		1.0
C22	0.1		0.1
C22:1	Trace		0.1

Sources: Masui et al., 2001

TABLE 18.2
Physicochemical Properties of DAG Oil and TAG Oil[a]

		DAG oil	TAG oil
Specific gravity (g/mL)	8.8°C	0.926	0.922
	20.0°C	0.923	0.914
	30.0°C	0.920	0.908
Viscosity (mPa·s)	8.8°C	84.8	74.5
	20.0°C	55.3	50.1
	30.0°C	37.9	35.3
Melting point (°C)		−2.0	−24.0
Smoke point (°C)		220	250
Flash point (°C)		298	344
Fire point (°C)		320	354
Auto-ignition point (°C)		416	435
Surface tension (mN/m at 25°C)		33.8 [33.9[b]]	33.7
Interfacial tension (mN/m at 25°C)		11.9 [14.6[b]]	23.8

[a]Sources: Masui et al., 2001.
[b]DAG oil without additives was measured.

Because migration occurs in cooking oil during both storage and the heating production process used for deodorization, the ratio of *sn*-1,2-DAG and *sn*-2,3-DAG may reach 1:1 in the trace amount of DAG contained in natural cooking oil and DAG cooking oil.

The higher structural specificity of the enantiomers of DAG in terms of their physiologic action as intracellular signal transmitters was investigated (Nomura et al., 1986; Rando & Young, 1984; Boni & Rando, 1985). However, the difference between

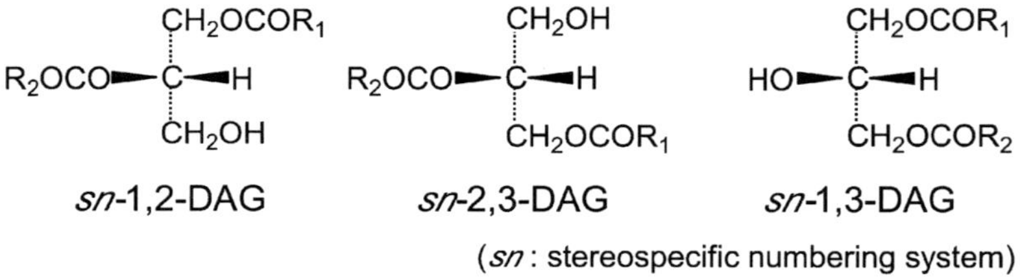

$$\begin{array}{ccc} CH_2OCOR_1 & CH_2OH & CH_2OCOR_1 \\ R_2OCO-C-H & R_2OCO-C-H & HO-C-H \\ CH_2OH & CH_2OCOR_1 & CH_2OCOR_2 \\ sn\text{-}1,2\text{-DAG} & sn\text{-}2,3\text{-DAG} & sn\text{-}1,3\text{-DAG} \end{array}$$

(*sn*: stereospecific numbering system)

FIG. 18.1. Structure of diacylglycerol.

enantiomers has no effect on the fat and oil chemistry or the dietetics. Thus, in this chapter, *sn*-1,2-DAG and *sn*-2,3-DAG are collectively referred to as 1,2-DAG.

Migration

The structural isomers of DAG, 1,2-DAG and 1,3-DAG, are readily equilibrated in the presence of acid, alkali, and heat (Sedarevich, 1967). A silica gel column causes little isomerization of DAG, but contact with Florisil leads to extensive isomerization due to its alkali property. This phenomenon is referred to as migration, and is common in MAG and DAG. At equilibrium, MAG is ~90% 1-MAG and ~10% 2-MAG. The mean isomerization equilibrium ratio of 1,2-DAG to 1,3-DAG was reported to be ~36:64 in branched-chain DAG, ~43:57 in diolein, and ~41:59 in dipalmitin. The ratio of 1,2-DAG and 1,3-DAG in DAG cooking oil (Table 18.1) and common cooking oils is nearly the same, ~3–4:7–6 (Takano & Itabashi, 2002).

The mechanism of isomerization is similar to that of MAG. The isomerization of DAG is likely caused by a steric effect due to the conversion to the more thermodynamically stable linear form of 1,3-DAG. The transition state results from the reaction of the adjacent hydroxyl group and an ester carbonyl group via a 5-membered ring intermediate (Fig. 18.2). In contrast to the equilibrium of MAG to 1-MAG (~90%), the ratio of the DAG 1,3-isomer is ~60–70%, and this is considered due to van der Waals forces between the two intramolecular fatty acid chains of 1,2-DAG, which slightly decreases the steric effect causing the transition to 1,3-DAG. Migration of DAG generally occurs by the intramolecular transition of fatty acids (Mank et al., 1976). However, it is also likely that the transition from the 1,2-isomer to the 1,3-isomer in the crystal state is caused by the intermolecular transition of a fatty acid (Dorset, 1987).

Because 1,2-DAG and 1,3-DAG are readily interconverted, as described above, to investigate the properties due to structural differences, careful handling of the samples is necessary. A method exists for producing highly pure 1,3-DAG by taking advantage of the equilibration property. The fats and oils are stored at a temperature slightly lower than the melting point of DAG (De Groot, 1972). Liquid 1,2-DAG can be equilibrated and 1,3-DAG can be sequentially separated as crystals, likely by solid-liquid separation. This method is able to increase the final purity of 1,3-DAG to >90%. Although this method is applicable to DAG that is a solid at room temperature, its use for liquid DAG may be difficult because the crystallizing temperature is low and the transition velocity is slow.

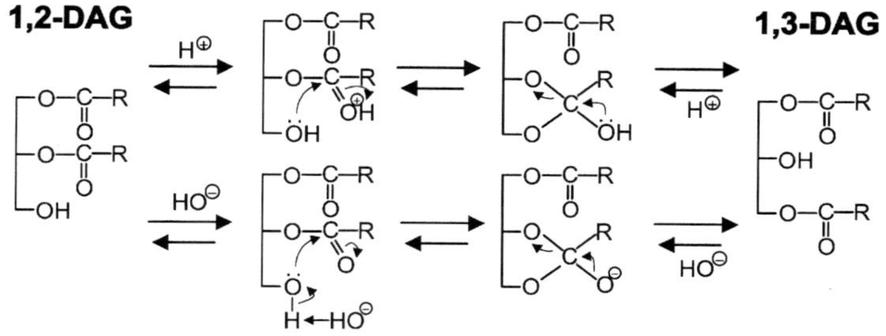

FIG. 18.2. Isomerization of DAG by acid and base catalysis.

Specific Gravity, Melting Point, and Viscosity

The melting point of glycerol-fatty acid esters increases in the order TAG < DAG < MAG, showing that the melting point increases as the degree of esterification decreases. Generally, when the fatty acid composition is the same, the melting points of 1,3-DAG and 1-MAG are ~10 and 20°C higher, respectively, than that of TAG. For the minor components, 1,2-DAG and 2-MAG, the melting point is ~10°C lower than those of its isomers (Benson, 1967; Formo, 1979; Bockish, 1998). The differences in melting points may be due to the ease of fatty acid chain arrangement and the strength of the hydrogen bonding of the hydroxyl group. The melting points of DAG-enriched oils prepared from various plant oils are shown in Table 18.3. To obtain a DAG-enriched oil that maintains a liquid state throughout the year, it is necessary to reduce the content of saturated fatty acids and increase the polyunsaturated fatty acid content compared with TAG (Yasukawa, 1996).

DAG and TAG are different, not only in melting point, but also in viscosity and specific gravity. The viscosity of DAG is slightly higher than that of TAG. The specific gravity of DAG is also higher than that of TAG. These differences may also be affected by the weak bonding of molecules by the hydrogen bonding of the hydroxyl group. Although the viscosity and specific gravity of DAG are slightly higher than those of TAG, these differences are not a constraint in the application of DAG oil for food products because the differences are small.

Smoke, Ignition, Flash, and Fire Point

The molecular weight of DAG is lower than that of TAG, and this affects some of the physicochemical properties. The smoke, flash, fire, and ignition points of DAG cooking oil are 30–40°C lower than those of TAG oil. The influence of hydrogen bonding in DAG decreases at a very high temperature, and the molecular motion of DAG, with a lower molecular weight than TAG, increases. This difference in molecular motion at high temperature may be reflected in differences in physical properties. These physiccal values are related to the handling of cooking oils, and the values for DAG are lower than those of TAG. However, differences in the physical values are not problematic, in practice, for cooking oils for family use (see Chapter 19). As described above, hydrogen

TABLE 18.3
Slip Melting Point of DAG-Enriched Oil Prepared from Various Oils[a]

	Origin of fatty acids of DAG oil					
	Canola	Safflower[b]	Corn	Soybean	Cottonseed	Olive
Slip melting point (°C)	9.3	2.6	16.9	15.6	20.2	20.7
Glyceride distribution (%)						
Triacylglycerol	13.5	19.8	18.8	13.1	27.9	22.3
Diacylglycerol	85.8	79.5	80.4	85.6	71.5	76.6
Monoacylglycerol	0.7	0.7	0.8	1.0	0.6	0.8
Fatty acid composition (%)						
C14	Trace	0.1	Trace	Trace	0.6	Trace
C16	4.1	6.3	10.5	10.1	18.1	9.8
C16:1	0.3	0.1	0.2	0.1	0.5	0.8
C18	1.9	2.7	1.9	3.7	2.6	3.5
C18:1	60.1	12.5	23.3	24.7	17.2	79.2
C18:2	21.7	77.1	56.9	53.9	60.0	5.4
C18:3 + C20:1	11.2	0.7	1.8	7.1	0.6	0.8
C20	0.7	0.5	0.5	0.5	0.4	0.4

[a]Source: Nakajima, 2004.
[b]High-linoleic safflower oil.

bonding of the hydroxyl group and molecular size clearly affect the physical properties of DAG.

Crystal Structure

Because this chapter discusses mainly liquid DAG, only points relating to the crystal structure of DAG are described. For details, the reader is referred to other papers (Lutton, 1972; Hernqvist, 1988).

Crystallization and Polymorphism. TAG exhibits crystal polymorphism, forming mainly 3 typical crystal structures, the α-, ß'-, and ß-forms. Asymmetrical DAG (1,2-DAG) exhibits the α- and ß'-forms. Asymmetric DAG adopts a hairpin conformation in which the hydrocarbon chains of the molecule are arranged in parallel in the same monolayer. The bilayer is constructed via a head-to-head orientation of molecules in adjacent monolayers. For the symmetrical DAG (1,3-DAG), the absence of the aform and the presence of two types of ß-form, ß1 and the more unstable ß2, were reported. Because the polar group is located in the center of the molecule and two hydrocarbon chains extend outward from the center, the resulting bilayer forms a V-shaped (herringbone type) conformation (Larsson, 1963; Hybe & Dorset, 1971).

Liquid Crystal-Mesomorphic State. MAG has a mesomorphic structure, which is another ordered lipid state called liquid crystal (Lutton, 1972). DAG is classified as a polar lipid, but DAG, similar to TAG, exhibits no mesomorphism alone or with water. DAG is only slightly soluble in the mesophase of MAG/water (Kelker & Hatz, 1980).

Solubilization

DAG oil solubilizes hydrophilic substances because of its hydrophilic nature. In addition, when used as a solvent, DAG oil exhibits a behavior different from that of TAG. These solvent properties of DAG oil offer great influences in terms of its application to food products, affecting the retention of flavor components (Yoshida et al., 1999), hydrolysis during heat-cooking (Shimizu et al., 2004), selection of the emulsifier to be used in the production of emulsified food products (Kawai et al., 2000), and the solubilization of functional substances (see Chapters 19-21). However, it is difficult to obtain clear-cut answers concerning the solvent effect and methods for using DAG oil in complex food product systems. Therefore, only the basic properties are described here.

Solubility of Water. The most familiar solvent property of DAG oil is its hygroscopicity. The water content in DAG cooking oil containing the maximal amount of solubilized water was reported to be 8900 ppm. The changes in moisture that occurred when DAG cooking oil was allowed to absorb environmental moisture at various relative humidities (RH) at 20°C are shown in Fig. 18.3. The water content of DAG cooking oil increased with an increase in environmental RH, and reached 5800 ppm at 80% RH. TAG solubilizes a maximum of only ~1000 ppm water, indicating that DAG is able to solubilize ~6–9 times more water than TAG. The solubility of water in DAG is thought to occur by the hydration of DAG molecules. This ability of DAG to solubilize hydrophilic substances is a positive property that specifically solubilizes functional substances in oil; it is also a negative property, i.e., its hydrolyzability (see Chapter 19).

Solubility of Phytosterols. Phytosterols are natural components of plant oils that act on bile acid micelles and inhibit cholesterol uptake by the body. Phytosterols are solubilized to only 0.5 g/100 g at 5°C in TAG oil, but this value is 4.2 g/100 g in DAG cooking oil. DAG oil containing 4% phytosterols significantly decreases blood cholesterol in humans, as evidenced by a clinical study involving the 4-wk ingestion of cooking oil (Goto et al., 1999a). On the basis of this information, DAG cooking oil containing 4% phytosterol (Goto et al., 1999b) was approved as a "Food for Specified Healthy Use" (FOSHU) by the Ministry of Health Welfare and Labor of Japan in 1999, and has been sold in Japan since 2001.

Solubility and Crystallizing Behavior of Solid TAG. The crystallizing behavior of tristearin in oil was investigated using DAG (diolein) and TAG (triolein) as a solvent (Masui et al., 2001). Under the same supersaturating conditions of tristearin, fewer crystal nuclei were generated (increase in nuclear generating energy) and the growth velocity was also slower in diolein than in triolein (Fig. 18.4). The crystal form of tristearin in triolein is a large and solid needle, whereas its form in diolein is small and thin (plate crystal) (Fig. 18.5). This difference in crystallizing behavior was reported to

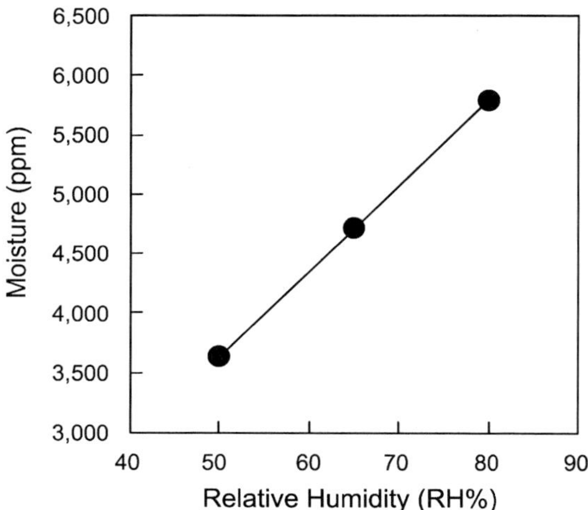

FIG. 18.3. Moisture absorption of DAG oil. DAG oil was spread on a stainless steel laboratory dish, and the moisture was measured after 1 wk at 20°C at relative humidities of 45, 65, and 80%. (*Source*: Nakajima, Y., *et al.*, unpublished data).

be due to a large difference in polarity between DAG as a solvent and solid TAG as a solute. Restraint of crystallization and inhibition on the growing surface of the crystals may be the cause.

It was reported that in common cooking oils, a low level of DAG affects the crystal structure of solid TAG. DAG delays the α-→β′-form and β-→β′-form transitions of TAG, and stabilizes the metastable polymorphism (Wright & Marangoni, 2002). In fat

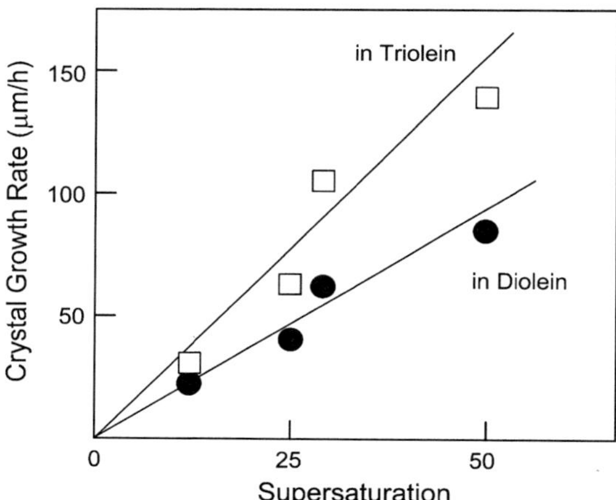

FIG. 18.4. Relationship between the growth rate of tristearin crystals and the degree of supersaturation of tristearin in triolein (□) and in diolein (●) (Source: Masui et al., 2001).

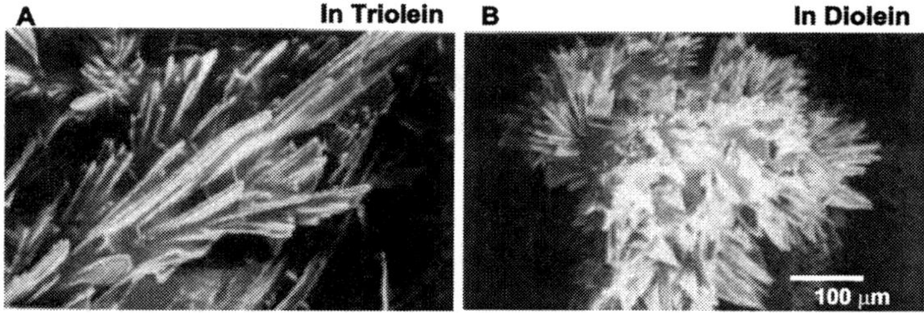

FIG. 18.5. Electron microphotograph of tristearin crystals in triolein (A) and in diolein (B) (Source: Masui et al., 2001).

and oil products such as margarine and shortening, physical properties such as consistency and plasticity are controlled by controlling the content of solid fat and the shape, size, and structure of crystals (Erickson, 1985). The β′-form crystals are small and highly capable of retaining liquid oil; thus, they are the crystal shape in the most demand. The most stable β-form crystals are large and cause graining, and are therefore not favorable. Therefore, DAG might be used as a crystal modifier to improve the quality of these fat and oil products (Hernqvist & Anjou, 1983).

Interfacial Phenomena

Because DAG contains a hydroxyl group, it would be expected to be different from TAG in terms of surface-chemical phenomena. Cooking oils are frequently used in the preparation of emulsified food products such as margarine, mayonnaise, and dressings, and information on the interfacial properties is very important for the manufacture of these food products. The properties of DAG as an emulsifier have been studied as a component of food emulsifiers, mono- and diglyceride, but only a few studies investigated the interfacial properties of DAG alone. These interfacial properties of DAG oil, mainly its emulsifying properties, are described here.

Surface and O/W Interfacial Tension. The surface tensions of DAG oil and TAG oil are nearly the same, ~35 mN/m (Shimada & Ohashi, 2003), which may be due to their constituents. Both DAG and TAG contain hydrocarbon chains in which the hydrophobic region consists of fatty acids. In contrast, oil/water interfacial tension differs greatly between TAG oil and DAG oil (Table 18.2). The oil/water interfacial tensions of purified TAG and purified DAG were 28.5 and 15.8 mN/m (25°C), respectively, and the interfacial tension for DAG/water was about half that of TAG/water. The DAG/water interfacial tension varies according to differences in the constituent fatty acids (Table 18.4). The lower interfacial tension of DAG/water compared with TAG/water may be due to the regular orientation generated in the oil/water interface by the hydroxyl group present in DAG.

Figure 18.6 shows the oil/water interfacial tension-concentration curves for DAG or MAG dissolved in TAG, and for MAG dissolved in DAG. At the TAG/water interface,

TABLE 18.4
Oil/Water Interfacial Tension of DAG with Various Fatty Acid Compositions at 25°C[a]

			Oil/water interfacial tension (mN/m)
Complex-chain DAG			
C_2	-C18:1	DAG	4.0
C_3	-C18:1	DAG	7.9
C_4	-C18:1	DAG	10.4
C_6	-C18:1	DAG	13.3
C_8	-C18:1	DAG	14.9
C_{10}	-C18:1	DAG	16.0
C_{12}	-C18:1	DAG	17.0[b]
Uni-chain DAG			
C_3	$-C_3$	DAG	6.4
C_4	$-C_4$	DAG	8.3
C_6	$-C_6$	DAG	12.4
C_8	$-C_8$	DAG	14.2
C18:1	-C18:1	DAG	17.2

[a]Source: Nakajima, 2004.
[b]Under supercooling.

MAG exhibited a Langmuir-type adsorption behavior. The interfacial tension began to decrease rapidly at 0.005% and broke at ~2%. When DAG was mixed in TAG, the oil/water interfacial tension decreased with the mixing ratio, indicating that DAG did not exhibit Langmuir-type adsorption at the TAG/water interface. In addition, in the oil/water interfacial tension-concentration curve for MAG dissolved in DAG, an ~10-fold or greater concentration was necessary to reduce the oil/water interface tension of DAG, compared with the concentration required to reduce the oil/water interfacial tension of TAG. DAG is probably a good solvent for MAG, and the repelling force of the oil phase against the hydrophilic region of MAG may be weak, suggesting that the potency of the emulsifier could be decreased when DAG is used for the oil phase rather than TAG.

Air/water Interface (Monolayer) The characteristics of DAG monomolecular films on the air/water interface were reported (Rodríguez Patino & Ruiz Domínguez, 2000). The behavior of a monomolecular film varies as a function of the DAG isomers present. 1,2-Distearin forms condensed layers, whereas 1,3-distearin forms a more expanded and unstable monolayer. In contrast, in an isomer mixture, distearin, which consists of linear fatty acids, forms a liquid- condensed structure, but diolein consisting of unsaturated fatty acids forms only a liquid-expanded structure. DAG is less likely to form a solid phase than MAG (Rodríguez Patino & Ruiz Domínguez, 1993). DAG may not readily form a closed packing structure at an air/liquid interface because of the presence of two alkyl chains in the molecule and the small size of the hydrophilic region. The adsorption of surfactants to the air/liquid interface is also an important phenomenon in food manufacturing. Advantage has been taken of the antifoaming action of MAG, but DAG also

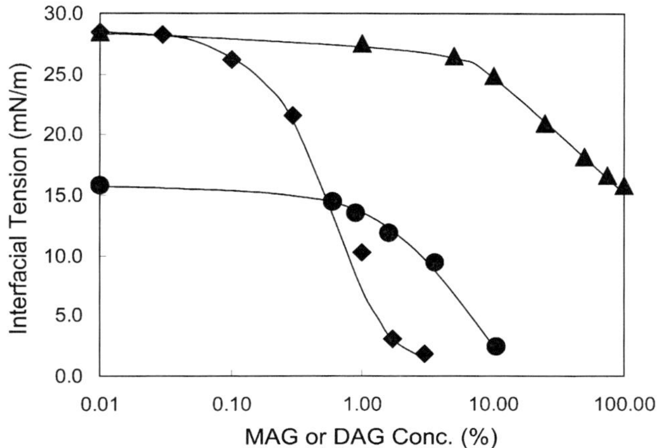

FIG. 18.6. The interfacial tension/concentration curve of MAG in TAG (◆), DAG in TAG (▲), and MAG in DAG (●) (Source: Shimada & Ohashi, 2003).

exhibits a high antifoaming effect. An antifoaming agent for tofu involving the adsorption of DAG to the air/liquid interface was proposed (Suzuki et al., 1999).

Emulsification Properties

W/O Emulsion. DAG oil is known to readily form a W/O emulsion, and a comparison of the basic emulsification properties between DAG oil and TAG oil was reported (Hernqvist & Anjou, 1983). When oil/water mixtures at different ratios were emulsified under the same conditions, TAG oil formed a W/O emulsion when the ratio of the oil phase was 60–90%, and an O/W emulsion layer and a small amount of W/O emulsion layer were formed at a lower ratio. In contrast, DAG oil formed a W/O emulsion when the ratio of the oil phase was 10–90%, and an O/W emulsion layer was not formed even at a low oil phase ratio (Fig. 18.7). The presence of a single hydroxyl group in the molecule may lead to these differences in emulsification properties between TAG and DAG. Generally, the issue of whether emulsifiers are more likely to form W/O or O/W emulsions can be predicted on the basis of their hydrophilic lipophilic balance (HLB). Emulsifiers with HLB values of 3–6 are reported to be appropriate for O/W emulsions (Rosen, 1989). The HLB value of DAG as obtained by Griffin's equation is ~2.8.

Water-Retaining Ability. DAG oil not only has highly emulsifying characteristics, but it is also able to retain a large volume of water (Shimada & Ohashi, 2003; Omura et al., 1996). One volume of TAG oil is able to emulsify only 0.97 volumes of water, whereas 1 volume of DAG oil is able to emulsify 5.83 volumes of water. In this emulsion, the ratio of the internal phase is (Φ) 85%, which exceeds the closest packing of spheres (Φ74%). The emulsion exceeding the closest packing has a structure like a liquid bubble, and water droplets in the internal phase are stabilized in a polygonal state (Fig. 18.8).

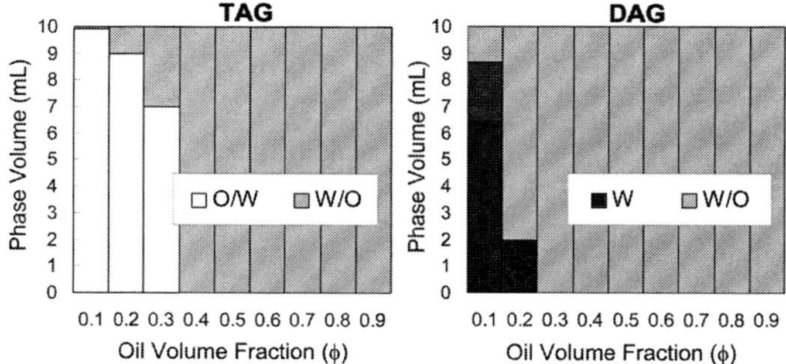

FIG. 18.7. Effects of the oil-volume fraction on the type of emulsion prepared using TAG & DAG (*Source*: Shimada and Ohashi, 2003).

Emulsions with such a high ratio of internal phase were investigated, and are denoted by various names such as high internal phase ratio emulsions (HIPRE) (Lissant, 1970), gel-emulsions (Kunieda et al., 1987), and highly concentrated emulsions (Aronson & Petko, 1993). However, in these previous studies, emulsions were prepared using mineral oil such as paraffin and emulsifiers, and only a few studies used glyceride such as DAG for the oil phase. In general, an emulsifying capability and a stable emulsion at a high internal phase volume are obtained when the interfacial film has a high elasticity resulting from the condensed surfactant layers via strong lateral intermolecular forces (van der Waals force) (Rosen, 1989; Aronson & Petko, 1993). Thus, DAG oil may form a highly elastic film in the oil/water interface. On the basis of the relationship between fatty acid composition and the water-retaining ability of DAG, at least one long alkyl chain of which is C_{12} greater, is necessary in the molecule for DAG to exhibit a high

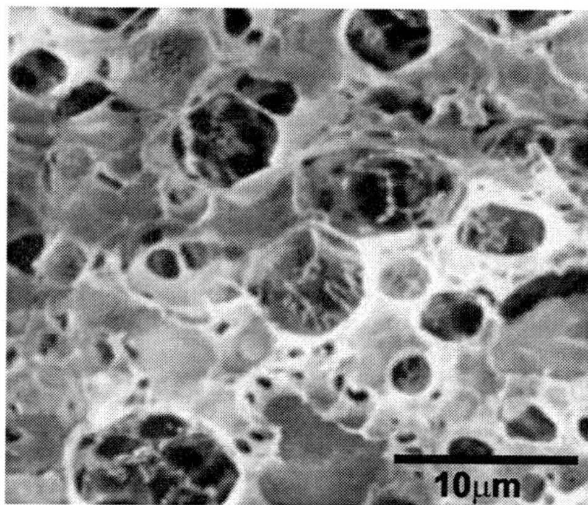

FIG. 18.8. Electron microphotograph of a water/DAG emulsion at a water volume fraction of 85% (*Source*: Nakajima, Y., unpublished data).

water- retaining ability (Nakajima, 2004). The stability of a high water content W/O emulsion increases in the following order: long chain-short chain DAG > long chain-long chain DAG > long chain-middle chain DAG (Aronson & Petko, 1993). These findings indicatethat intramolecular interactions between the two fatty acids in the DAG molecule may play important roles in the high water-retaining ability of a long chain-long chain DAG such as DAG oil, in addition to the intermolecular interactions between DAG molecules.

Stabilization of W/O Emulsion. A W/O emulsion using DAG oil will separate after standing for a long period. However, solubilization of polyols such as sugar (Mizukoshi et al., 1995a), organic acids (Mizukoshi et al., 1995b), and electrolytes such as sodium chloride (Shimada & Ohashi, 2003; Nakajima et al., 1995) in the aqueous phase markedly increase the stability of a W/O emulsion. Stabilization of a high water content W/O emulsion by the addition of electrolytes such as sodium chloride in other systems was reported (Kunieda et al., 1987; Kunieda et al., 1989). Stabilization by salt is likely the result of the increased adsorption density of the surfactant to the oil/water interface. In studies of MAG monolayer at the air/water interface, the behavior of the film after solubilizing polyols in the aqueous phase (230) was similar to that after solubilizing salts (Rodríguez Patino & Ruiz Domínguez, 1996). On the basis of these findings, the stabilization of a DAG emulsion by polyols may also be based on the same mechanism involving salts. High water content margarine, produced by taking advantage of the high water-retaining ability of DAG and the W/O emulsion stabilizing effect of salts has been proposed (Maeda et al., 1997). Phospholipids such as soybean lecithin also markedly improve the stability of a W/O emulsion of DAG oil (Kasuga et al., 1995). The stabilization mechanism of lecithin has not been clarified. Because DAG and lecithin have two long-chain molecules in their structure, and lecithin contains a large hydrophilic group, it may form a stable packing structure at the oil/water interface, similar to the relationship between SPAN and TWEEN (Rosen, 1989).

O/W Emulsion. To prepare an O/W emulsion using DAG oil, it is necessary to use higher amounts of an emulsifier and a higher HLB than that used for TAG. The amounts and HLB of an emulsifier (polyglycerin-fatty acid ester) necessary for the preparation of an O/W emulsion using DAG oil and TAG were reported. Using TAG, emulsifiers with an HLB of 4.5–14.5 form an O/W emulsion at 0.25% or higher. In contrast, using DAG oil, emulsifiers with an HLB of 14.5, 13, and 10–4.5 require a concentration of 0.25, 0.5, and 1%, respectively (Shimada & Ohashi, 2003). Because DAG has a higher polarity than TAG, repelling by water may decrease the affinity between the hydrophobic region of the emulsifier adsorbed to the DAG/water interface and the oil phase and may not allow the emulsifier to form a closed orientation.

In the food industry, protein, in addition to polyglycerine-fatty acid esters and sucrose-fatty acid esters, is also an important component for O/W emulsification and emulsion stabilization. When protein is used as an emulsifier, an O/W emulsion is efficiently formed, even if DAG oil is used (Fig. 18.9). The O/W emulsifying ability of a protein that emulsifies by spreading and adsorbing to the oil/water interface (Phillips et al., 1994) may be markedly stronger than that of DAG which emulsifies by orientation toward the interface caused by the minute molecular balance.

A functional agent using two properties, the DAG W/O emulsion-stabilizing effect of salt and the W/O emulsion-destabilizing effect of protein, was proposed (Suzuki et al., 1999). This agent is a W/O emulsion-type bittern (magnesium chloride) added to soybean milk during tofu production. Bittern is a solidifier with a high ability to rapidly solidify soybean milk. To prepare smooth, fine-grained tofu, use of glucono-d-lactone with a slow solidifying speed or the addition of bittern after cooling soybean milk is necessary. When this agent (bittern/DAG emulsion) is dispersed into soybean milk while it is hot, the protein slowly destroys the emulsion, and the soybean milk is slowly solidified by the magnesium chloride released during the destruction of the emulsion. This method can be used to produce a smooth tofu without altering the original taste.

Summary

The physicochemical properties of DAG (DAG oil) as a liquid oil are almost the same as those of TAG. However, because DAG has a hydroxyl group in the molecule, the physicochemical properties related to hydrogen bonding are different from those of TAG. Therefore, the use of DAG oil for food production may require different handling than that for TAG. However, the DAG-specific properties that are not associated with TAG may be used for the development of new products. DAG is an attractive new food material not only for its nutritional properties but also its physicochemical properties.

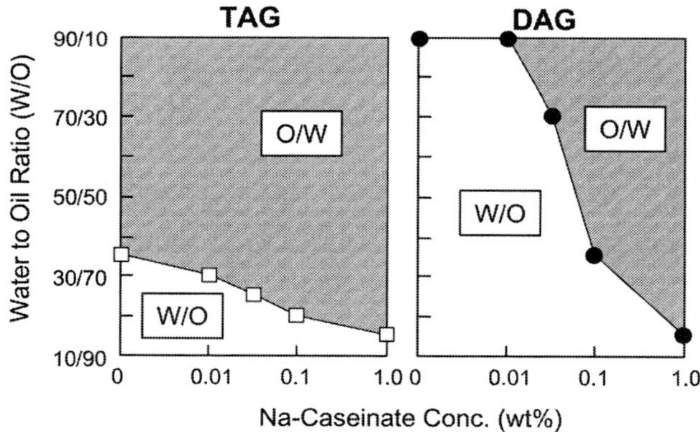

FIG. 18.9. Effects of Na-caseinate on the phase transition of an oil/water system. To prepare emulsions, TAG or DAG and Na-caseinate solution (20 mL) were emulsified at various ratios by means of an ultradisperser (5 min) (Source: Masui, K., unpublished data).

References

Aronson, M.P., and Petko, M.F. Highly Concentrated Water-in-Oil Emulsion:Influence of Electrolyte in Their Properties and Stability. *J. Colloid Interface Sci.* **1993,** *159,* 134–149.

Benson, F.R. Polyol Surfactants. in *Nonionic Surfactants* (Shick, M.J., ed.) **1967**; pp. 247–299, Marcel Dekker, New York.

Bockish, M.Composition, Structure, Physical Data, and Chemical Reactions of Fats and Oils, Their Derivatives, and Their Associates. in *Fats and Oils Handbook*, **1998**; pp. 53–120, AOCS Press, Champaign, IL.

Boni, L.T., and Rando, R.R. The Nature of Protein Kinase C Activation by Physically Defined Phospholipid Vesicles and Diacylglycerols. *J. Biol. Chem.* **1985,** *260,* 10819–10825.

De Groot, W.T.H.M. Acyl Migration Solid Phase Isomerization of 1,2-Diaglycerides to 1,3-Isomers. *Lipids* **1972,** *7,* 626-628.

Dorset, D.L. Is the Initial Solid State Acyl Shift of 1,2-Diglycerides Intermolecular? *Chem. Phys. Lipids* **1987,** *43,* 179–191.

Erickson, D.R. Table Spreads and Shortenings. in *Baily's Industrial Oil and Fat Products*, **1985**; Vol.3, 4th edn. (Applewhite, T.H., ed.) pp. 41–126, John Wiley & Sons, New York.

Formo, M.W. Physical Properties of Fats and Fatty Acids. in *Bailey's Industrial Oil and Fat Products*. **1979**; Vol.1, 4th edn., (Swern, D., ed.) pp. 177–232, John Wiley & Sons, New York.

Goto, N., Mori, H., Katsuragi, Y., Toi, T., Yasukawa, T., and Shimasaki, H. Effects of Diacylglycerol Containing Phytosterol on Reducing Blood Cholesterol Level. *Nippon Yukagaku Kaishi* **1999a,** *48,* 235–240.

Goto, N., Nishide, T., Tanaka, Y., Yasukawa, T., and Masui, K., WO **1999b**; Patent 99/48378.

Hernqvist, L.Crystal Structures of Fats and Fatty Acids. in *Crystallization and Polymorphism of Fats and Fatty Acids* (Garti, N., and Sato, K., eds.) **1988**; pp. 97–137, Marcel Dekker, New York.

Hernqvist, L., and Anjou, K. Diglycerides as a Stabilizer of the ß'-Crystal Form in Margarines and Fats. *Fette Seifen Anstrichm.* **1983,** *85,* 64–66.

Hyble, A., and Dorset, D. The Crystal Structure of the 1,3-Diglyceride of 11-Bromoundecanoic Acid. *Acta Cryst.* **1971,** *27,* 977–986.

Kasuga, Y., Maeda, H., and Inaoka, S. **1995**; JP Patent 1954773.

Kawai, S., Konishi, Y., Masui, K., and Yasukawa, T., **2000**; PCT WO 78162.

Kelker, H., and Hatz, R. Liquid Crystals in Living Systems. in *Handbook of Liquid Crystals* (Giesler, G., ed.) **1980**; pp. 553–591. Verlag Chemie, Deerfield Beach, FL.

Kunieda, H., Solans, C., Shida, N., and Parra, J.L. The Formation of Gel-Emulsions in a Water/ Nonionic Surfactant/Oil System. *Colloid Surf.* **1987,** *24,* 225–237.

Kunieda, H., Yano, N., and Solans, C. The Stability of Gel-Emulsions in a Water/Non-Ionic Surfactant/Oil System. *Colloid Surf.* **1989,** *36,* 313–322.

Larsson, K. The Crystal Structure of the 1,3-Diglyceride of 3-Thiadodecanoic Acid. *Acta Cryst.* **1963,** *16,* 741–748.

Lehner, T., Kuksis, A., and Itabashi, Y. Stereospecificity of Monoacylglycerol and Diacylglycerol Acyltransferases from Rat Intestine as Determined by Chiral Phase High-Performance Liquid Chromatography. *Lipids* **1993,** *28,* 29–34.

Lissant, K.J., Geometry of Emulsions. *J. Soc. Cosmet. Chem.* **1970,** *21,* 141–154.

Lutton, E.S. Lipid Structures. *J. Am. Oil Chem. Soc.* **1972,** *49,* 1–9.

Maeda, H., Kasuga, Y., Inaoka, S., and Nakajima, Y. **1997**; JP Patent 2632900.

Mank, A.P., Ward, J.P., and Van Drop, D.A. A Versatile, Flexible Synthesis of 1,3-Diglycerides and Triglycerides. *Chem. Phys. Lipids* **1976,** *16,* 107–114.

Masui, K., Yasunaga, K., Nishide, T., Nakajima, Y., and Yasukawa, T. Physicochemical and Frying Characteristics of Oil High in Diacylglycerol. *Proceedings of the 2001 PIPOC International Palm Oil Congress (Chemistry and Technonology).* **2001**; pp 65-73.

Mizukoshi, M., Nakajima, Y., and Kokumai, K. **1995a**; JP Patent 1966846.

Mizukoshi, M., Nakajima, Y., and Kokumai, K. **1995b**; JP Patent 2003513.

Nakajima, Y., Maeda, H., Kasuga, Y., Inaoka, S., and Mizukoshi, M. **1995**; JP Patent 1995194.

Nakajima, Y. **2001**; JP Patent 3235842.

Nakajima, Y. Water-Retaining Ability of Diacylglycerol. *J. Am. Oil Chem. Soc.* **2004,** *81,*907-912.

National Archives and Records Administration, Washington. §184.1505 Mono- and Diglycerides. in *21 Code of Federal Regulations*, **1998**; pp. 498–499.

Nomura, H., Ase, K., Sekiguchi, K., Kikkawa, U., and Nishizuka, Y. Stereospecificity of Diacylglycerol for Stimulus-Response Coupling in Platelets. *Biochem. Biophys. Res. Commun.* **1986,** *140,* 1143–1151.

Omura, H., Katada, M., and Tanaka, Y. **1996**; JP Patent 2087454.

Phillips, L.G., Whitehead, D.M., and Kinsella, J. Emulsions. in *Structure-Function Properties of Food Proteins* (Taylor, S.L., ed.) **1994**; pp. 153–169, Academic Press, San Diego.

Rando, R.R., and Young, N. The Stereospecific Activation of Protein Kinase C. *Biochem. Biophys. Res. Commun.* **1984,** *122,* 818–823.

Rodríguez Patino, J.M., and Ruiz Domínguez, M. Surface Properties of Monoglyceride Monolayers Spread on Aqueous Glycerol Solutions. *Colloids Surf.* **1993,** *A 75,* 217–228.

Rodríguez Patino, J.M., and Ruíz Domínguez, M. Study of Monostearin Films in the Presence of Electrolytes. *Colloids Surf.* **1996,** *A. 114,* 287–296.

Rodríguez Patino, J.M., and Ruiz Domínguez, M. Interfacial Characteristics of Diglyceride Monolayers at the Air/Aqueous Phase Interface. *Colloids Surf.* **2000,** *A 168,* 35–43.

Rosen, M.J. Emulsification by Surfactants. in *Surfactants and Interfacial Phenomena* **1989**; 2nd edn., pp. 304–336, John Wiley & Sons, New York.

Sedarevich, B.Glyceride Isomerization in Lipid Chemistry. *J. Am. Oil Chem. Soc.* **1967,** *44,* 381–393.

Shimada, A., and Ohashi, K. Interfacial and Emulsifying Properties of Diacylglycerol. *Food Sci. Technol. Res.* **2003,** *9,* 142–147.

Shimizu, M., Moriwaki, J., Nishide, T., and Nakajima, Y. Thermal Deterioration of Diacylglycerol and Triacylglycerol Oils During Deep-Frying. *J. Am. Oil Chem. Soc.* **2004,** *81,* 571-576.

Suzuki, K., Yoshida, H., and Kanai, K. **1999**; JP Patent 2908633.

Suzuki, K., Yoshida, H., Tamaishi, R., Komikado, M. **1999**; JP Patent 2880871.

Takano, H., and Itabashi, Y. Molecular Species Analysis of 1,3-Diacylglycerols in Edible Oils by HPLC/ESI-MS. *Bunseki Kagaku* **2002,** *51,* 437–442.

Wright. A.J., and Marangoni, A.G. Effect of DAG on Milk Fat TAG Crystallization. *J. Am. Oil. Chem. Soc.* **2002,** *79,* 395–402.

Yasukawa, T. **1996**; JP Patent 2010558.

Yoshida, H., Suzuki, K., and Komikado, M. **1999**; JP Patent 2938701.

Young, F.V.K. Refining and Fractionation of Palm Oil. in *Critical Reports on Applied Chem: Palm Oil* (Gunstone, F.D., ed.) **1987**; pp. 39–70. John Wiley & Sons, Chichester.

Chapter 19

Cooking Oil: Cooking Properties and Sensory Evaluation

Tsutomu Nishide[a], Masao Shimizu[a], Thomas R. Tiffany[b], Hisae Ogawa[c]

[a]Health Care Food Research Laboratories, Kao Corporation, Sumida-ku, Tokyo, Japan; [b]James R. Randall Research Center, Archer Daniels Midland Company, Decatur, IL; and [c]Junior College of Kagawa Nutrition University, Toshima-ku, Tokyo, Japan

Introduction

Cooking oil is used as a heating medium, but it also affects the taste, flavor, and texture of foods. This chapter describes the stability of diacylglycerol-enriched liquid oil as a heating medium in cooking, and its influence on these food qualities. In this study, diacylglycerol-enriched oil that consists of 80% (w/w) or higher diacylglycerol (DAG) without additives is defined as DAG oil. Triacylglycerol (TAG), which has almost the same fatty acid composition and tocopherol content as DAG oil, is defined as TAG oil. In addition, a DAG oil product that contains a small amount of additives is defined as DAG cooking oil. The formulation of DAG cooking oil is shown in Table 19.1. The salad oil product that is the most popularly used in Japan is defined as TAG cooking oil. Typical glyceride compositions, fatty acid compositions, and the tocopherol contents of these oils are shown in Table 19.2. First, the basic properties of DAG oil regarding stability are compared with TAG oil. Next, the cooking characteristics of DAG cooking oil currently on the market are summarized. Finally, consumer test results regarding the use of DAG cooking oil for home use are described.

Stability of DAG Oil

The stability of DAG oil was compared with that of TAG oil, which has almost the same fatty acid composition and tocopherol content as DAG oil.

TABLE 19.1
DAG Cooking Oil Formulation

DAG oil	99.42–99.83%
(≥80% diacylglycerol)	
Emulsifier	0.07–0.35%
Antioxidants	0.1–0.23%
Tocopherols	
L-Ascorbyl palmitate	

TABLE 19.2
Glyceride Composition, Fatty Acid Composition, and Tocopherol Content of the Test Oils[a]

	DAG oil	TAG oil	DAG cooking oil	TAG cooking oil
Glyceride composition (%)				
MAG	1.2	0.0	0.5	0.0
DAG	81.4	1.7	86.6	1.5
TAG	17.3	98.3	12.9	98.5
Fatty acid composition (%)				
C16:0	3.1	5.7	3.1	7.0
C18:0	1.2	2.1	1.2	3.0
C18:1	39.8	35.2	39.8	46.0
C18:2	47.1	48.2	47.0	33.9
C18:3	8.4	8.1	8.5	9.2
C20:0	0.3	0.5	0.3	0.6
C20:1	0.2	0.3	0.2	0.4
Tocopherol content (%)				
α-	0.031	0.032	0.023	0.017
β-	0.003	0.003	0.001	0.012
γ-	0.050	0.044	0.098	0.039
δ-	0.011	0.015	0.041	0.003
Total	0.096	0.094	0.163	0.072

[a]Source: Shimizu et al., 2004

Autoxidation Stability

The autoxidation of fats and oils is promoted by heat, light, and prooxidant trace metals as well as by increased unsaturation levels of the fatty acids constituting the oil. The influence of differences in the glyceride structure on autoxidation was evaluated using the oil stability index (OSI) (Shimizu et al., 2004). The OSI value was measured at 120°C by a Rancimat test (Model 679, Metrohm AG, Herisau, Switzerland), and the values for DAG oil and TAG oil (3.2 and 2.9 h, respectively) did not differ.

Nakatsugawa and co-workers (Nakatsugawa et al., 2001) evaluated the stability of DAG oil with respect to autoxidation in the presence of ultraviolet light at 40°C, and found that DAG oil was more stable than TAG oil against autoxidation irrespective of the presence or absence of ultraviolet light. This suggests that the presence of a free hydroxyl group in the structure of DAG is a factor in its stability. They proposed that the hydroxy group of DAG acts as an antioxidant, similar to a sugar alcohol, or as a donor for the chelation of trace metals. In conclusion, the autoxidation stability of DAG oil is greater than or equal to that of TAG oil.

Heating Stability (Deep-Frying Test)

Deep-frying tests were conducted in a temperature range of 160–180°C. The deterioration of a cooking oil at high temperature is caused mainly by hydrolysis and thermal oxidation, leading to decomposition and polymerization (Fritsch, 1981). The characteristics of the thermal oxidation and hydrolysis of DAG oil are summarized below.

Thermal Oxidation Stability. Continuous deep-frying tests were conducted under severe conditions as a model for the industrial use of cooking oil, i.e., 8 h at 180°C without replenishment with fresh oil. Details of the cooking conditions are shown in Table 19.3. The specification of frying oil differs among countries, as shown in Table 19.4, which summarizes the frying oil regulations in Japan and countries in the EU (Firestone, 1996). In the United States, the Food and Drug Administration (FDA) does not regulate the quality of frying media. [Note: the U.S. Department of Agriculture (USDA)/Food Safety and Inspection Service (FSIS) contains only guidelines for processing of meat and poultry.] We determined the following major indices of deterioration of thermal oxidation during frying: the anisidine value, polymerized glyceride content, polar compounds, which are frequently used in European countries, such as France, Germany, and Italy (Billek et al., 1978), and petroleum ether-insoluble oxidized fatty acids.

Changes in the indices regarding oxidation are summarized in Table 19.5. No substantial differences in the anisidine value or amounts of petroleum ether-insoluble oxidized fatty acids were found between DAG oil and TAG oil. The polymerized glycerides from DAG oil tended to be lower than those from TAG oil. However, the polymerized glycerides, representing the wt% of polymerized TAG or DAG, do not reflect the polymerization of the fatty acid chains because of the difference in molecular weights of the polymers from DAG vs. TAG . Thus, the samples were hydrolyzed and the polymerized fatty acids were determined; no significant difference was observed between DAG oil and TAG oil. The term polar compounds is inappropriate as a deterioration index in the case of DAG oil because all of the DAG molecules are also measured as polar materials. Nakatsugawa and co-workers (Nakatsugawa et al., 2001) conducted a heating test at 180°C for 12 h to evaluate the thermal oxidation stability of TAG oil and DAG oil, and concluded that almost no difference existed between the oils.

TABLE 19.3
Frying Test Conditions[a]

Frying oil	DAG and TAG with the same fatty acid and antioxidant composition
Fryer	Electric domestic fryer (National, NT-110)
Frying temperature	SV[b] = 180°C, PV[c] = 150–185°C, mean = 170°C
Oil amount	1 kg
Frying period	8 h
Frying time	2.5 min/batch, 8 batches/h
Food	Frozen sliced potato (5 mm thick)
Frying amount	Approximately 10 wt% of oil per batch (0–2 h: 100 g potato; 2–4 h: 80 g; 4–6 h: 70 g; 6–8 h: 60 g
Fresh oil replenishment	None

[a]Source: Shimizu et al., 2004
[b]SV, setting value.
[c]PV, process value.

TABLE 19.4
Worldwide Regulations for Frying Oils[a]

	Acid value	Smoke point	Total polar compounds	Dimeric and polymeric triglyceride	Oxidized FA insoluble in petroleum ether	Other
Austria	<2.5	>170°C	<27%	—	<1%	Frying temperature <180°C
Belgium	FFA <2.5%	>170°C	<25%	<10%	—	Viscosity at 50°C <37 mPa·s (fats), Viscosity at 20°C <27 mPa·s (oils) C18:3 ≤ 2%
Czech	—	—	<25%	<10%	—	
Finland	<2.5 (fats) < 2.0 (vegetable oil)	>170°C (fats) >180°C (vegetable oil)	<25%	—	—	
France	—	—	<25%	—	—	C18:3 ≤ 2%
Germany	<2.0	>170°C	<24%	—	<0.7%	Smoke point difference <50°C
Italy	—	—	<25%	—	—	Frying temperature <180°C
Japan	<2.5	>170°C	—	—	—	COV <50°C
Netherlands	<4.5	—	—	<16%	—	
Spain	—	—	<25%	—	—	
Switzerland	—	>170°C	<27%	—	—	
USA[b]	FFA <2.0%	—	—	—	—	

[a]*Source*: Firestone, 1996.
[b]USDA/FSIS Meat and Poultry Inspection Manual Guidelines.

During high temperature cooking, emissions of volatile aldehydes which are secondary oxidation products including acrolein worsen the cooking environment. Katsuta and co-workers (Katsuta et al., in press) compared the emission of volatile aldehydes from TAG oil and DAG oil with different degrees of unsaturation of fatty acid moieties during the deep-frying of sliced potatoes. Both the amount and the composition of volatile aldehydes from DAG oil and TAG oil were almost identical. On the contrary, the patterns of aldehydes were completely different among the oils with different fatty acid compositions. This result suggests that fatty acid composition has a strong influence on the pattern of volatile aldehydes emissions. On the basis of the above findings, there appears to be no significant difference in thermal oxidation stability between DAG oil and TAG oil.

They also examined the production of acrylamide generated during deep-frying of potato chips (Katsuta et al., in press). The content of acrylamide in potato chips cooked with DAG oil and TAG oil were comparable levels (0.7 ± 0.3 and 1.3 ± 0.4 ppm, respectively). Acrylamide in both oils after a frying test were not detected. The production of acrylamide is not affected by glyceride structure.

Hydrolysis Stability. Hydrolysis is an important reaction in cooking oil, as well as thermal oxidation. Deep-frying tests were conducted under the severe cooking conditions that were used in the previous evaluation regarding oxidation. As indices of hydrolysis, the DAG content and acid value (AV) were measured, as shown in Table 19.5. The DAG content decreased from the baseline value; however, the rate of decrease was slight, i.e., only 1.4% after an 8-h period of frying. The AV increased more rapidly with frying time in DAG oil than in TAG oil, although it was within the specifications of AV, as shown in Table 19.4. This indicates that DAG oil may be more susceptible to hydrolysis than TAG

TABLE 19.5
Changes in the Indices of Oxidation and Hydrolysis During Heating[a]

	Test oil	Initial	2 h	4 h	6 h	8 h
Oxidation						
p-Anisidine value	TAG	4.0	84	122	178	214
	DAG	1.2	85	123	175	214
Oxidized fatty acids (%)	TAG	0.02	—	—	—	0.99
	DAG	0.01	—	—	—	0.87
Polymerized glycerides (%)	TAG	1.0	4.7	7.3	11.0	15.6
	DAG	1.7	4.2	6.0	8.4	11.7
Polymerized fatty acids (%)	TAG	0.8	1.8	3.9	6.5	8.0
	DAG	1.1	1.5	3.9	5.8	7.8
Polar compounds (%)	TAG	3.2	—	—	—	24.8
	DAG	85.2	—	—	—	90.9
Hydrolysis						
Acid value	TAG	0.04	0.10	0.15	0.23	0.30
	DAG	0.16	0.30	0.48	0.96	0.76
DAG content (%)	DAG	81.4	80.7	80.5	80.0	80.0

[a]Source: Shimizu et al., 2004

evaluated. The characteristics of both oils as a heating medium are similar because no changes in the temperature at the core of the foodstuff between the oils were observed. Overall, the sensory evaluation results were better for the dishes cooked with TAG cooking oil than for those with DAG cooking oil.

Ohno conducted various frying tests (Ohno, 2002a; Ohno, 2002b) and reported that DAG cooking oil was more resistant to autoxidation than TAG cooking oil. The deterioration indices of DAG cooking oil during deep-frying or stir-frying were similar to those of TAG cooking oil. Ogawa and co-workers (Ogawa et al., 2001a) reported on the results of sensory evaluations of home-made confectionaries cooked with DAG and TAG cooking oils. In the evaluation of chiffon cake, financier, chou, cookies, and doughnuts (used as frying oil), no significant difference was observed in flavor or texture between the

TABLE 19.6
Sensory Evaluation of Dishes Cooked with DAG and TAG Cooking Oils

		Immediately after cooking				
	Test oil	Appearance	Flavor	Taste	Texture	Overall impression
Fried chicken without flour	DAG	0.00 ± 0.86	0.00 ± 0.73	0.05 ± 0.89	0.05 ± 1.00	−0.05 ± 0.83
	TAG	−0.30 ± 0.66	−0.20 ± 0.70	−0.25 ± 0.85	−0.20 ± 0.77	−0.15 ± 0.81
		NS	NS	NS	NS	NS
Fried potato without flour	DAG	−0.95 ± 1.08	−0.74 ± 0.87	−0.79 ± 0.85	−0.89 ± 0.88	−0.89 ± 0.88
	TAG	−0.42 ± 1.22	−0.05 ± 0.91	−0.47 ± 1.12	−0.63 ± 1.16	−0.47 ± 1.17
		NS	NS	$P < 0.05$	NS	NS
Potato chips	DAG	−0.50 ± 0.76	−0.35 ± 0.75	−0.55 ± 1.10	0.60 ± 1.14	−0.10 ± 1.02
	TAG	0.10 ± 0.91	0.10 ± 0.91	0.00 ± 0.92	0.35 ± 1.31	0.20 ± 1.11
		$P < 0.05$	NS	NS	NS	NS
Fried chicken	DAG	0.50 ± 0.69	0.10 ± 0.72	0.15 ± 1.04	0.35 ± 0.88	0.20 ± 0.83
	TAG	0.25 ± 0.79	0.20 ± 0.80	0.30 ± 0.79	0.55 ± 0.73	0.50 ± 0.82
		$P < 0.05$	NS	NS	NS	NS
Breaded fried chicken	DAG	0.10 ± 0.72	−0.20 ± 0.62	−0.30 ± 0.92	0.40 ± 0.82	0.00 ± 0.86
	TAG	0.00 ± 0.56	−0.05 ± 0.51	−0.10 ± 0.72	0.50 ± 1.00	0.20 ± 0.83
		NS	NS	NS	NS	NS
Breaded fried potato	DAG	−0.40 ± 0.82	−0.50 ± 1.05	−0.50 ± 1.00	0.35 ± 1.14	−0.05 ± 1.28
	TAG	−0.30 ± 0.92	−0.10 ± 0.97	−0.20 ± 0.95	0.85 ± 0.93	0.10 ± 1.07
		NS	NS	NS	NS	NS
Fried chicken with batter (Tempura)	DAG	−0.05 ± 1.00	0.15 ± 1.14	−0.25 ± 1.25	0.45 ± 1.23	0.15 ± 0.99
	TAG	0.00 ± 0.79	0.00 ± 0.92	−0.20 ± 0.89	−0.10 ± 1.07	−0.10 ± 1.02
		NS	NS	NS	NS	NS
Fried potato with batter (Tempura)	DAG	−0.20 ± 0.89	−0.10 ± 0.91	−0.25 ± 1.07	0.30 ± 1.34	0.05 ± 1.32
	TAG	−0.25 ± 0.72	−0.35 ± 0.88	−0.55 ± 1.19	−0.15 ± 1.18	−0.30 ± 1.13
		NS	NS	NS	NS	NS
Chicken sauté	DAG	−0.05 ± 0.83	−0.45 ± 0.89	−0.30 ± 0.92	−0.25 ± 1.02	−0.25 ± 0.91
	TAG	−0.40 ± 0.68	−0.15 ± 0.67	−0.30 ± 0.80	−0.45 ± 0.76	−0.40 ± 0.82
		NS	NS	NS	NS	NS
Vegetable stir–fry with cabbage	DAG	−0.35 ± 0.87	−0.30 ± 0.92	−0.50 ± 0.94	−0.50 ± 0.95	−0.50 ± 0.83
	TAG	−0.30 ± 0.73	−0.20 ± 0.69	−0.35 ± 0.74	−0.40 ± 0.68	−0.30 ± 0.68
		NS	NS	NS	NS	NS

Source: Ogawa et al., 2001a. NS, not significant.

oils, with the exception of chiffon cake. The cross section of the chiffon cake prepared with DAG used as a cooking oil revealed the presence of large cavities and a lack of homogeneity compared with that prepared with TAG cooking oil. The evaluation regarding the texture of the chiffon cake prepared with DAG cooking oil was significantly lower than that of TAG cooking oil. It was reported that DAG easily adsorbs to the interface between air and water compared with TAG (see Chapter 18). The antifoaming effect of DAG caused by adsorption to the interface between air and water might influence the homogeneity of chiffon cake, resulting in a poorer texture than that with a TAG cooking oil. The addition of ~3% baking powder with the flour in conventional recipes for chiffon cake improved the homogeneity and texture of the chiffon prepared with DAG cooking oil. Details of the application technologies for DAG in baked products, meal

ABLE 19.6
ontinued)

30 min after cooking				
Appearance	Flavor	Taste	Texture	Overall impression
0.25 ± 0.64	0.00 ± 0.79	−0.40 ± 0.82	−0.50 ± 0.95	−0.50 ± 0.76
0.15 ± 0.67	0.05 ± 0.60	−0.10 ± 0.79	−0.30 ± 0.80	−0.20 ± 0.77
NS	NS	NS	NS	NS
−0.63 ± 1.07	−0.63 ± 0.96	−0.89 ± 0.94	−0.89 ± 0.88	−0.84 ± 0.90
−0.68 ± 0.89	−0.25 ± 0.81	−0.37 ± 0.90	−0.84 ± 0.96	−0.63 ± 0.90
NS	NS	NS	NS	NS
−0.45 ± 1.00	−0.90 ± 0.79	−0.95 ± 1.00	−0.10 ± 0.91	−0.80 ± 1.01
0.10 ± 0.72	0.10 ± 0.64	−0.10 ± 0.79	0.80 ± 1.01	0.25 ± 0.79
NS	$P < 0.01$	$P < 0.01$	$P < 0.01$	$P < 0.01$
0.30 ± 0.47	−0.05 ± 0.51	−0.05 ± 0.51	0.10 ± 0.85	−0.20 ± 0.70
0.25 ± 0.64	−0.05 ± 0.69	−0.30 ± 0.73	−0.30 ± 0.66	−0.35 ± 0.67
NS	NS	NS	NS	NS
0.20 ± 0.70	−0.25 ± 0.72	−0.45 ± 0.83	0.30 ± 1.13	−0.10 ± 0.97
0.10 ± 0.45	0.10 ± 0.64	−0.05 ± 1.00	0.45 ± 0.94	0.30 ± 0.98
NS	NS	NS	NS	NS
−0.05 ± 0.69	−0.70 ± 0.80	−0.70 ± 0.86	0.15 ± 1.31	−0.55 ± 0.94
−0.15 ± 0.81	−0.15 ± 0.88	−0.60 ± 0.99	0.45 ± 1.32	−0.15 ± 1.18
NS	NS	NS	NS	NS
−0.15 ± 0.49	−0.25 ± 0.64	−0.50 ± 1.10	−0.50 ± 0.76	−0.35 ± 0.81
0.05 ± 1.00	−0.05 ± 0.60	−0.45 ± 0.83	−0.70 ± 0.92	−0.50 ± 0.83
NS	NS	NS	NS	NS
−0.55 ± 1.10	−0.65 ± 0.81	−1.30 ± 0.86	−0.75 ± 1.37	−1.15 ± 1.14
−0.25 ± 0.79	−0.45 ± 0.83	−0.45 ± 1.05	−0.25 ± 1.33	−0.40 ± 1.23
NS	NS	NS	NS	NS
0.00 ± 0.97	−0.25 ± 0.91	−0.40 ± 0.75	−0.35 ± 0.75	−0.35 ± 0.81
−0.20 ± 0.95	0.00 ± 0.56	−0.20 ± 0.77	−0.20 ± 0.70	−0.30 ± 0.80
NS	NS	NS	NS	NS
−0.25 ± 0.55	−0.25 ± 0.55	−0.30 ± 0.65	−0.25 ± 0.63	−0.40 ± 0.68
−0.30 ± 0.57	−0.15 ± 0.59	−0.40 ± 0.68	−0.35 ± 0.59	−0.35 ± 0.59
NS	NS	NS	NS	NS

oil under severe conditions. A recent study revealed that DAG dissolves more moisture than TAG because of the presence of a hydroxyl group that is not present in TAG (see Chapter 18). In the measurement of moisture content during frying, DAG oil contained 1.8-fold more moisture than TAG oil, which may be a factor in promoting the hydrolysis of DAG oil. Further investigations to elucidate the mechanism of AV increment will be forthcoming. Sakai and coworkers (Sakai et al., 2003) recently reported that the increase in AV in DAG oil can be controlled by the addition of an organic acid, an interesting finding.

Under in-home cooking conditions, it was confirmed that the increasing rate of AV of DAG cooking oil is similar to that of TAG cooking oil as described below. Moreover, no significant change in the AV of DAG cooking oil was observed in a 1-y storage test at room temperature (Nakajima, 2004).

Characteristics of DAG Cooking Oil

Recent studies reported on the cooking properties of DAG cooking oil on the market fortified with antioxidants and an emulsifier as shown in Table 19.1. Ogawa and coworkers (Ogawa et al., 2001b) conducted sensory evaluations for fried foods, a stir-fry, and a sauté prepared using DAG and TAG cooking oils. This was a single-blind test involving 10 women aged 20–50 y who were experienced in sensory evaluation. The test oil and control oil were DAG cooking oil and TAG cooking oil. (In this study, the fatty acid composition of DAG and TAG cooking oils was adjusted to be approximately equal.) Subjects were instructed to mark one among 7 levels from -3 (very bad) to +3 (very good) in a questionnaire after they tasted each dish. Detailed results are shown in Table 19.6. According to the sensory evaluation conducted immediately after cooking, no significant differences between the oils were observed in the overall evaluation of dishes such as fried foods with and without flour, breaded fried food, tempura and sauté as shown in Table 19.6. In particular, tempura cooked with DAG cooking oil was given a higher evaluation than that with TAG cooking oil, but the difference was not significant. When the sensory evaluation was conducted 30 min after cooking, evaluations of potato chips cooked with DAG cooking oil were significantly lower than those with TAG cooking oil. In general, potato chips easily absorb moisture. Hence, a dish cooked with DAG cooking oil may absorb more moisture than that with TAG cooking oil as shown in Chapter 18. No significant differences between the oils were observed for other foods.

Fukuda and co-workers (Fukuda et al., 2002) compared the deterioration indices of DAG cooking oil with those of TAG cooking oil after a deep-frying test. An elevation in AV value of DAG cooking oil was observed under frying conditions for a total of 244 min compared with TAG cooking oil. The initial carbonyl value of DAG cooking oil was slightly higher than that of TAG cooking oil; however, the rate increases in this value for the two oils during frying for a total of 244 min did not differ. No significant difference in peroxide or iodine value was observed. A sensory evaluation of fried foods was also conducted (Fukuda et al., 2002). Vegetable tempuras, a shrimp tempura, deep-fried chicken, and a fried shrimp prepared with DAG cooking oil and TAG cooking oil were

TABLE 19.7
Changes in the Acid Value of Test Oils Used for Repeated Deep-Frying[a]

	Initial	1st Tempura	2nd Tempura	3rd Cutlets	4th Cutlets	5th Fritter
DAG cooking oil	0.46	0.46	0.45	0.51	0.55	0.54
TAG cooking oil	0.05	0.07	0.10	0.15	0.19	0.15

[a]Source: Masui et. al. 2001.

replacement, sauces, gravies, and frozen entrées are described in Chapter 22.

To investigate changes in the quality of DAG cooking oil for home use, deep-frying tests were conducted. AV values were used as an index of oil deterioration. Cooking oil is usually used for deep-frying several times with fresh oil replenishment and the removal of frying sediment in Japan (Kajimoto, 2000). Tempura (twice), cutlets (twice), and fritters (once) were prepared 5 times over an interval of 1 wk to simulate home use. The initial AV of the DAG cooking oil was higher than that of the TAG cooking oil because of the acidic additives used. Although the AV increased slightly with cooking, the changes were similar for both products as shown in Table 19.7 (Masui et al., 2001). An elevation in AV sometimes leads to a decrease in the smoke point (Kochhar, 2001). However, no significant occurrences of smoke or irritating odor were observed in this test, nor in any other frying tests. In conclusion, DAG cooking oil is sufficiently stable for repeated use at home.

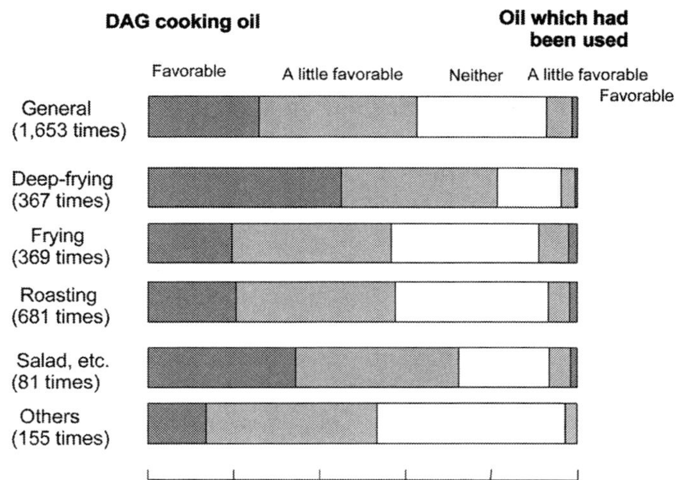

FIG. 19.1. Results of consumer evaluations comparing DAG cooking oil with their ordinary cooking oil for 3 wk in the home (Source: Masui et al., 2001).

Consumer Evaluation

The in-home use test was conducted to investigate the practical use of DAG cooking oil. Women (n = 111; 20–60 y old) participated in the test, and were instructed to compare DAG cooking oil with their ordinary cooking oil for 3 wk in the home. The evaluation results, which were summed with respect to each dish, showed that DAG cooking oil was superior to their ordinary cooking oil for all dishes, as shown in Figure 19.1. No one reported a difference such as a low smoke point in physical properties from TAG cooking oil. Many women communicated that dishes prepared with DAG cooking oil were not oily and had a light taste. The absence of a heavy stomach feeling was reported by some subjects. This supports the relationship with fast gastric emptying, as observed in the gastric retention study (Chapter 3).

Summary

The oxidation stability of DAG was comparable to that of conventional cooking oil. DAG showed a greater increase in AV than TAG under severe heating conditions, analogous to the industrial use of cooking oil; however, it was within the regulations for frying oil. On the other hand, no difference in the rate of increase in AV between the two oils was observed in the in-home use test. The results of the sensory evaluation in which DAG cooking oil was compared with TAG cooking oil indicated that no significant differences between the oils were observed in the evaluation of fried foods, except for potato chips. The results of consumer research indicated that dishes cooked with DAG cooking oil were superior to those prepared with conventional cooking oil. It can be concluded that DAG cooking oil has sufficient quality for use as a cooking oil in the home. DAG cooking oil was approved as "Foods for Specified Health Use" by the Ministry of Health, Labor and Welfare in Japan in 1999, and is currently marketed in Japan and in the United States.

References

Billek, G., Guhr, G., and Wailbel, J. Quality Assessment of Used Frying Fats: A Comparison of Four Methods. *J. Am. Oil Chem. Soc.* **1978,** *55*, 728-733.

Fritsch, C. W. Measurements of Frying Fat Deterioration: A Brief Review. *J. Am. Oil. Chem. Soc.* **1981,** *58*, 272-274.

Firestone, D. Regulation of frying fats and oils. In *Deep frying: chemistry, nutrition and practical applications.* Perkins, E.G., Erickson, M.G., Eds. AOCS Press; Champaign, IL, **1996**; 323–34.

Fukuda, H., Kimura, T., Sugimoto, T., and Iritani, N. Studies of Triacylglycerols and Diacylglycerols for Cookery. *J. Cookery Sci. Jpn.* **2002,** *35*, 172-179.

Kajimoto, G. Use of Edible Oil in Household and Disposal of Used Frying Oil. *J. Cookery Sci. Jpn.* **2000,** *33*, 387-391.

Katsuta, I. Shimizu, M., Yamaguchi, T., Nakajima, Y. Emission of volatile aldehydes from DAG-rich and TAG-rich oils with different degrees of unsaturation during deep-frying. *J. Am. Oil Chem. Soc.* (in press).

Kochhar, S.P. The composition of frying oils. *Frying, Improving quality,* Rossell, J.B. Ed.; CRC Press; LLC, New York, **2001**; pp. 87-114.

Masui, K., Yasunaga, K., Nishide, T., Nakajima, Y., and Yasukawa, T. Physicochemical and Frying Characteristics of Oil High in Diacylglycerol. in *Proceedings of the 2001 PIPOC International Palm Oil Congress*, Kuala Lumpur, **2001,** pp. 65-73.

Nakajima, Y. Water-Retaining Ability of Diacylglycerol. *J. Am. Oil Chem. Soc.* **2004,** *81*, 907-912.

Nakatsugawa, K., Ohashi, K., and Shimada, A. Comparison of Oxidation Stability of Diacylglycerol and Triacylglycerol. *Nippon Shokuhin Kagaku Kogaku Kaishi,* **2001,** *48*, 429-436.

Ogawa, H., Okushima, S., and Kodama, H. The Cooking Characteristics of Diacylglycerol Rich Cooking Oils when Preparing Confectionery. *J. Integr. Study Diet. Habits* **2001a,** *12*, 36-42.

Ogawa, H., Okushima, S., and Kodama, H. The Cooking Characteristics of Diacylglycerol Rich Cooking Oils When Preparing Dishes. *J. Integr. Study Diet. Habits* **2001b,** *12*, 100-108.

Ohno, Y. Deep-Frying Oil Properties of Diacylglycerol-Rich Cooking Oil. *J. Oleo Sci.* **2002a,** *51*, 275-279.

Ohno, Y. Thermal Oxidative Stability of Diacylglycerol-Rich Cooking Oil. *J. Cookery Sci. Jpn.* **2002b,** *36*, 397-402.

Sakai, H., Katada, M., and Ishibashi, M., U.S. Patent Application Pub. No. 2003/0026886.

Shimizu, M. Moriwaki, J. Nishide, T. Nakajima, Y. Thermal Deterioration of Diacylglycerol and Triacylglycerol Oils During Deep-Frying. *J . Am. Oil Chem. Soc.* **2004,** *81*, 571-576.

Chapter 20

Oil in Water Foods: Mayonnaise and Salad Dressing

Shigeru Kawai

Health Care Food Research Laboratories, Kao Corporation, Sumida-ku, Tokyo, Japan

Introduction

Mayonnaise and salad dressings are typical oil in water (O/W) food products in which liquid oils are used. DAG has specific emulsification properties and physical characteristics that are not associated with TAG, due to the presence of a free hydroxyl group in the former. Therefore, in some cases, different handling is required to prepare O/W food products using DAG. In this chapter, methods for the production of mayonnaise utilizing DAG oil are reported as a typical example of O/W emulsion food products. In addition, the preparation of several salad dressings using DAG oil is described.

Mayonnaise

Mayonnaise typically contains the following ingredients: oil (usually >65%), egg yolk, vinegar, and seasonings (e.g., sugar, salt, mustard) (Table 20.1). Mayonnaise is an acidic O/W emulsion product that contains a high amount of oil. In mayonnaise, the emulsifier originates in the egg yolk (Mizutani & Nakamura, 1984). In this section, a method for the production of DAG mayonnaise and the interaction between DAG and egg yolk lipoprotein are described.

Characteristics of DAG Oil Mayonnaise

When TAG and DAG mayonnaise samples (TAG-M, DAG-M) are prepared with normal egg yolk and TAG oil or DAG oil (formula: Table 20.1), the viscosity of DAG-M is higher, and the emulsified droplets are smaller than those of TAG-M. Ohashi and Shimada reported the same tendency in mayonnaise models (Ohashi & Shimada, 2002). They suggested that the higher viscosity of the DAG emulsion does not correlate directly with the diameter of the emulsion droplets.

DAG mayonnaise prepared with normal egg yolk is comparable to TAG-M in taste, flavor, and color. Although it gives the mayonnaise a slightly different texture (viscosity), DAG oil can be used for preparing homemade mayonnaise. However, when DAG mayonnaise is under accelerated preservation conditions (40°C), many cracks appear and the aqueous phase is released into the cracked region (Fig. 20.1). This phenomenon occurs not only in the laminated tubes generally used for commercial products in Japan, but also in the glass containers that are used mainly in Europe and the United States.

TABLE 20.1
Typical Composition of Mayonnaise

	Western countries Real mayonnaise	Japan Egg yolk type
Oil	Vegetable oil(s)	Vegetable oil(s)
Emulsifier (egg)	Whole eggs and egg yolks	Egg yolks
Acidifying ingredients	Vinegar	Vinegar
	Lemon juice	
Seasoning	Salt	Salt
	Sugar	Sugar
		Amino acid(s)
	Spice	Spice
	Natural flavors	
Sequestrant	Calcium disodium EDTA	

DAG mayonnaise, prepared using phospholipase A_2 (PLA2, EC 3.1.1.4)-treated egg yolks (DAG-PLM) does not develop cracks (Kawai & Konishi, 2003). Treatment of the egg yolk with PLA2 selectively hydrolyzed the ester bond at the *sn*-2 position of 1, 2-diacyl-*sn*-glycero-3-phosphatide; as a result, the polarity of the phospholipid molecules in the egg yolk is increased. DAG oil is more polar than TAG oil. Therefore, the stability of mayonnaise prepared from DAG oil stored under accelerated preservation conditions arises from the interactions between the oil and phospholipids. For this reason, analyses of the chemical state of the phospholipids were performed.

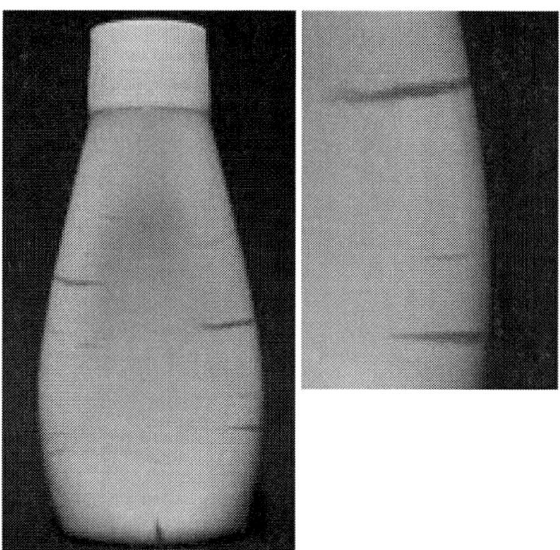

FIG. 20.1. Cracks in the DAG-M sample stored at 40°C for 4 wk. Source: Kawai, 2004.

Interaction between DAG Oil and Phospholipids Located in the Egg Yolk Lipoprotein Structure

^{31}P nuclear magnetic resonance (NMR) analyses of phospholipid molecules located in the lipoprotein structure of egg yolk are useful for the analysis of phospholipids. In mayonnaise, phosphorus atoms are contained in phospholipid molecules as well as in the protein of egg yolks. However, ^{31}P NMR measurements generally do not detect phosphorus atoms in protein because their mobility is too low. In general, the spectral widths of the NMR signals are a function of the mobility of the molecules, i.e., a low mobility results in a broad signal and a high mobility in a sharp signal.

In the spectrum of TAG-M, only a broad signal was observed. This indicates that the mobility of the phospholipid molecules is restricted and that the phospholipid molecules are located in the lipoprotein structure and/or at the surface of the emulsion droplet. In the case of DAG-M and DAG-PLM, other sharp signals and a broad signal similar to that of TAG-M were observed. These results suggest the presence of phospholipid molecules with a high mobility in DAG-M and DAG-PLM. The sharp signals may

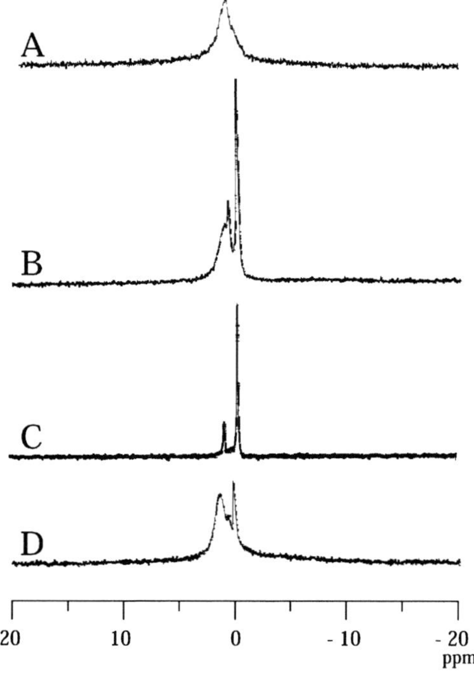

FIG. 20.2. ^{31}P NMR spectra of TAG-M (A), DAG-M (B), egg yolk lecithin dissolved in DAG oil (C), and DAG-PLM (D). ^{31}P NMR measurements were carried out under the following conditions: 109.25 MHz; 45° plus (6.5 µs); 16K data points; 20,000 Hz spectral width; 0.41-s acquisition time and a 1.59-s relaxation delay time; one-pulse sequence applied proton decoupling. Source: Kawai, 2004.

arise in the case in which the phospholipid is dissolved in the DAG oil droplets, based on comparisons with authentic samples. In the spectrum of DAG-PLM, the intensity of the sharp signals was lower than that of those in the DAG-M spectrum (Fig. 20.2).

The phospholipid concentrations in the oil phase were analyzed for TAG-M, DAG-M, and DAG-PLM (Fig. 20.3). Little phospholipid dissolution was observed for TAG-M. In the case of DAG-M, ~30% of the total phospholipid in the egg yolk was dissolved in the oil phase, whereas in the case of DAG-PLM, only 3% of the total phospholipids was dissolved.

^{31}P NMR and quantitative analyses of phospholipid suggest the following: (i) in DAG-M, in which crack formation was observed, a fairly large amount of phospholipids in egg yolk lipoprotein is dissolved in the DAG oil droplets; and (ii) in TAG-M and DAG-PLM, in which no crack formation was observed, the phospholipids in egg yolk lipoprotein are only slightly dissolved in oil droplets.

The mechanism of crack formation for DAG-M and the improvement of DAG-PLM can be explained as follows. These different behaviors of phospholipids may be due to differences in the polarity of TAG oil vis-à-vis DAG oil; the high polarity of DAG oil can be attributed to the presence of a hydroxyl group in the molecule, which increases the solubility of the phospholipids in the oil droplets. The molecular polarity of phospholipids in the egg yolk is increased as the result of the enzyme treatment. This effect greatly reduces the amount of phospholipids that can be dissolved in the DAG oil, which is the primary reason for the high stability of the DAG-PLM.

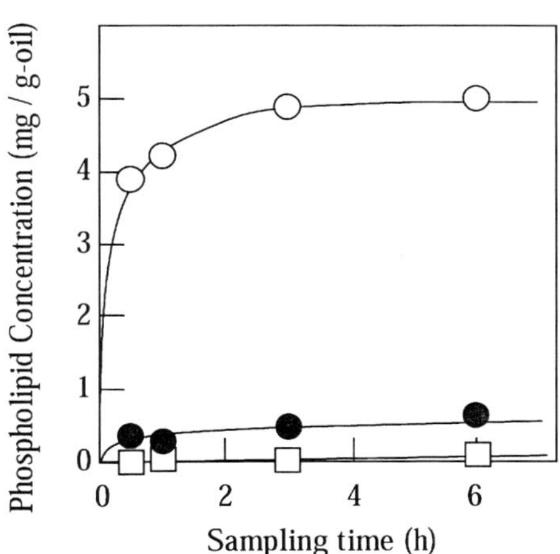

FIG. 20.3. Phospholipid concentrations [except for phosphatidylethanolamine (PE)] of TAG-M (□), DAG-M (○), and DAG-PLM (●) in the oil phase. Zero sampling time corresponds to the point at which the oil phase was added to the vacuum mixer in the premix process. The mayonnaise samples were centrifuged to separate the oil phase. PE can not be detected by this technique; therefore, the analyzed phospholipids were estimated to be ~85% of all the phospholipids in the egg yolk. Source: Kawai, 2004.

An Industrial Production Method for DAG Oil Mayonnaise Using PLA2-Treated Egg Yolk

Formulation. In this section, a production method using a Japanese egg yolk formulation (Table 20.1) is described. PLA2-treated egg yolk is used. It is preferable to optimize the reaction rate of PLA2-treated egg yolk by adjusting the amounts of oil and egg yolk. Some thickeners are useful for increasing the viscosity and improving the shear resistance in the final product during the production process (Kawai & Konishi, 2003).

Production Method. The same production method as the conventional TAG mayonnaise can be applied to mayonnaise using DAG oil. The materials are mixed using a vacuum mixer and then emulsified using a colloid mill to produce the final product. Mechanical shear rate and time for producing an aqueous phase containing egg yolk are important in terms of the viscosity and stability of final product (Kawai & Konishi, 2003).

Stability and Physical Properties. The shelf life of industrially produced DAG-PLM is equivalent to conventional TAG mayonnaise, and the physical properties are maintained during that time. The crystallization temperature of DAG oil is higher than that of TAG oil (see Chapter 18). Therefore, when DAG oil is used for the oil phase, the ability of the mayonnaise to withstand low temperatures declines compared with TAG-based mayonnaise. To improve the low-temperature resistance, the addition of a crystallization inhibitor such as polyglycerol fatty acid esters, sucrose fatty acid esters, and sorbitan fatty acid esters is effective (Shiiba et al., 2002).

Salad Dressing

Although the definitions and regulations vary among countries, the difference between salad dressing and mayonnaise depends mainly on the oil content. The typical oil contents of mayonnaise and salad dressing are ~65–85 and ≤60%, respectively. Salad dressings are classified as pourable and spoonable. Pourable dressings include the two-phase type, in which the oil and water phase are separate, and the emulsion type which has the same emulsion properties as mayonnaise. Several salad dressings that utilize DAG oil are described here.

Pourable Salad Dressing

Five types of pourable salad dressings containing DAG oil, i.e., Sesame taste, Japanese taste, Italian taste, Chinese roasted sesame taste, and Herb French taste, are sold in Japan. The Sesame taste product is an emulsion type, and the others are the two-phase type, which is easy to make using DAG oil. The production method is the almost same as that used for the conventional process. Emulsion types such as Sesame taste can be prepared by emulsification of an ~40% oil phase with egg, with the addition of vinegar and seasonings such as salt, sugar, sesame, miso, or soy sauce. In this case, however, PLA2-

treated egg yolk is not necessary to achieve a stable emulsion because of the low content of the oil phase.

Spoonable Salad Dressing

Table 20.2 shows an example of forumula of spoonable salad dressings. This formula allows the production of a spoonable salad dressing without the need to use PLA2-treated egg yolk as in the preparation of an emulsion-type pourable salad dressing.

Summary

Typical O/W food products, mayonnaise and salad dressings, prepared using DAG oil as the oil phase can be summarized as follows:

1. In the case of products with a high oil content (>65%) that use egg yolk as an emulsifier, a product such as mayonnaise would be produced using PLA2-treated egg yolk to achieve sufficient stability for long-term preservation.
2. In the case of products with a low oil content (<60%), products such as dressings would be produced via the same procedure used for TAG oil.

The application of DAG oil to O/W food products occasionally warrants particular materials and/or procedures. An understanding of the physicochemical properties of DAG will permit a wide variety of O/W food products to be produced.

TABLE 20.2
Formula for Spoonable Salad Dressing[a]

Ingredient	(%)
Water	36.75
DAG Oil	33.00
Sugar	9.50
Vinegar (100 grain)	8.50
Egg yolk	7.00
Modified food starch	2.50
Salt	1.75
Mustard powder	0.50
Xanthan gum	0.24
Lemon juice	0.10
Spices	0.06
Potassium sorbate	0.05
EDTA	0.05

[a]*Source*: Sikorski, D., unpublished data.

References

Kawai, S., and Konishi, Y., **2003**; U.S. Patent 6, 635, 777 B1.

Kawai, S. Characterization of Diacylglycerol Oil Mayonnaise Emulsified Using Phospholipase A2-Treated Egg Yolk. *J. Am. Oil Chem.* **2004,** *81*, 993-998.

Kudo, N., Nakajima, Y., Satou, M., Sugiura, M., Yamaguchi, H., and Miyatani, T. **2003**; U.S. Ser. 10/608030.

Mizutani, R., and Nakamura, R. Emulsifying Properties of Egg Yolk Low Density Lipoprotein (LDL): Comparison with Bovine Serum Albumin and Egg Lecithin. *Lebensm.-Wiss. Technol.* **1984,** *17*, 213-216.

Ohashi, K., and Shimada, A. Emulsifying Properties of Diacylglycerol in a Concentrated Oil-in-water Emulsion System. *J. Cookery Sci. Jpn.* **2002,** *35*, 132-138.

Shiiba, D., Asabu, Y., Kawai, S., and Nakajima, Y. **2002**; U.S. Patent 0,119,239 A.

Chapter 21

Water-in-Oil Type of Emulsion Foods: Margarine, Spreads, and Butter Cream

Kenji Masui

Health Care Food Research Laboratories, Kao Corporation, Sumida-ku, Tokyo, Japan

Introduction

Typical water-in-oil (W/O) emulsion foods include margarine and fat spreads that are not only consumed at the table but are also widely used in bakery and confectionery products as an essential material for food preparation. Butter cream fillings and icings are also used in confectionary and bakery products. In this chapter, the physical properties of W/O emulsion foods, such as margarine, spreads, and butter cream fillings, using diacylglycerol are described. The typical composition of glyceride and fatty acids of DAG are shown in Table 21.1.

Margarines and Spreads

"Margarines" and "spreads" are classified on the basis of their fat/oil content (Bumbalough, 2000; Bockisch, 1998). The fat/oil content of margarine should generally be ≥ 80%, whereas that of a spread should be <80%. However, the purpose of this study is to describe the physical properties of W/O emulsion foods that contain DAG. Therefore, this distinction between margarines and spreads does not apply here. The properties of a DAG spread, whose fat/oil content is 70%, are described below.

Effects of DAG on Emulsion Stability. DAG is a polar lipid molecule that contains a hydroxyl group (OH). As a result, DAG has excellent emulsibility and stability in W/O emulsions (see Chapter 18). A typical formula and preparation methods of DAG spread are shown in Table 21.2. The DAG spread consisted of 30% solid fat using partially hydrogenated palm oil [iodine value (IV), 40; melting point (mp), 46°C] and 70% liquid oil using DAG. The control TAG spread contained 70% soybean oil instead of DAG. Both DAG and TAG spread emulsions were stable for 6 mo at 5°C. In particular, the DAG spread does not require the use of emulsifiers such as monoacylglycerol and lecithin, which must be added to a TAG spread.

Effects of DAG on Taste and Flavor Release. A DAG spread melts in the mouth more quickly than a TAG spread; however, the salty taste of the DAG sample was perceived later than that for TAG. As a result, the DAG spread had an unsatisfactory flavor release, which was related directly to the degree of pleasant taste. To perceive saltiness, fat crys-

TABLE 21.1
Glyceride and Fatty Acid Composition of DAG[a]

	DAG
Glyceride	
Triacylglycerol	13.7
Diacylglycerol	85.6
Monoacylglycerol	0.7
Fatty acid	
C16	3.0
C18	1.3
C18:1	38.9
C18:2	47.3
C18:3	8.0
C20	0.1
C20:1	0.2
C22	0.1
C22:1	—

[a]Source: Masui et al., 2001.

TABLE 21.2
Typical Formula for Margarine[a]

	DAG	TAG
Oil phase		
Oil and fat	68.6	68.0
MAG	—	0.5
Lecithin	—	0.1
Flavor	0.1	0.1
Water phase		
Water	29.7	29.7
Powdered milk	0.3	0.3
Salt	1.3	1.3

[a]The manufacturing process for spread in the laboratory is as follows: While the temperature of the oil phase was maintained at 60–70°C, the water phase was gradually added to the oil while mixing with a homo-mixer (TK. Homo Mixer Mark II: Tokusyu Kika Kogyo). After the addition of water to the oil, premixing was conducted with a homo-mixer at 8000 rpm for 5 min. Rapid cooling and crystallization was achieved with a closed chilling machine (Chiller: Tama Seiki Kogyo) immediately after premixing. The product was allowed to crystallize for ~15 min with cooling at a rate of 3°C/min until the temperature of the product reached 14–16°C. After crystallization, a tempering treatment was carried out for 12–15 h at 15°C, and the product was stored at 5°C.

tals should melt in mouth, i.e., the W/O emulsion should be disrupted and converted to an O/W type of emulsion. In the case of a DAG spread, demulsifying in the mouth might be delayed due to its high stability. To promote demulsification during crystal melting, various types of demulsifiers, such as polyglycerine fatty acid esters and hydrolyzed protein, are useful (Masui & Konishi, 2001).

The effects of demulsifiers on a DAG spread are shown in Figure 21.1. In the case of a DAG spread without a demulsifier, demulsification and salt release were delayed.

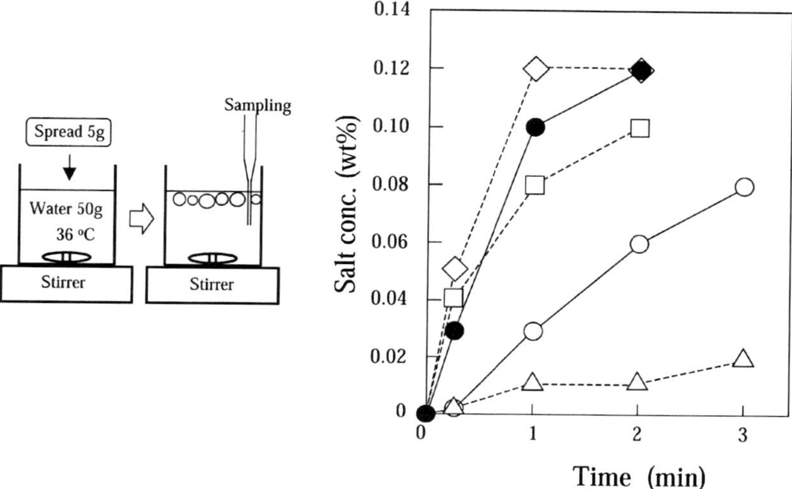

FIG. 21.1. Effects of various polyglycerine fatty acid esters on taste development. The method used is as follows: 5 g of spread was placed in 50 g of distilled water at 36°C. The salt concentrations in the water were measured with a salt meter as a function of time. HLB 4.5 (△), HLB 11 (□), HLB 15 (◇), DAG (○), and TAG (●). *Source*: Masui, K., unpublished data.

When various types of polyglycerine fatty acid esters with different hydrophilic-lipophilic balance (HLB) values were added to the spreads at a concentration of 0.5%, those with HLB ≥7 had an improved salt release rate; in the case of those with HLB ≥11, the rate increased to the same level as that of the TAG spread. A sensory evaluation was simultaneously conducted and the results were correlated with the rates of salt release. On the other hand, a polyglycerine fatty acid ester with a low HLB (HLB 4.5), suitable for a W/O emulsion, reduced the transfer rate, making it less tasty. Concerning the protein agents, hydrolyzed whey protein had significant effects on improving the salt release rate and taste development (Masui & Konishi, 2001).

These results indicate that emulsifiers with a high HLB, which are not normally used for W/O emulsions, and hydrolyzed protein improved the demulsification of the DAG spread. Furthermore, the addition of emulsifiers with a high HLB and hydrolyzed protein had no effect on the stability of the products during storage; demulsification was not observed after storage at 5°C for 6 mo.

Effects of DAG on the Consistency of a Spread. The consistency of a spread is controlled by the solid fat content (SFC), and the shape and form of the crystals. The amount of crystals tends to increase as a function of the storage time due to crystallization and further growth of crystals (Bockisch, 1998; Erickson, 1985). The change in consistency of the spreads is shown in Figure 21.2. The consistency of the DAG spread is slightly lower than that of TAG immediately after its preparation, but this relationship is reversed

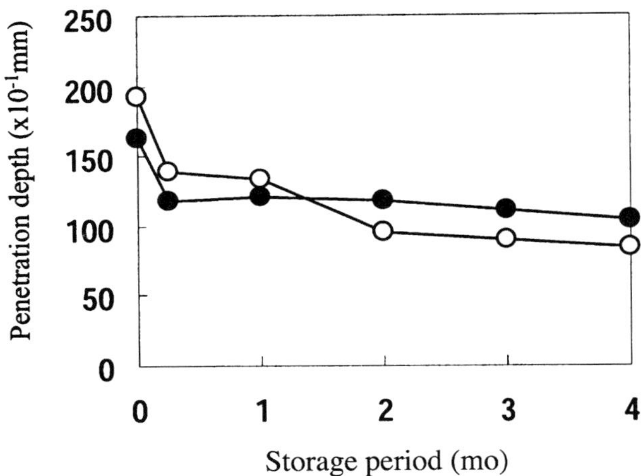

FIG. 21.2. Consistency of DAG and TAG spread stored at 5°C. The spreads had a 70% fat/oil content; the oil phase consisted of 20% partially hydrogenated soybean oil (IV, 64; mp, 42°C) and DAG or TAG (soybean oil). The spreads were prepared and stored for 4 mo at 5°C. Consistency was measured with a metal cone (Type PNR10: Sommer & Runge KG) through the macropenetration method, and represented as the amount of cone penetration depth (mm) × 10. DAG (○) and TAG (●). *Source*: Masui, K., unpublished data.

after 2 mo of storage at 5°C. The consistency of the TAG spread became constant in 1 wk; on the other hand, the DAG spread continued to gradually harden, and became slightly harder than the TAG sample after 2 mo of storage. The amount of solid fat added (hydrogenated oil content) in the oil phase was the same in both spreads; therefore, the changes in consistency might be caused by differences in the amount and shape of crystals. It was reported that when solid TAG is crystallized in DAG, the rate of crystal growth is lowered and smaller crystals are formed, probably as the result of the effect of DAG (Masui et al., 2001) Actually, the SFC of the oil phase in DAG spreads are slightly lower than those of TAG (Table 21.3.).

In examining the reasons why a DAG spread becomes harder than a TAG spread with time, it is likely that there are more deposited solid TAG crystals in DAG, i.e., low saturated solubility (Masui, K. et al., unpublished data). Therefore, the SFC of the oil phase in a DAG/TAG spread was monitored; however, the amount of crystals did not differ between the two even after 4 mo (data not shown). Other reasons might include more minute crystals of TAG in the DAG spread. They might create a closer and finer crystal network, resulting in a harder DAG spread.

Butter Cream Filling/Icing Shortenings

Butter cream fillings and icings are fat/oil products that contain sugar and are intended for confectionery use. They are comprised primarily of syrup, taste and flavor agents,

TABLE 21.3
Solid Fat Contents (SFC) of the Oil Phase in Margarine[a]

Temperature (°C)	DAG	TAG
5	13.7	14.9
10	12.6	14.0
15	11.1	11.2
20	7.9	9.2
25	2.3	7.7
30	2.0	4.0
35	0.9	0.9
40	0.6	1.2

[a]SFC was measured with SFC-900 (Praxis Co.) according to standard methods of fat analysis (AOCS official method Cd16-81). The oil phase consisted of 20% partially hydrogenated soybean oil (iodine value, 64; melting point, 42°C) and DAG or TAG (soybean oil). *Source*: Masui, K., unpublished data.

and shortening; they should have whipping characteristics, emulsion stability, heat resistance, and the ability to melt in the mouth as essential functions. Because DAG has excellent W/O emulsion stability and the effect of controlling the properties of TAG crystals (see Chapter 18), it may be useful for fillings and icings with good whipping ability and emulsion stability (Omura, 1995). In this chapter, the whipping ability and emulsion stability of shortenings containing DAG are evaluated especially with respect to a shortening for butter cream filling.

DAG Shortening for Butter Cream Filling. The DAG shortening contained 40% partially hydrogenated Canola oil (IV, 69; mp, 43.7°C) as the solid fat and 60% DAG. As a control, TAG shortening that is commonly sold on the U.S. market for all purposes was used for a comparison with DAG. The SFC of DAG and TAG (commercial) shortening showed nearly the same amount of crystals (data not shown). The specific volume (mL/g) and syrup absorption ability after whipping are shown in Figures 21.3 and 21.4, respectively. Both the whipping ability and syrup retainability of the DAG shortening were higher than those of the TAG shortening. In examining the whipping ability of the DAG shortening, the specific volume reached 4 mL/g in 5 min after the start of mixing, the whipping rate was also high, and the whipped product remained stable even under continuous mixing. The syrup retainability of DAG shortening was ~1.5 times as large as that of TAG shortening, indicative of the high W/O emulsion stability of DAG shortening. In general, a shortening containing minute crystals has a high whipping ability, and the crystals of DAG shortening may be smaller than those of TAG as the result of the effect of DAG.

Use of Fully Hydrogenated Oil for DAG Shortening. To examine the effect of DAG more clearly, two types of shortening containing 20% fully hydrogenated soybean oil (IV, 1.1; mp, 67.6°C) and 80% DAG or TAG (soybean oil) were prepared to compare their whipping ability and emulsion stability. The DAG and TAG shortenings contained the same amount of solid fat; however, large differences in both whipping ability and syrup retainability were found (Figs. 21.5 and 21.6).

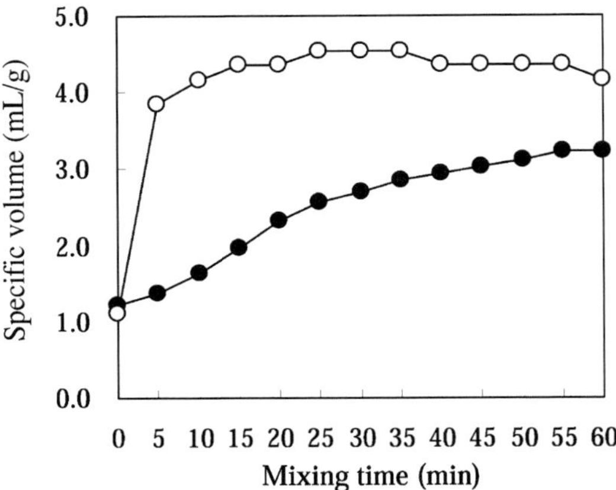

FIG. 21.3. Whipping properties of DAG shortening. The manufacturing process for shortening in the laboratory is as follows: After 0.1% of lecithin was added and fully dissolved at 60–70°C, it was rapidly cooled and crystallized using a chiller. After crystallization, a tempering treatment was carried out for 2 d at 25°C, and the product was stored at 5°C. Whipping ability was measured at 25°C as follows: 300 g of shortening was mixed at medium speed (No. 2) with a Hobart Mixer (Model N-50; Hobart). The specific volume (mL/g) was measured at 5-min intervals. DAG (○) and commercial shortening (●). *Source*: Masui, K., unpublished data.

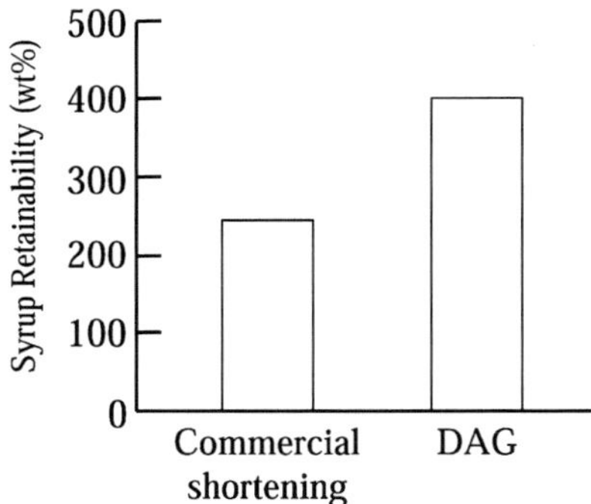

FIG. 21.4. Emulsion stability of DAG shortening. The emulsion stability was measured at 25°C as follows: After whipping the shortening and adding a high maltose syrup (Brix70), it was continuously mixed. The amount of absorbed syrup was then measured immediately before the separation of the shortening and syrup. *Source*: Masui, K., unpublished data.

Concerning TAG shortening, the whipping ability and syrup retainability were slightly lower than those of the commercial shortening. This might result from the fully hydrogenated oil used as the solid fat, i.e., it was likely due to the effects of the properties of these crystals. Because the size and shape of the crystals absorbed to the interface

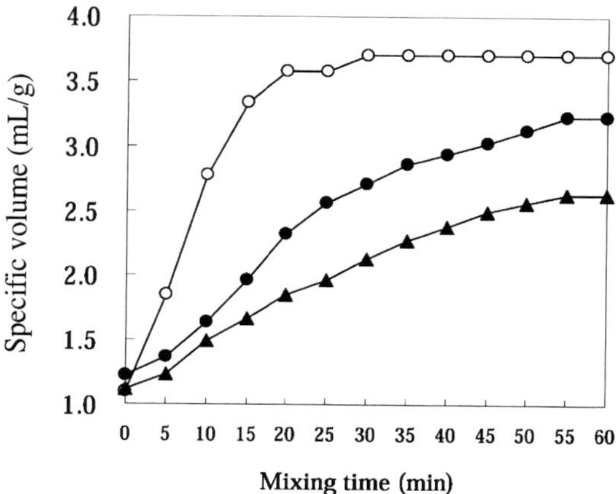

FIG. 21.5. Whipping properties of DAG shortening using fully hydrogenated soybean oil; DAG (○) , TAG (▲), and commercial shortening (●). *Source*: Masui, K., unpublished data.

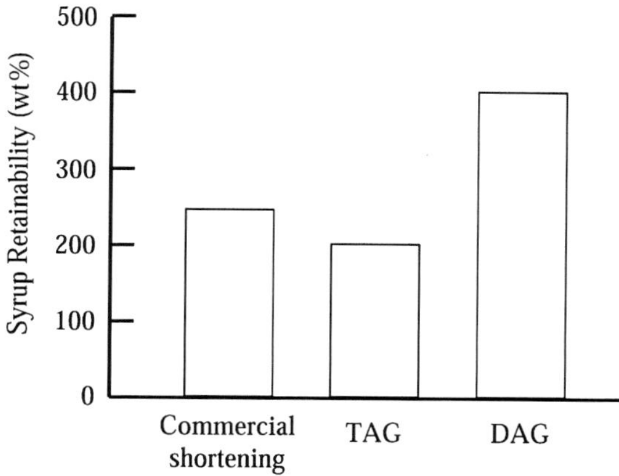

FIG. 21.6. Emulsion stability of DAG shortening using fully hydrogenated soybean oil. *Source*: Masui, K., unpublished data.

have significant effects on the stability of emulsions and foams, the foam stability of a shortening using partially hydrogenated oil including *trans* fatty acids is generally higher than that of a fully hydrogenated oil. The commercial shortening used in this study consisted of partially hydrogenated soybean oil, including 26% saturated fatty acids and 16% *trans* fatty acids.

On the other hand, both the whipping ability and syrup retainability of the DAG shortening were significantly improved in spite of the use of fully hydrogenated oil for the solid fat. This might be because minute and plate-shaped crystals formed by crystallization of fully hydrogenated oil in DAG were able to stabilize both interfaces, oil/water (syrup) and oil/air (foam).

Summary

The physical characteristics of DAG used in W/O foods such as margarine, spreads, and butter cream were described in this chapter. DAG has some surface activity due to the presence of a hydroxyl group (OH), resulting in excellent W/O emulsion ability. In margarines and spreads, DAG can replace the liquid TAG currently in use, and it might also be useful for low-fat spreads. Although its high emulsion stability may cause a delay in taste development in the mouth, it was demonstrated that this effect can be adjusted by the addition of emulsifiers with a high HLB.

Due to the solvent properties conferred by the polar group, DAG decreases the crystal growth rate of solid TAG, leading to very small TAG crystals. Through such changes in the properties of solid TAG crystals, butter cream fillings and icings with a light taste can be developed and manufactured by improving the emulsion stability and whipping ability of shortenings that contain DAG. Furthermore, DAG can improve the emulsion stability and whipping ability even of non-*trans* fat, such as fully hydrogenated oil and tropical oil.

From the above results, DAG may be a useful material for developing emulsion foods with new functions and properties.

References

Bockisch, M. Fats as, or in, Food, in *Fats and Oils Handbook,* **1998**; pp. 719, 02, AOCS Press, Champaign, IL.

Bumbalough, J. Margarine Types and Preparation Technology, in *Introduction to Fats and Oils Technology,* **2000**; 2nd ed., pp. 452, 62, AOCS Press, Champaign, IL.

Erickson, D.R. Table Spreads and Shortenings, in *Bailey's Industrial Oil and Fat Products,* **1985**; *Vol. 3,* 4th ed. (Applewhite, T.H., ed.) pp. 41, 26, John Wiley & Sons, New York.

Masui, K., and Konishi, Y., PCT **2001**; Patent WO 0101787 A3.

Masui, K., Yasunaga, K., Nishide, T., Nakajima, Y., and Yusakawa, T. Physicochemical and Frying Characteristics of Oil High in Diacylglycerol, Proceedings of the 2001 PIPOC International Palm Oil Congress (Chemistry and Technology), **2001**; pp. 65, 3.

Omura, H., Katada, M., and Tanaka, Y., 1995; JP Patent 2010513.

Chapter 22

Application of Diacylglycerol Oil in Baked Goods, Nutritional Beverages/Bars, Sauces, and Gravies

Dawn Sikorski

James R. Randall Research Center, Archer Daniels Midland Company, Decatur, IL 62521–1656

Use of Diacylglycerol Oil in Food Products

Because consumption of diacylglycerol (DAG) oil can have a positive effect on weight management (see Chapters 9–11), visceral fat (see Chapter 10), postprandial (see Chapters 6–8) and fasting triglycerides (see Chapters 11 and 12), and appetite suppression (see Chapter 5), there has been great interest in incorporating this oil into a variety of food products. Not only would the use of DAG oil provide clinical benefits through consumption of 1,3-DAG, but in many cases, it would also reduce the intake of saturated and *trans* fats through partial replacement of vegetable shortening or dairy fat. By replacing vegetable shortening or dairy fat with DAG oil, the levels of polyunsaturated and monounsaturated fats in the food product increase, which subsequently increases the overall healthful value and nutritional benefit of the product. The level of DAG oil substitution is influenced by the application or product category in which diacylglycerol oil is used to replace triacylglycerol (TAG). In this chapter, the utilization of DAG oil in various baked goods, nutritional supplements, meal replacement beverages, sauces, gravies, and frozen entrées will be discussed.

Use of DAG Oil in Cookies

To understand the effects of DAG oil in cookies and how best to incorporate it into the dough, one must first understand the factors that have the greatest effect on cookie quality. The type of fat, sugar, and flour selected will each have a significant effect on the texture and appearance of the finished product, regardless of cookie type.

Fat imparts a variety of beneficial properties to the cookie dough and to the finished cookie. Fat lubricates the dough and aids in aeration of the dough before baking (Pyler, 1973a). It also aids in flavor release and helps to provide the desired texture and mouthfeel characteristics in the finished product (Hoseney, 1994). In addition, the type of fat selected can have a great effect on dough viscosity, directly influencing the type of manufacturing equipment that can be used to process the dough (Matz, 1987). Typically, some type of plastic fat (such as shortening, margarine, butter, or lard) is used for most cookies (Kulp, 1994). The solids contributed by the plastic fat provide a framework for air entrapment during the creaming stage, which increases the volume in the finished cookie and provides the desired textural attributes (Kulp, 1994). Because liquid fats and oils do not contain the same level of solids as plastic fats, they are not able to incorporate volume to the same extent as plastic fats; consequently, their application in cookies has been somewhat limited up to this point (Pyler, 1973a).

Cookies using plastic fat will tend to be more crisp and have better snap, whereas cookies using liquid oils or fats will tend to be softer, more tender, and less crisp than their plastic fat counterparts. Depending on how much liquid oil is incorporated into the formula, manufacturing problems may arise. If a high portion of liquid oil is added, the dough will become very flowable and sticky, making it difficult to process using conventional wire-cutting or rotary die equipment (Kulp, 1994). Consequently, using a deposit or extrusion set-up may be necessary when producing products containing low levels of solid fat (Matz, 1987).

Sugar has a large effect on the functional properties and appearance of cookies. The type of sugar used can affect the spread, texture, and appearance of the finished cookie (Kulp, 1994). The most common type of sugar used in cookies is granulated sugar; however, corn syrups are used in applications such as soft-batch cookies in which extra water binding and softness are required in the finished product (Kulp, 1994). In a typical cookie, granulated sugar is partially solubilized by the formula water present due to moisture constraints (Hoseney, 1994). During baking, the sugar melts, resulting in a more fluid dough, which allows the cookie to spread (Hoseney, 1994). Coarsely ground sugar absorbs less water, yielding less spread, a higher degree of surface cracking, and a crisp, hard texture in the finished cookie, whereas finely ground sugar absorbs more water, yielding greater spread, less surface cracking, and a softer texture in the finished cookie (Hoseney, 1994. Kulp, 1994).

The type of protein/flour used can also have a significant effect on cookie performance. Most flours used in cookie manufacture are derived from soft wheat (Hoseney, 1994), which has a lower protein/gluten content than hard wheat. Higher protein contents in hard-wheat varieties generally yield tougher textures, reduced spread, and less surface cracking in cookies than their soft-wheat counterparts (Kulp, 1994). Because both protein and sugar fractions compete for the available water in the system, the gluten fraction in cookies is typically not hydrated to a great extent (Hoseney, 1994). Lack of gluten development is an advantage in cookie dough because development of the gluten would make the dough more difficult to process through wire-cut, rotary, deposit, or extrusion set-ups and would give the finished cookie a tough, undesirable texture (Kulp, 1994). The only exception with respect to development of gluten proteins occurs when sugar syrups are used in the formulation. Because the sugar syrups provide a higher level of available moisture in the formula, the gluten is able to more fully hydrate and develop (gluten cannot develop if it is not hydrated) (Kulp, 1994). However, due to the humectant properties imparted by the sugar syrups, the texture of the cookies is not negatively affected, and softness of the final product is maintained.

Incorporation of DAG oil into cookies is fairly straightforward; it can be added as either a partial or complete replacement of the shortening typically used in cookies. It is added at the same stage as the shortening to ensure proper mixing with the shortening and sugar. To determine the optimum level of DAG oil in the cookie, a model system should be used. Once the desired level has been determined, the effect on the flavor system should be investigated. By using a liquid oil to replace part or all of the solid fat originally in the system, perception and release of flavor compounds may be altered. Consequently, minor changes in the flavor system may be required to maintain similar flavor and aroma characteristics to the original cookie. Depending on how much liquid oil is incorporated into the system, it may be necessary to alter packaging, antioxidants, bulking agents, preservatives, or crumb softeners to obtain similar shelf-life characteristics (Kulp, 1994). Because functional attributes and desired eating quality vary depending on the type of cookie baked, similar practices should be used to determine the optimum inclusion level, flavor profile, and storage considerations for other cookie types utilizing DAG oil.

Work was done using a model sugar cookie formula to examine how DAG oil affected the functional properties, flavor profile, and shelf life of the cookie. The formula contained 40% fat, 63% sugar, and 9% protein (based on flour weight). The dough was prepared using a three-stage process. In the first stage, shortening (or oil, if applicable) was creamed with sugar to facilitate aeration. Eggs were added in the second stage to provide emulsification of the shortening and/or oil with the dry ingredients; flavor (vanilla and lemon extracts) was also added at this stage to achieve uniform dispersion before the addition of the flour. Flour and water were added in the final stage to minimize gluten development. After the dough was made, it was chilled for ~30 min, sheeted, cut, and baked. After the cookies were completely cooled, they were packaged in foil pouches and stored at room temperature to evaluate shelf life. Dough rheology and ease of machining were evaluated during make-up and manufacture, whereas spread, texture (gforce, in g, by TA-XT plus texture analyzer), water activity (by Aqualab Series 3 TE water activity meter), and moisture content (by Mettler LP 16 drying oven equipped with Mettler PM 100 balance) were evaluated at various time points over the shelf life of the cookies.

Comparison of Functional Properties in Cookies Utilizing DAG Oil at Various Inclusion Levels Relative to Shortening. To determine the effect of DAG oil in cookies, various inclusion levels relative to shortening were examined. Cookies were made in which DAG oil replaced shortening at 25, 50, 75, and 100% of the amount of shortening originally contained in the formula. Little difference was seen in make-up, dough viscosity, chill time, or sheeting properties of the dough when 25% DAG oil:75% shortening or 50% DAG oil:50% shortening blends were used. Moderate differences were seen when the 75% DAG oil:25% shortening blend was used. Although the dough was not as fluid or sensitive to temperature as the complete shortening replacement (see discussion below), it did not have sufficient structure to be easily sheeted and repeatedly worked. Consequently, a depositor or extrusion system would be recommended over a wirecut system for commercial production of cookies with this shortening:oil composition (Matz, 1987).

Major differences were seen when DAG oil was used as a complete replacement for shortening. Because sufficient solids were not present in the oil to provide structure, the dough was significantly more fluid than when shortening was incorporated into the dough at higher levels. Consequently, the time required to chill the dough to enable it to be machined increased from 30 to 45 min. Scrap dough was also more difficult to rework using the manufacturing procedure described; the dough was more sensitive to increases in temperature and became more sticky and difficult to handle as its temperature increased. Therefore, to produce cookies utilizing DAG oil commercially as a complete replacement for shortening, it would be necessary to use a depositor or extrusion setup to work and machine the dough efficiently.

In addition to the effects seen in dough rheology and machinability, effects in spread, texture, water activity, and moisture of the finished cookie were also compared. The control, made with 100% shortening as the fat source, had a spread (Anonymous, 1998) of 57.8 (Fig. 22.1). Cookies made with 75% shortening:25% DAG oil, 50% shortening:50% DAG oil, and 25% shortening:75% DAG oil had comparable spread results (Anonymous, 1998) of 64.2, 62.3, and 60.2, respectively (Fig. 22.1). Although the spread of these cookies was a bit higher than that of the control, they were similar in appearance to the control. In contrast, cookies made with 100% DAG oil as the fat source had considerably less spread than the control (Fig. 22.1), averaging 50.0. Moreover, the cookies made with 100% DAG oil had an appearance and texture more similar to a soft-batch cookie as opposed to a snap cookie. The texture of cookies made with 100% shortening, 75% shortening:25% DAG oil, and 50% shortening:50% DAG oil was similar (all were snap type). The texture of the 25% shortening:75% DAG oil cookie was intermediate between a soft-batch and snap cookie, but favored the soft-batch type.

Through examination of the physical and chemical changes occurring during make-up and baking of traditional cookie dough formulas, one can devise probable explanations to describe what happens when DAG oil is added to the dough at various levels. At partial replacements of up to 50% shortening, DAG oil provides lubricity and increases flowability of the dough, yielding increased spread. When DAG oil is used to replace

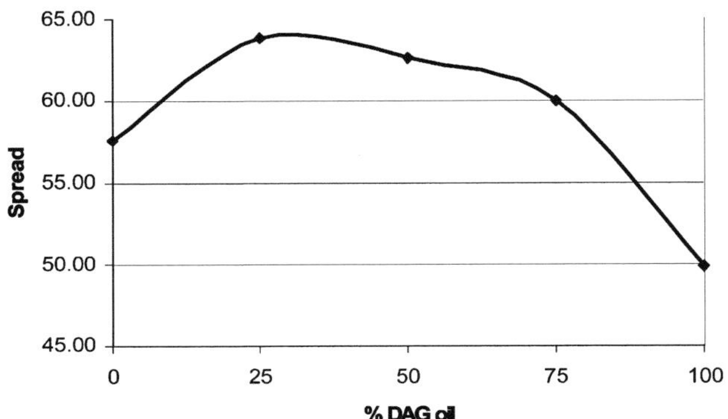

FIG. 22.1. Effect of diacylglycerol oil utilization on sugar cookie spread. DAG oil was used to replace 0, 25, 50, 75, and 100% of the all-purpose shortening typically contained in the formula.

75% of the shortening, additional lubricity is imparted to the dough; however, the increased polarity of DAG relative to TAG allows more water to be retained during baking, which allows more gluten to be developed before the cookie is completely baked. Gluten development decreases cookie spread, changing the texture from a snap type to more of a soft-batch type. Differences in spread compared with formulations containing 25 and 50% DAG oil are minimized because the added lubricity has a positive effect on spread and increased gluten development has a negative effect on spread. When DAG oil is used to completely replace shortening in cookies, the marked increase in fluidity of the dough combined with the increased polarity of DAG relative to TAG allows even more water to be retained and a higher level of gluten to be developed. As a result, spread is further decreased and texture changes from an intermediate between a snap cookie and soft-batch cookie to one that solely exhibits characteristics of a soft-batch cookie.

Texture readings (by TA-XT plus, Texture Technologies, Scarsdale, NY), water activity, and moisture results support the hypothesis described above. No major differences were seen in texture (as measured by TA-XT plus or as described by informal sensory analysis), moisture, or water activity of cookies made with 100% shortening, 75% shortening:25% DAG oil, or 50% shortening:50% DAG oil after 4 wk (Fig. 22.2). The initial texture of cookies made with 25% shortening:75% DAG oil was similar to that of cookies made with 0–50% DAG oil; however, after 4 wk, these cookies became firmer and had a slight hard/stale texture relative to cookies made with 0–50% DAG oil (Fig. 22.2). Water activity was higher (~0.50 vs. 0.45) although moisture was about the same (4.7 vs. 4.5%) in cookies containing 75% DAG oil compared with cookies containing 0–50% DAG oil. Cookies made with 100% DAG oil had texture scores similar to those of other cookies for the first week after manufacture; however, the texture became progressively firmer and increasingly stale in subsequent weeks (Fig. 22.2). After 2 wk, the cookies made with 100% DAG oil were judged to be too stale for consumer acceptance. Both water activity (0.54 vs. 0.45) and moisture scores (6.3 vs. 4.5%) were notably higher in 100% DAG oil cookies compared with cookies containing shortening or blends of shortening and DAG oil. To compensate for these increases in water activity and moisture, it may be necessary to add preservatives to maintain a shelf life similar to that of shortening-based cookies.

Investigation of Crumb Softeners

Effect of High Fructose Corn Syrup (HFCS). Additional methods were investigated to improve the texture and keeping qualities of cookies containing 100% DAG oil. One of the ingredients investigated was high-fructose corn syrup (HFCS). HFCS is used as a humectant in soft-batch cookies (Kulp, 1994); because the cookies with 100% DAG oil had a soft-batch quality, it was reasoned that HFCS might prolong this character and improve shelf life through improved management of moisture. Formulations using HFCS to replace 15% of the granulated sugar in the formula were unsuccessful because the dough was extremely fluid and too sticky to work with using available equipment. Therefore, HFCS was abandoned in favor of more conventional crumb softeners having less effect on dough rheology.

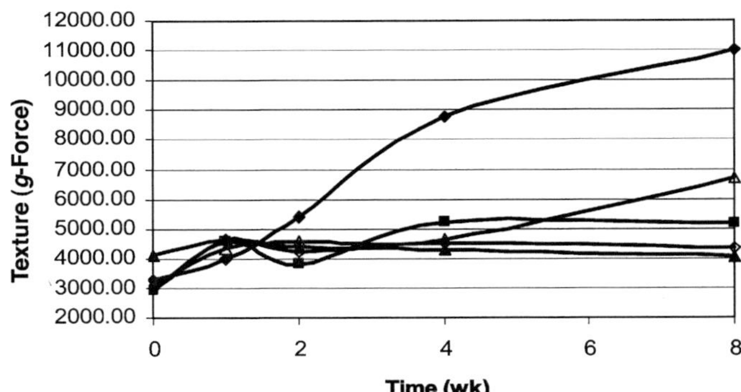

FIG. 22.2. Effect of diacylglycerol oil utilization on texture of sugar cookies. DAG oil was used to replace 0 (△), 25 (▲), 50 (◇), 75 (■), and 100% (◆) of the all-purpose shortening typically contained in the formula.

Effect of Traditional Crumb Softening Agents. Additional crumb softeners investigated included distilled monoglycerides (DMG), deoiled lecithin, polyglycerol esters (PGE), and sodium stearoyl lactylate (SSL). The formula and manufacturing procedure were the same as previously described; equivalent levels of softener (1%, flour weight basis) were used in each treatment. Although it would be necessary to adjust the level of some softeners in practice (due to limitations based on flavor, permitted use in the category, for example), each was compared at the same level to compare performance properties on an equivalent basis. In addition to comparing performance properties of shortening and DAG oil, performance properties of TAG oil (with oils selected to match the fatty acid composition of the DAG oil) were also examined. All oils tested were tested as pure systems; in other words, no blending of shortening, DAG oil, or TAG oil was done in this study.

Formulations prepared with distilled monoglycerides [iodine value (IV) = 20] showed improvement in hardness values in all three treatments in the initial and 1-wk results (Figs. 22.3 and 22.4). However, results showed little improvement compared with the control after 1 wk (Figs. 22.3 and 22.4). These results support existing work in the literature which shows that there is little, if any, gelatinization of starches in cookies during baking (Kulp, 1994). As a result, crumb softeners that function by interfering with starch retrogradation will most likely have a minimal effect on improving the shelf life of cookies.

Formulations prepared with deoiled lecithin showed large increases in spread in all treatments (Fig. 22.5). Spread (Anonymous, 1998) increased from 57.6 to 66.8 in cookies made with shortening, from 50.0 to 57.8 in cookies made with DAG oil, and from 60.7 to 69.0 in cookies made with TAG oil (Fig 22.5). Increases in spread resulted from modification of dough rheology before baking, which was noted in each of the three doughs immediately after make-up. Because the doughs made with liquid oils (either DAG or TAG oils) were inherently more fluid, the addition of deoiled lecithin caused

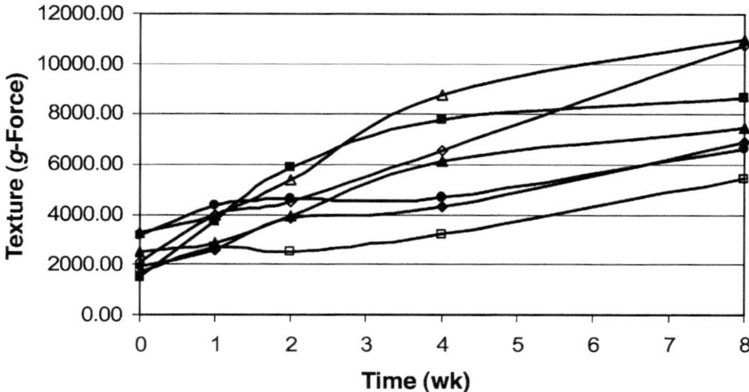

FIG. 22.3. Texture of 100% DAG oil cookies made with various crumb softeners. Crumb softeners used were as follows: DMG (■), 3-PGE (◇), 10-PGE (◆), deoiled lecithin (▲), and SSL (□). In addition, results for the diacylglycerol oil control [100% DAG as fat source, no crumb softener] (Δ) and overall control for the process [100% shortening as fat source, no crumb softener] (●) are shown.

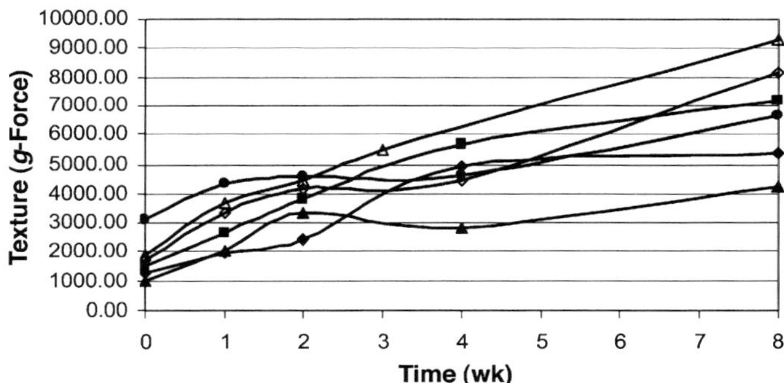

FIG. 22.4. Texture of 100% TAG oil cookies made with various crumb softeners. Crumb softeners used were as follows: DMG (■), 3-PGE (◇), 10-PGE (◆), and deoiled lecithin (▲). In addition, results for the triacylglycerol oil control [100% TAG as fat source, no crumb softener] (△) and overall control for the process [100% shortening as fat source, no crumb softener] (●) are shown.

them to be stickier and more difficult to handle than when they were processed without additives. Because the dough made with shortening had solid fat present to help build structure and provide the desired handling characteristics, the addition of deoiled lecithin increased fluidity of this dough without negatively affecting its handling or machining properties.

Improvements in texture were noted in all three treatments when deoiled lecithin was used; however, more pronounced and lasting effects were seen in shortening and TAG oil treatments than in the DAG oil treatment (Figs. 22.3 and 22.4). When deoiled

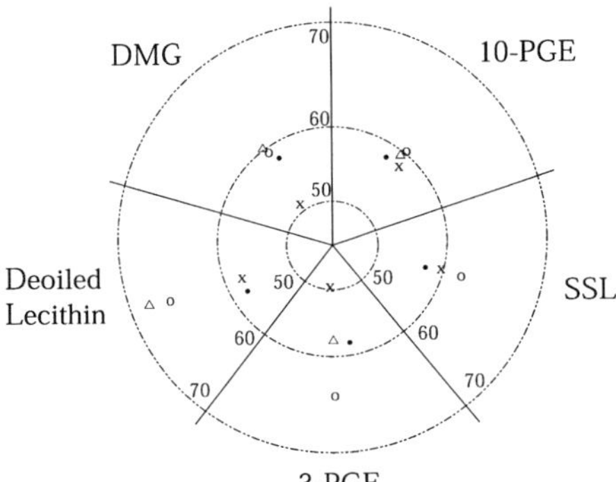

FIG. 22.5. Effect of fat/oil source and crumb softener type on sugar cookie spread. Cookies were made with 100% shortening (○), 100% DAG oil (x), or 100% TAG oil (△) as the fat source. In addition, the control for the process [100% shortening as fat source, no crumb softener] (•) is shown.

lecithin was used as a crumb softener in cookies made with shortening or TAG oil, texture readings recorded at 4 wk in the deoiled lecithin treatments were similar to those recorded initially in the control (Fig. 22.4). In contrast, the deoiled lecithin treatment made with DAG oil had texture readings at 2 wk similar to those recorded initially in the control (Fig. 22.3); at 4 wk, a crumbly, stale texture began to emerge. Therefore, it was reasoned that deoiled lecithin alone would not be suitable to extend the shelf life of cookies using DAG oil as a complete replacement for shortening. Deoiled lecithin may be suitable for cookies using TAG oil as a complete replacement for shortening; however, the level would most likely have be reduced because there were very slight, but notable, off-flavors and aromas present at the 1% inclusion level.

In addition to distilled monoglycerides and deoiled lecithin, two different types of polyglycerol esters [triglycerol monostearin (3-PGE) and decaglycerol monostearin (10-PGE)] were investigated to determine their utility as crumb softeners in cookies. The different PGE were selected to determine whether the degree of polymerization of the polyglycerol chains had an effect on crumb softening power. Minor improvements were seen when 3-PGE was used (Figs. 22.3 and 22.4); however, because the improvements were not significant and would not have the necessary effect on shelf life, no further exploration was done with this additive. Notable improvements were seen when 10-PGE was used to extend the shelf life of cookies made with DAG oil. Spread increased (50.0 to 57.1) (Anonymous, 1998) to values seen in shortening-based cookies (57.6) without affecting dough rheology (Fig. 22.5). In addition, DAG oil–based cookies made with 10-PGE had a softer texture for a longer period of time than DAG oil–based cookies made with the previously discussed crumb softeners (Fig. 22.3). Despite improvements in texture, spread, and dough rheology, improvements were not sufficient to yield more than a 4-wk shelf life; thus, the use of 10-PGE as the sole crumb softening agent in DAG oil-based cookies was ruled out.

Of the crumb softeners tested, SSL appeared to be the most promising in DAG oil. Use of SSL increased spread (Anonymous, 1998) in DAG oil–based cookies from 50.0 to 58.5, making the spread similar to that of the shortening-based control (Fig. 22.5). Use of SSL did not change the rheological properties of the dough before baking, enabling the dough to be more easily handled and machined compared with other crumb softeners. In addition to the positive change in spread, positive changes were also noted in texture. In fact, when SSL was used as a crumb softener in cookies containing DAG oil, texture readings recorded at 4 wk in the SSL treatments were similar to those recorded initially in the control (Fig. 22.3).

Cookies made with shortening displayed similar trends with respect to spread (increasing from 57.6 to 61.6) and texture when SSL was used (Fig. 22.5). As with DAG oil, no changes were seen in dough functionality or handling when SSL was used in cookies made with shortening. Although use of SSL provides some distinct advantages over other crumb softeners tested, it is important to note that this study was designed to compare all crumb softeners equally; thus, SSL was used at a level exceeding the level permitted in cookies by the Code of Federal Regulations (CFR). Therefore, to fully elucidate the benefit of SSL in cookies for commercial manufacture, additional testing at levels within the permitted use requirements would be required.

Use of DAG Oil in Cakes

Cakes and cookies are similar in the fact that they both have high levels of sugar incorporated into their respective formulations; however, because the water content in cake is so much higher than that of cookies, the sugar in cake completely dissolves in the batter, leaving a tender product devoid of sugar crystals (Hoseney, 1994). Because the amount of available water in the batter is high, starches can gelatinize during baking, locking incorporated air into place and setting the structure in the finished cake (Hoseney, 1994). High amounts of available water also allow more thorough development of the gluten, which, in turn, affects the structure of the finished cake (Pyler, 1973b). The type of fat, emulsifier, leavening agents, and flour selected each has a significant effect on volume, texture, and appearance of the finished product, regardless of cake type.

Effect of Formula Ingredients on Cake Quality

Arguably, the ingredients that have the most significant effect on cake volume and texture are fat and emulsifiers. Fats aid in air entrapment, providing volume in the finished cake (Hoseney, 1994). They also lubricate protein and starch fractions, disrupting structure formation, yielding increased tenderness and fine crumb grain (Pyler, 1973b). Use of emulsifiers improves the functionality of fat by reducing the surface tension between fat and the aqueous phase, allowing the fat to be more completely dispersed, which enhances the formation of smaller air cells and improves air incorporation and crumb grain (Pyler, 1973b; Zelch, 2001). Because the fat is dispersed into small particles over a large surface area, lubrication of dough and water-holding capacity of the batter are improved, giving a more tender crumb and longer shelf life (Zelch, 2001).

Fats selected can be either shortening, emulsified shortening, or fluid shortening (Pyler, 1973a; Pyler, 1973b). The type of fat selected and level of fat used will affect cake tenderness and volume (Pyler, 1973a). Comparisons of cakes made with emulsified and fluid shortenings showed that fluid shortenings could provide greater volume, finer grain, greater water-holding capacity, improved batter flow, and greater mixing tolerance than cakes made with emulsified shortening (Pyler, 1973a). Not only could improvements in performance be gained through the use of fluid shortenings, but healthier consumption alternatives could also be realized by replacing more of the solid fat with liquid oil.

In addition to the type of fat used, the selection of appropriate leavening agents is important for proper air incorporation and growth of air cells during baking. Typically, a baking powder that contains a mixture of slow- and fast-acting leavening acids is used in cakes (Zelch, 2001). The fast-acting type aids in nucleation of air cells in the batter, increasing the amount of air incorporated within the solid fat (Hoseney, 1994). The slow-acting type works in the oven through expansion of the batter, aiding in growth of air cells formed during the creaming stage (Hoseney, 1994). Proper selection of fast- and slow-acting leavening acids will be influenced by the type of cake to be produced, as well as the process and equipment being utilized.

Due to the large amount of moisture present in cake batters, gluten development takes place to a higher extent in cakes than in cookies (Pyler, 1973b). To allow sufficient gluten development to occur without creating excessive toughness, the protein content of cake flour should be ~6–8% compared with 8–10% for pastry (cookie) flour (ADM Catalog). In addition to a lower protein content, it is also important to use a bleached cake flour rather than an untreated cake flour (Hoseney, 1994). It is thought that the bleaching aids in the swelling of the starch, which increases the viscosity of the cake batter (Hoseney, 1994). By increasing viscosity of the cake batter and providing a more ordered structure for air entrapment, coalescence and subsequent collapse of air cells is reduced, yielding increased volume in the finished cake (Hoseney, 1994).

Although used at very low levels in cakes, pregelatinized starches and hydrocolloids can have a significant effect on dough rheology and moisture retention. Typically, pregelatinized starches are used to provide increased viscosity in the batter before baking (Zelch, 2001), which allows more air to be retained within the batter, giving the finished cake increased uniformity and finer grain (Pyler, 1973b). Hydrocolloids, such as carboxymethylcellulose (CMC), guar gum, or xanthan gum, not only provide increased batter viscosity, but also increase moisture retention and softness and improve the shelf life of the finished cakes (Pyler, 1973b. Zelch, 2001).

Effect of the Mixing Processes. The two mixing processes most widely used in cake manufacture are multistage mixing and single-stage mixing. Multistage mixing is used in the production of scratch cakes (typical in commercial production), whereas single-stage mixing is used to produce cakes from box mixes (typical in home use) (Hoseney, 1994). In multistage mixing, shortening (either emulsified or fluid) and sugar are creamed together to disperse fat and facilitate development of small air cells (Hoseney, 1994). The addition of liquids (such as water and eggs) and flour takes place in subsequent mixing stages.

During baking, the shortening melts, releasing air cells into the aqueous phase, allowing them to combine with the leavening acids to expand and provide volume in the finished cake (Hoseney, 1994). Because a separate creaming stage is used to nucleate the batter, air incorporation is quite efficient, yielding cakes with high volume, tender crumb, and fine grain; consequently, cakes made by multistage mixing techniques tend to be more resilient and better able to withstand the abusive conditions encountered in shipping and distribution (Hoseney, 1994).

In single-stage mixing, shortening and emulsifiers are melted together and plated onto the dry ingredients, allowing the fat system to be thoroughly dispersed within the mix (Hoseney, 1994). Liquids, such as water, milk, eggs, and oil, are then added to the dry mix. Because all dry ingredients and liquids are present at the same time, a creaming stage is not possible. Thus, nucleation of air cells takes place directly in the aqueous phase as opposed to being encapsulated in the fat phase (Hoseney, 1994). Because air cells are not encapsulated in fat, they are more mobile in the aqueous phase, which leads to coalescence and formation of larger air cells as well as increased air loss (Hoseney, 1994). As a result, cakes made from single-stage mixing are generally lower in volume, have coarser, more open grain, and are less resilient than cakes made from multistage mixing.

Incorporation of DAG oil in cakes is fairly straightforward; it can be added either as a partial replacement for shortening or as the oil phase component in a fluid shortening system. If the DAG oil is added as a partial replacement for shortening, it is added at the same stage as the shortening to ensure proper mixing with the shortening and sugar. If the DAG oil is added as the oil component in a fluid shortening, it is typically added in with the sugar to facilitate creaming and subsequent air incorporation before the addition of dry ingredients.

To determine the optimum inclusion rate of DAG oil in cake, a model system should be used. Once the desired inclusion rate has been determined, the effect on flavor and shelf life should be considered. By using a liquid oil to replace part or all of the solid fat originally in the system, perception and release of flavor compounds may be altered; consequently, minor changes may be required to maintain flavor and aroma characteristics. Depending on the level of liquid oil in the system, it may be necessary to alter the preservatives/mold inhibitors, antioxidants, or packaging used to obtain the desired shelf-life characteristics (Pyler, 1973a; Pyler,1973b). Because functional attributes and eating qualities vary depending on the type of cake baked, similar practices would have be used to determine the inclusion level, flavor profile, and storage considerations for other cakes made with DAG oil. To test the concepts illustrated above, work was done using four model cake formulas to examine how DAG oil affected functional properties and flavor profile in each cake type. Model systems were based on both scratch and box-mix formulas to determine the effect of DAG oil on nucleation and formation of air cells in multistage and single-stage mixing processes.

Use of Diacylglycerol Oil in Multi-Stage Mixing (Scratch Formulas)

To determine performance of DAG oil under multistage mixing conditions, a yellow cake formula containing 110% sugar, 45% fat (shortening + oil + emulsifiers), and 100%

cake flour was selected as a model system. Because all scratch formulas use the yellow cake formula as a base, it was reasoned that performance differences seen between DAG and TAG oils in 110% sugar yellow cakes would be representative of performance differences seen in other 110% sugar cakes (e.g., white, chocolate). All-purpose shortening was used as the plastic shortening in all trials, whereas either DAG or TAG oil was used as the liquid oil in the trials. Plastic shortening accounted for 50–70% of the fat source used, whereas liquid oil accounted for 30–50% of the fat source used, depending on the trial. To form the fluid shortening for each trial, the liquid oil, plastic shortening, and emulsifiers were creamed together before the addition of the dry ingredients. The remaining mixing steps were identical to those typically used in multistage mixing processes.

Results showed slight differences in volume and texture when DAG oil was used to replace TAG oil in a modified fluid shortening system (Table 22.1). In general, cakes made with DAG oil had lower volume, but had softer texture and were less gummy than cakes made with TAG oil (Table 22.1). Cakes made utilizing 30, 40, or 50% DAG oil to replace all-purpose shortening were 5.5, 5.9, and 3.0% lower in volume, respectively, than cakes made with TAG oil. Slight differences were seen in softness and gumminess scores when 30% DAG oil was used, whereas differences in softness and gumminess scores were more marked at the 50% use levels (Table 22.1).

Although lower volumes usually lead to increased hardness and gumminess values, the difference between expected and actual results was most likely due to differences in polarity between DAG and TAG oil. Increased polarity of DAG oil increased its interaction in the aqueous phase; consequently, when used to encapsulate air cells, segregation of the air cells from the aqueous phase was not as complete as when TAG oil was used. As a result, the air had greater mobility within the aqueous phase before gelatinization of the starch, leading to the formation of larger air cells, resulting in lower volume (Hoseney, 1994). Although interaction with the aqueous phase resulted in slightly lower cake volumes, it improved the water-holding capacity of the batter, yielding increased softness and decreased gumminess in the finished cake (Table 22.1). At higher inclusion levels of oil, less difference was seen in finished cake volumes between DAG and TAG; in addition, the texture was proportionally softer and less gummy in cakes made with DAG oil (Table 22.1). Improvements in volume are most likely related to higher batter viscosities in the DAG oil treatment, which reduced mobility and subsequent loss of air cells; improvements in texture are most likely related to the improved water-holding capacity of DAG oil relative to TAG oil. Although it is believed that differences in volume between cakes made with DAG and TAG oils would be within the acceptable range of variation for a manufacturer, one could restore volume in cakes made with DAG to levels seen in cakes made with TAG by applying one of the following options: modification of the leavening acid (Hoseney, 1994), increased level of leavening agents (Hoseney, 1994. Matz, 1987), small additions of hydrocolloid gums (Pyler, 1973b), modification of the emulsification system (Hoseney, 1994), or modification of the mixing conditions (Matz, 1987; Pyler, 1973b).

TABLE 22.1
110% Sugar Scratch Cakes: Mean and SD for Textural Attributes[a]

% Shortening/	Surface volume index units[b]		Hardness (g)		Gumminess		Springiness		Cohesiveness		Resilience	
% Oil ratio	Mean	SD	Mean	SD	Mean	SD	Mean	SD	Mean	SD	Mean	SD
70/30 DAG	746.9	14.98	610.85	55.03	475.20	40.43	0.855	0.046	0.778	0.005	0.473	0.006
70/30 TAG	790.7	9.28	699.54	51.72	551.98	39.12	0.908	0.020	0.789	0.013	0.484	0.011
60/40 DAG	780.0	9.30	548.01	81.51	425.96	57.24	0.846	0.130	0.779	0.021	0.466	0.023
60/40 TAG	828.8	6.43	627.76	18.08	477.17	18.69	0.855	0.043	0.760	0.017	0.452	0.019
50/50 DAG	777.1	40.01	509.12	13.31	400.81	12.50	0.899	0.035	0.787	0.006	0.465	0.004
50/50 TAG	801.5	8.68	675.36	24.68	530.32	24.66	0.915	0.036	0.785	0.016	0.483	0.021

[a]For the shortening/oil blends, all-purpose shortening was used as the shortening fraction and either DAG or TAG oil was used as the oil fraction.
[b]Based on AACC Method 10-91. Method was modified to more precisely calculate surface area and account for asymmetry in the product. One surface volume index unit = 0.1 mm^2.

Use of DAG Oil in Single-Stage Mixing (Box Mixes)

Box-mix formulas tested were white, yellow, and devil's food cake. Due to differences in polarity between DAG and TAG oil, white cake was tested to determine whether functionality differed with use of egg whites, yellow cake was tested to determine whether functionality differed with use of whole eggs, and devil's food cake was tested to determine whether functionality differed with addition of cocoa and a lower mixing time. Independent triplicate trials were conducted on each cake type to have sufficient data to determine statistical significance between treatments.

Results obtained for white cake made from single-stage mixing procedures showed no differences in volume, hardness, gumminess, springiness, or cohesiveness between cakes made with DAG and TAG oils (Table 22.2). However, the cakes made with DAG oil were significantly more resilient than the cakes made with TAG oil (Table 22.2). Results indicate that both liquid oils have similar interaction with the emulsifiers and egg whites used in the formula.

Although results were similar when DAG oil was used to replace TAG oil in white cakes, more differences were seen when it was used to replace TAG oil in yellow cakes. Yellow cakes made with DAG oil were significantly harder, more gummy, more cohesive, and more resilient than those made with TAG oil; volume and springiness did not differ in the two treatments (Table 22.2). Because the main difference between white and yellow cake formulas was the use of egg whites vs. whole eggs, it was reasoned that differences in hardness, gumminess, and cohesiveness between the two oils were most likely due to differences in their interaction with egg yolk phospholipids. Because DAG oil has a higher polarity than TAG oil, it can solubilize more of the egg yolk phospholipids, reducing their ability to act as emulsifying and tenderizing agents (Kawai & Konishi, 2000). Consequently, hardness of the cake made with DAG oil increases. Because fewer of the egg yolk phospholipids orient themselves at the interface between water and oil to reduce surface tension and provide emulsification, more are available to participate in interactions with free water, increasing water-holding capacity and gumminess of the cake. Increased gumminess makes the cake more cohesive, causing it to break down in the mouth more slowly. Despite the differences seen in mechanical texture analysis, no significant differences were seen between cakes made with DAG or TAG oils when presented to consumers in a triangle test. Therefore, although some significant differences were detected in the texture of cakes made with DAG and TAG oil, differences were minor as judged by actual consumers of the two products.

Finally, results obtained for devil's food cake made from box mixes showed no differences in volume, hardness, gumminess, springiness, or cohesiveness between cakes made with DAG and TAG oils (Table 22.2). However, cakes made with DAG oil were significantly more resilient than cakes made with TAG oil (Table 22.2). Results indicate that both liquid oils have similar interactions with the emulsifiers and alkalized cocoa used in the formula. Similarity in texture was unexpected, considering the results obtained for DAG and TAG oils in yellow cake. One possible explanation for the observations is that reductions in both mixing speed and time reduced gluten development in devil's food cake compared with yellow cake. Consequently, less tenderization from the egg

TABLE 22.2
Box-Mix Cakes: Least Significant Difference Means for Textural Attributes[a]

Cake type	Surface volume index units[b]		Hardness (g)		Gumminess		Springiness		Cohesiveness		Resilience	
	DAG	TAG	DAG	TAG	DAG	TAG	DAG	TAG	DAG	TAG	DAG	TAG
White	718.7	731.2	454.42	429.8	337.83	308.32	0.947	0.962	0.743	0.717	0.419[a]	0.379[b]
Yellow	860.9	867.3	483.66[a]	373.73[b]	383.72[a]	286.49[b]	0.961	0.959	0.794[a]	0.767[b]	0.472[a]	0.429[b]
Devil's food	748.7	755.6	559.69	544.80	457.71	426.13	0.988	0.975	0.823	0.782	0.503[a]	0.468[b]

[a]Means with different letters for an attribute differ at the 95% confidence level.

[b]Based on AACC Method 10-91. Method was modified to more precisely calculate surface area and account for asymmetry in the product. One surface volume index unit = 0.1 mm^2.

yolk phospholipids was required; therefore, differences in hardness between devil's food cakes made with DAG or TAG oil were minimal. Because no differences were seen in gumminess or cohesiveness between the DAG oil and TAG oil treatments, the egg yolk phospholipids were most likely interacting with the cocoa particles in the formula. Interaction with the cocoa reduced the amount of phospholipid in the DAG oil; consequently, the water-holding capacity of the DAG oil was reduced to a level similar to that seen in TAG oil, thus minimizing differences in gumminess between the two treatments. Because no difference was observed in gumminess, the cakes disintegrated in the mouth in a similar fashion; therefore, no differences were seen in cohesiveness between the two treatments.

Although some differences between the performance of DAG oil and TAG oil in cakes appear to be dependent upon the particular formulation tested, one parameter that was consistent among all formulation types was resilience. White, yellow, and devil's food cakes made with DAG oil were all found to be significantly more resilient than the same cakes made with TAG oil (Table 22.2). Improved resilience is particularly important for box-mix cakes because these cakes are usually more delicate and susceptible to damage than cakes made from multistage mixing processes (Hoseney, 1994). Therefore, using DAG oil to replace TAG oil could be advantageous in box-mix cakes.

Use of DAG Oil in Muffins

Muffin formulations are similar to high-ratio cakes; however, they are typically less sweet, use a higher protein flour, and have lower levels of fat than are commonly found in most high-ratio cakes (Willyard, 2000). In addition, muffins are more dense and have a chewier texture than their high-ratio cake counterparts (Matz, 1987). Ingredients affecting volume, texture, and eating quality of muffins to the greatest extent are the fat source, flour, eggs, sugar, leavening agents, gums, and starches used (Willyard, 2000). Mixing conditions are also important. Typically, muffins use less vigorous mixing conditions for less time than those used in cakes to avoid excessive gluten development from the use of a higher protein flour (Benson, 1998). As in cake, either single-stage or multistage mixing can be used to make muffins (Willyard, 2000). Multistage mixing is preferred in commercial settings to obtain maximum air incorporation, resilience, and resistance to breakage. Because it produces muffins with less structural stability than multistage mixing, single-stage mixing is typically used in food service, in-store, or consumer settings in which production conditions are less abusive.

To test the functional properties of DAG oil in muffins, both apple streusel and banana muffins were selected. Both were made utilizing box-mix formulas and single-stage mixing procedures. The apple streusel formula was selected to determine whether differences in polarity between DAG oil and TAG oil affected the suspension of inclusions, whereas the banana formula was selected to determine whether functional properties were altered by high inclusion levels of oil. Independent duplicate trials on each mix were done to have sufficient data to determine statistical significance between treatments. Results obtained for both muffin types showed no differences in volume, hardness, gumminess, springiness, cohesiveness, or resilience between DAG oil and TAG oil treatments (Table 22.3). Therefore, use of DAG oil does not appear to affect suspension of inclusions, texture, or volume differently from TAG oil, even at high inclusion levels of oil. Actual product testing by a consumer panel (triangle test) confirmed that no significant differences were perceived between DAG oil and TAG oil treatments.

Use of DAG Oil in Brownies

Brownie formulations have characteristics similar to both cakes and cookies (Smith, 2001). They are similar to cakes with respect to sugar and egg content, but resemble cookies with respect to shortening and water content. Brownies can be divided into three main types: fudge, cake, and bar (Smith, 2001). For the purpose of this discussion, we will focus on fudge and cake types. Cake brownies have less sugar, more cocoa, more shortening, and less egg than fudge brownies (Smith, 2001). The reduction in sugar allows the gluten to be more highly developed, which gives the brownies light texture, increased volume, and improved tolerance to changes in mixing time (Smith, 2001). In contrast, fudge brownies have more sugar, less cocoa, less shortening, and more egg (Smith, 2001). Corn syrup is added to bind water, creating the characteristic "fudgy" texture (Smith, 2001). Increased sugar and the presence of corn syrup increase water binding by the sugar fraction, which minimizes gluten development and gives the brownies a dense texture and low volume (Smith, 2001). Because minimal air incorporation and gluten development are desired, the batter cannot be mixed for extended periods of time and is not tolerant to changes in mixing time or conditions (Smith, 2001).

Evaluation of Brownie Formulas

Box-mix formulas were tested to determine whether there were differences in texture or flavor between fudge brownies made with DAG oil and TAG oil. Independent triplicate trials were done to have sufficient data to evaluate statistical significance between treatments. Results showed no differences in hardness, resilience, springiness, or cohesiveness between treatments (Table 22.4). However, brownies made with DAG oil were significantly more chewy than those made with TAG oil (Table 22.4). Results can be explained by examining differences in polarity between DAG oil and TAG oils. Increased polarity of DAG oil relative to TAG oil increases its water-holding capacity, which increases gluten development during mixing, thereby changing the texture of the brownie.

Although texture measurements indicated that brownies made with DAG oil were chewier than those made with TAG oil, these results were not confirmed in consumer testing. When brownies made with DAG oil and TAG oil were compared in a triangle test, consumers thought brownies made with DAG oil were less moist, chewy, and flavorful than brownies made with TAG oil. Further analysis of their comments revealed that brownies made with DAG oil were perceived as being slightly more cake-like in texture whereas brownies made with TAG oil were perceived as being slightly more fudge-like. Because fudge brownies are more dense and bind more water than cake brownies, they are likely perceived by consumers as being more moist and chewy (Smith, 2001).

TABLE 22.3
Box-Mix Muffins: Least Significant Difference Means for Textural Attributes[a]

	Surface volume index units[b]		Hardness (g)		Gumminess		Springiness		Cohesiveness		Resilience	
Cake type	DAG	TAG	DAG	TAG	DAG	TAG	DAG	TAG	DAG	TAG	DAG	TAG
Apple	250.0	244.5	402.77	390.92	306.88	302.84	0.820	0.856	0.762	0.775	0.436	0.442
Banana	259.0	258.5	499.08	479.37	383.28	365.00	0.922	0.924	0.769	0.762	0.435	0.420

[a]No significant differences were noted in any of the attributes tested.
[b]Based on AACC Method 10-91. Method was modified to more precisely calculate surface area and account for asymmetry in the product. One surface volume index unit = 0.1 mm^2.

TABLE 22.4
Box-Mix Brownies: Least Significant Difference Means for Textural Attributes[a]

Hardness (g)		Chewiness		Springiness		Cohesiveness		Resilience	
DAG	TAG	DAG	TAG	DAG	TAG	DAG	TAG	DAG	TAG
967.89	780.72	232.37[a]	209.60[b]	0.530	0.561	0.458	0.481	0.136	0.144

[a]Means with different letters for an attribute differ at the 95% confidence level.

The differences described above were not observed consistently between DAG oil and TAG oil treatments in repeated evaluations. Additional trials comparing the same box mix (from different lots/manufactured at different times) rated DAG oil brownies higher in chocolate flavor than TAG oil brownies. Though DAG oil brownies were rated higher in overall chocolate flavor, the perception of flavor release occurred later than the perception of chocolate flavor in TAG oil brownies. Therefore, although release of chocolate flavor was consistent between treatments, perception of flavor intensity was not. Differences were also seen in texture between different lots of mix tested. Some evaluations rated brownies made with DAG oil as being more chewy and fudge-like, whereas other evaluations rated brownies made with DAG oil as being less chewy and more cake-like. Because differences between texture and flavor seemed to vary with the lot of mix being tested, it was reasoned that lot-to-lot variability between mixes is greater than texture and flavor differences based on the type of oil used.

Use of DAG Oil in Nutritional Supplements

Nutritional supplements encompass a wide variety of products. Because guidelines for this product segment are not clearly defined, for the purpose of discussion in this chapter, we will define nutritional supplements as products specifically designed to meet the health needs of a particular class of patients. Typically, nutritional supplements can be consumed either orally or enterally, depending on the health status of the patient. Because the formulations have been balanced to meet the patients' daily needs for protein, carbohydrate, fat, vitamins, and minerals, they may consume nutritional supplements as a partial or complete replacement for food.

Consumption of nutritional supplements can yield a variety of benefits for patients. Clinical studies in cancer patients have shown that consumption of nutritional supplements can reduce weight loss, preserve lean muscle mass, and improve quality of life (Barber et al., 1999). Benefits are derived from multiple components working in synergy within the formula. Inclusion of eicosapentaenoic acid (EPA) enables metabolism to be more closely regulated, reducing weight loss resulting from metabolic changes caused by the tumor (Barber et al., 1999). Protein helps the body to build and maintain muscle mass, and multiple fat sources provide a more consistent supply of energy for the body (Barber et al., 1999). Medium-chain triglycerides (MCT) provide a more readily available energy source for the body, whereas vegetable oils such as safflower and canola provide a more slowly digested fat source; overall digestion/release of energy is more controlled, which enables the product to be digested more thoroughly and to be better tolerated by patients (Bach & Babayan, 1982), increasing their energy levels and improving their quality of life (Barber et al., 1999).

Dialysis patients can also benefit from nutritional supplementation. Because they are tightly restricted in their fluid intake, their supplements are more concentrated . In addition, levels of protein, fat, and carbohydrate are adjusted to meet their specific needs. For example, lower protein levels compensate for reduced ability to process high protein loads, whereas higher fat levels offset decreased protein intake. Both readily digestible and slower digesting fats are used to provide a controlled release of energy into the body. Carbohydrate levels are reduced to avoid complications in hypertriglyceridemic dialysis patients (Abbott Nutrition Website).

Another major category of nutritional supplements is for patients with moderate-to-extreme breathing difficulties. In the mildest cases, patients may require periodic assistance in breathing from mechanical ventilation devices; in the most severe cases, patients would require 24-h breathing assistance from mechanical ventilation devices. In mild cases, either oral or enteral feeding would be permitted; in severe cases, feeding would be entirely by enteral means. Formulations are concentrated to provide a nutrient-dense product because patients have a restricted fluid intake. Levels of protein, fat, and carbohydrate are adjusted to provide a low respiratory quotient (RQ) to minimize the amount of carbon dioxide gas retained in the bloodstream (Das & Roberts, 2001). Carbohydrate has the most dramatic effect on RQ (raises carbon dioxide level the most), followed by protein, then fat (Das & Roberts, 2001). To provide a product with lower RQ to ease patient breathing, lower levels of protein and carbohydrate and higher levels of fat must be used. Higher RQ/levels of carbon dioxide gas in the blood would exacerbate breathing difficulties in patients with milder cases (Askanazi et al., 1980) and may result in respiratory failure in patients with more severe cases (Covelli et al., 1981).

Using DAG oil in nutritional supplements is equivalent to using vegetable oils from a processing standpoint. The only issue that may arise by using DAG oil is related to its polarity. Because DAG oil is more polar than TAG oil, it will be more reactive with oxygen if stored in tanks without an inert headspace. Therefore, to avoid oxidation before processing and to ensure the DAG oil is of optimal quality, it will either have to be stored under an inert headspace or at room temperature and used quickly if a modified headspace is not available.

To test the utility of DAG oil in nutritional supplements, soy-based formulas containing either DAG or TAG oils were processed by ultra-high temperature (UHT) pasteurization and stored for 1 wk before descriptive evaluation by trained panelists. The fatty acid profiles of both oils were matched to ensure that differences noted between the two beverages were due to structural differences between DAG and TAG instead of differences in their fatty acid composition. Soy-based formulas were chosen to investigate differences in the ability of the oils to mask soy notes or to reduce astringency. Protein, fat, and carbohydrate represented ~23, 26, and 51% of energy, respectively. This composition most closely resembled beverages used for patients recovering from injury or surgery (Abbott Nutrition Website). Results from the descriptive panel showed no major differences in flavor or mouthfeel between the beverages; thus, based on flavor and mouthfeel, DAG oil could be used to be a complete replacement for TAG oil in nutritional supplements. However, depending upon the needs of patients consuming the beverage, partial replacement may be preferred to provide the most desirable fatty acid profile (e.g., supplementation with EPA or γ-linolenic acid).

Use of DAG Oil in Meal Replacement Beverages

Like nutritional supplements, many products and formulations exist for meal replacement beverages. The main differences between nutritional supplements and meal replacement beverages are the customer base for which they are designed and their intended use. Nutritional supplements are designed for hospital patients with specific nutritional needs, whereas meal replacement beverages, with the exception of products designed for type II diabetics, are designed for consumption by the general population. Although most nutritional supplements are consumed to gain or maintain weight, most meal replacement beverages are consumed to lose or maintain weight. Because meal replacement beverages do not generally contain as much protein or fat as nutritional supplements, they are not recommended as sole-source nutrition items. In fact, they are designed to replace only 1–2 meals/d.

Although the primary benefits associated with meal replacement beverages are weight loss and weight maintenance, a number of secondary benefits can arise through their consumption. By losing weight (or maintaining a healthy weight), a person typically feels better, both physically and mentally. In addition, the risk for developing obesity-related cancers (Calle et al., 2003), non-insulin dependent diabetes (type II) (National Institutes of Health, 2000), and coronary heart disease (Calle et al., 1999) can be reduced through proper weight maintenance. Because numerous benefits can be derived from weight loss and/or weight maintenance, a number of clinical studies have been done to evaluate the use of meal replacement beverages as a weight loss aid. Studies have been done in overweight individuals, obese individuals, and individuals with type II diabetes. Overweight individuals are classified as those having a body mass index (BMI) of 25.02–29.9 kg/m^2, whereas obese individuals are classified as those having a BMI ≥ 30.0 kg/m^2 (Garrow & Webster, 1985). Because individuals with type II diabetes have impaired glucose control and may require a product more closely matched to their individual needs, they were evaluated among both diabetic and nondiabetic treatment groups.

On the basis of the results obtained from the studies surveyed (Ditschuneit et al., 1999; Flechtner-Mors et al., 2000; Heymsfield et al., 2003), it appears that meal replacement beverages can be used as an effective tool for weight loss and weight maintenance in both diabetic and nondiabetic populations. Further, diets incorporating meal replacement beverages and snacks at the start of their weight loss/weight maintenance program appear to have positive effects on blood pressure, serum triglycerides, blood glucose, and insulin levels in overweight and obese individuals. The reduction in insulin levels during consumption of meal replacement beverages and nutrition bars indicates that these products may also be useful to type II diabetics as dietary adjuncts to improve glucose control.

Using DAG oil in meal replacement beverages is equivalent to using vegetable oils from a processing standpoint. As discussed earlier, extra care should be taken when storing DAG oil because it is more reactive with oxygen than TAG oil. To maximize its keeping quality, the DAG oil should either be stored in tanks with an inert headspace or used quickly if an inert headspace is not available. To test the utility of DAG oil in meal replacement beverages, soy-based products containing either DAG or TAG oils were processed by UHT pasteurization and stored for 1 wk before evaluation by trained panelists. The fatty acid profiles of both oils were matched to ensure that differences seen between the beverages were due to structural differences between DAG and TAG rather than differences in their fatty acid composition. Soy-based formulas were chosen to investigate differences in the ability of the oils to mask soy notes or to reduce astringency. Protein, fat, and carbohydrate accounted for ~15, 10, and 75% of energy, respectively. This composition is similar to commercially available soy-based and dairy-based meal replacement formulations. Results from the descriptive panel showed no major differences in flavor or mouthfeel between the beverages; thus, based on flavor and mouthfeel, DAG oil could be used as a complete replacement for TAG oil in these products. However, partial replacement may be preferred to complete replacement to provide the fatty acid profile that best suits the needs of the intended population (Garg et al., 1988).

Although the study described above showed no major differences between beverages made with TAG or DAG oil, a previous study conducted on a soy milk beverage did show some differences between the oils. The soy milk beverage was less viscous than typical nutritional supplements or meal replacement beverages and used HTST (high temperature, short time) instead of UHT pasteurization. Although descriptive panelists thought the basic tastes of the soy milk beverages (sweet, sour, salty, bitter) were similar, they thought the sample made with DAG oil had direction-ally lower soy notes and astringency than the sample made with TAG oil. By examining descriptive results from the beverages tested, one can conclude that differences in flavor between products made with DAG and TAG oils will be more strongly influenced by the type of formulation and processing conditions used than by the oil source selected.

Use of DAG Oil in Nutritional Bars

Nutritional bars appeal to a wide variety of demographic and lifestyle groups. Application areas in which they are used include sports nutrition, body building, athletic endurance, women's health, meal replacements, and specialized diets (e.g., high-protein, low-carbohydrate, and balanced nutrition bars). A newly emerging category of nutritional bars are those that have eating qualities that are similar to commercially produced candy bars, but provide some functional health benefit (e.g., weight loss).

The majority of nutrition bars have a fat content of <7 g/serving, with the most popular bars having a fat content of ~5 g/serving. Much of the fat content in these bars is attributed to the chocolate and yogurt coatings covering the bars; thus the use of DAG oil in these bars may be limited to levels of 1–4 g/serving. DAG oil could either be incorporated into the bar center in typical bars before the forming process or into the caramel and nougat fractions of nutritional bars that mimic commercially produced candy bars.

DAG oil could provide several beneficial effects in nutritional bars, from improvements in the nutritional properties of the bar, to improvement in the bar manufacturing process, or improvement in the textural properties of the bars during storage. The biggest issue related to producing acceptable nutritional bars concerns the drying effect that the dry ingredients, especially proteins, have on the bars during storage. This drying effect can typically be minimized through use of proteins that have low water-binding properties in combination with modification of the water absorption properties of the proteins through coating them with a hydrophobic compound such as DAG or TAG oil.

Use of DAG Oil in Sauces, Gravies, and Frozen Entrées

Sauces and gravies are used in a wide variety of products and applications. They add flavor and elegance to almost any dish, transforming them from bland to spectacular in just minutes. They can be used as toppings for pasta, rice, meat, or vegetables and can be either acidic or neutral in pH. Consumers and restaurants have moved toward the use of processed sauces and gravies, often in the form of frozen entrées and prepared meals, to prepare great tasting meals more quickly and conveniently (Przybyla, 1983). By selecting the appropriate gravy or sauce, the same basic entrée can be tailored to suit a wide variety of tastes and occasions. In addition, gravies or sauces can be formulated to meet specific dietary needs or constraints for kosher, diabetic, or vegetarian meals (Przybyla, 1983).

Many ingredients are available to provide the desired stability in sauces, gravies, and frozen entrées. Perhaps the most critical ingredient to a successful gravy or sauce formulation is the type of starch selected. Typically, a starch high in amylopectin is preferred in applications in which a long shelf life or freeze/thaw stability is required (Baldwin et al., 1972; National Starch and Chemical Website). Although amylose is beneficial from a standpoint of binding water, it tends to have a much higher level of retrogradation over time than amylopectin (Przybyl et al., 1983).

In addition, freezing and thawing procedures can dramatically affect the appearance and texture of the finished product (Przybyl et al., 1983). Fast freezing causes the formation of many small ice crystals within the product, avoiding the destructive effects caused by large ice crystal formation. Fast thawing allows the ice to change to water as quickly as possible, reducing retrogradation and syneresis (Hoseney, 1994). The selection of appropriate emulsifiers and stabilizers can also improve freeze/thaw stability in frozen products (Prendergast, 1980). Because they act to improve freeze/thaw stability by different mechanisms, using a combination of these ingredients in conjunction with starch would most likely provide the greatest protection against syneresis. Emulsifiers enable smaller oil droplets to be formed, facilitating an even dispersion of the oil throughout the water phase of the product and allowing a more stable emulsion to be formed (Prendergast, 1980), whereas stabilizers, such as hydrocolloid gums, bind free water and provide viscosity in the gravy or sauce.

Processing conditions also influence the appearance and texture of gravies and sauces. The most common method for processing gravies and sauces is to heat the ingredients to 95°C, homogenize the slurry, and hot fill into cans or jars (Prendergast, 1980). Because the gravy or sauce will be subjected to high heat for long periods of time, a modified cook-up starch is typically used to achieve the desired stability for products made by this process (National Starch and Chemical Website). After the retort cycle is complete, the products may either be stored at room temperature in the original cans or jars or used in frozen-food products. Although this process is straightforward and easily done from a manufacturing perspective, subjecting the gravy or sauce to high heat for extended periods of time decreases the contribution in viscosity from the starch (Prendergast, 1980).

To avoid such losses in viscosity, newer manufacturing facilities process gravies and sauces at lower temperatures. Initial processing takes place at 50°C instead of 95°C; the product is then homogenized in traditional fashion, but sterilized and cooled through the use of heat exchangers before packaging or freezing (Prendergast, 1980). Heat exchangers heat and cool the product much more rapidly than bulk processing, preserving peak viscosity of the starch. Consequently, sauces and gravies made using this technology are more stable and have a longer shelf life than products made at 95°C (Prendergast, 1980).

Using DAG oil in gravies and sauces is fairly straightforward; it can be added as a partial, or in some cases, a complete replacement of the shortening, dairy fat, or oil typically used in these products. It is added at the same stage as the fat source typically used to make the product; thus, no specialized processing equipment or handling is required to use DAG oil in these systems. To determine the optimum inclusion rate of DAG oil in the gravy or sauce, a model system should be used. Once the desired inclusion rate has been determined, the effect on viscosity, texture, and mouthfeel should be investigated. By using a liquid oil to replace part or all of the solid fat originally in the system, it may be necessary to use additional stabilizing and thickening agents to produce a product with a viscosity, mouthfeel, and texture similar to the control product. In addition, the effect on shelf life should be considered. Depending on how much liquid oil is incorporated into the system, it may be necessary to use additional/different antioxidants or preservatives to obtain similar shelf-life characteristics. Because functional attributes and desired texture and mouthfeel vary depending on the product being made, similar practices would have to be used to determine the optimum inclusion level, type and level of stabilizing/ thickening agents, and storage considerations for other sauces or gravies made with DAG oil.

To test the concepts illustrated above, work was done using the model systems described:

1. Traditional white sauce made using milk and cream as the fat source. The white sauce contained 7.3% fat and 2.5% modified food starch in the control formula.

2. Modified white sauce made using butter as the fat source. The white sauce contained 7.3% fat, 2.5% modified food starch, and 0.10% xanthan gum in the control formula. In addition, the use and necessity of emulsifiers were investigated using this model system.
3. Traditional brown gravy made using partially hydrogenated vegetable shortening as the fat source. The brown gravy contained 6.1% fat, 3.5% modified food starch, and 0.17% xanthan gum in the modified control formula.
4. Traditional barbecue sauce made using vegetable oil as the fat source. The control formula for the process contained 5.0% fat and 2.0% modified food starch.
5. Meat marinade with 18.6% vegetable oil and 0.20% xanthan gum in the control product.

In all of the systems described, DAG oil was used as a complete replacement for the fat source ordinarily contained within the system. Fatty acid profiles of DAG and TAG oils were matched to ensure that differences seen between the performance of the two oils was due to their structure rather than differences in their fatty acid profiles. Model systems 1–4 were heated to 95°C and cooked for 5 min under continuous agitation. After cooling the samples in the refrigerator for approximately 24 h, viscosity readings were taken at 25 and 50°C to determine the viscosity profile at room temperature and consumption temperature, respectively. Model system 5 was processed at room temperature using moderate, continuous agitation for the entire make-up procedure of the product. Samples were allowed to equilibrate for approximately 24 h before viewing separation behavior in control and test products.

Traditional White Sauce Using Milk and Cream. White sauces were chosen as the first model system to be investigated due to their wide range of use across many different applications. White sauces are typically used as a base to make Alfredo sauce, cheese sauce, curry sauce, and mustard sauce, among others. Because these applications would transcend both home and industrial use, the functionality of DAG oil was investigated from the perspectives of both home and industrial users. Nonfat dry milk and either DAG or TAG oils were added to test formulas to replace the protein and fat ordinarily contributed by milk and cream. Initial formulations containing either DAG oil or TAG oil were less viscous and had a less creamy mouthfeel than the control formulation containing solid fat; however, formulations with DAG oil and TAG oil were comparable to the control when either extra starch or a starch/hydrocolloid blend were used to provide added thickening.

Modified White Sauce Using Nonfat Dry Milk and Butter. Due to variations in viscosity and texture between batches of white sauce controls made with milk and cream, a model system using nonfat dry milk and butter was explored. Because the fat source was the only variable in both control and test formulations, differences in texture, viscosity, and mouthfeel between each fat source could be more clearly identified. In addition, the effect of emulsifier use on the fat sources was tested. Because emulsifiers improve product stability and shelf life, their use would most likely be preferred in commercial applications in which the target shelf life of a gravy or sauce can range from 9 mo to 1 y.

When the white sauces were tested without emulsifiers, equivalent viscosity was measured in products made using either DAG oil or TAG oil at both 22 and 50°C (Figs. 22.6 and 22.7). Both sauces were lower in viscosity than the control (made with butter) (Figs. 22.6 and 22.7), indicating additional starch or a starch/hydrocolloid blend would be necessary to achieve equivalent viscosity and mouthfeel from a liquid oil compared with a solid fat source. When emulsifiers were added at a level of 0.2% (total formula weight basis), viscosity decreased in all cases (Figs. 22.6 and 22.7). When tested at 22°C, comparable drops in viscosity were observed for emulsified sauces made with butter and DAG oil but much higher drops occurred in emulsified sauces made with TAG oil (Fig. 22.6). In contrast, when the viscosity of emulsified and unemulsified sauces was compared at 50°C, little difference was noted between sauces made with butter, similar reductions were noted between sauces made with DAG oil, and the greatest reductions (even more than those seen at 22°C) were noted between sauces made with TAG oil (Fig. 22.7). These results suggest that either DAG or TAG oils could be used to make an acceptable white sauce in applications in which emulsifier usage is not required. However, DAG oil may offer a performance advantage over TAG oil in formulations containing emulsifiers because the viscosity of white sauces made with DAG oil appears to be less affected by emulsifiers than the viscosity of white sauces made with TAG oil. Further, fewer added stabilizers and thickeners would be required to make a product comparable to the control when DAG oil is used, allowing the flavor of the sauce made with DAG oil to be more closely matched to the flavor of the control product.

Traditional Brown Gravy Made with Shortening. Brown gravies were chosen as the second model system to be investigated because they are widely used as a base to make such sauces as Bordelaise sauce, mushroom sauce, Piquant sauce, and Provencale sauce. In these formulations, either DAG oil or TAG oil was added to test formulas to replace the fat ordinarily contributed by shortening in the control formula. Xanthan gum was added to all formulations to provide body and cling to enable the formulation to meet the most diversified set of needs for both home and industrial users.

Results on brown gravies evaluated at room temperature indicated there were no major differences in viscosity between gravies made with shortening, DAG oil, or TAG oil. However, when these same gravies were evaluated at 50°C, the viscosity of the gravy containing shortening was considerably less than that of gravies made with either DAG oil or TAG oil. These differences were most likely caused by the shortening changing from a solid phase to a liquid phase. Because DAG and TAG oils contain very little saturated fat, very little saturated fat was available to undergo a phase change from solid to liquid in either one of these oils; consequently, differences in viscosity were less obvious for the gravies made with liquid oil than the gravy made with shortening. Because 50°C simulates the consumption temperature of the gravy, thicker gravies made using liquid oils may be perceived as being richer, more premium products than gravies made

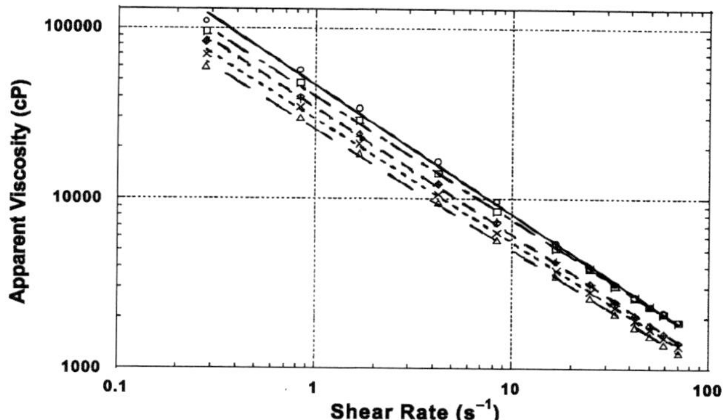

FIG. 22.6. Viscosity profiles of white sauces made with a nonfat dry milk/butter model system. Evaluations were made at room temperature (22°C) on the following fat source/emulsifier combinations: butter, no emulsifier (○); butter, with 0.2% emulsifier (□); DAG oil, no emulsifier (◇); DAG oil, with 0.2% emulsifier (x); TAG oil, no emulsifier (+); and TAG oil, with 0.2% emulsifier (△). Equations for the lines and *r*-squared values are as follows: (○) = 46,597·*x*(–0.75836), *R* = 0.9959; (□) = 40,071·*x*(–0.72404), *R* = 0.99877; (◇) = 34,416·*x*(–0.74449), *R* = 0.99926; (x) = 29,497·*x*(–0.72362), *R* = 0.9989; (+) = 33,891·*x*(–0.74173), *R* = 0.99966; and (△) = 25,667·*x*(–0.70888), *R* = 0.99902.

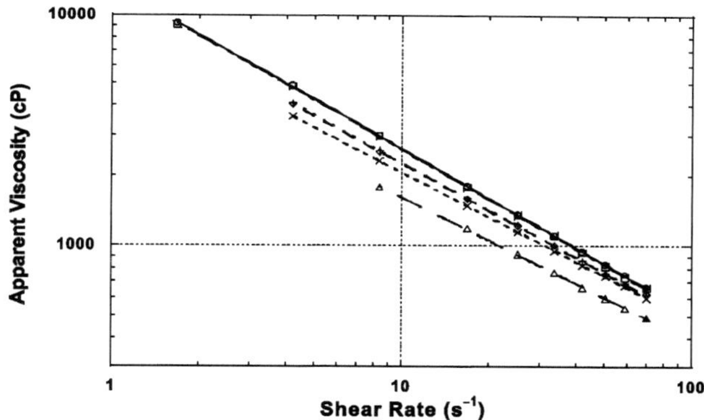

FIG. 22.7. Viscosity profiles of white sauces made with a nonfat dry milk/butter model system. Evaluations were made at consumption temperature (50°C) on the following fat source/emulsifier combinations: butter, no emulsifier (○); butter, with 0.2% emulsifier (□); DAG oil, no emulsifier (◇); DAG oil, with 0.2% emulsifier (x); TAG oil, no emulsifier (+); and TAG oil, with 0.2% emulsifier (△). Equations for the lines and *r*-squared values are as follows: (○) = 13,353·*x*(–0.70738), *R* = 0.99993; (□) = 13,257·*x*(–0.70777), *R* = 0.9999; (◇) = 10,583·*x*(–0.67004), *R* = 0.99998; (x) = 8,965.4·*x*(–0.6362), *R* = 0.99998; (+) = 10,787·*x*(–0.67446), *R* = 0.99999; and (△) = 6648.1·*x*(–0.61378), *R* = 0.99992.

with shortening. These results indicate that either DAG or TAG oils could be used as a complete replacement for shortening with little to no change in processing conditions in the model system tested. By using DAG oil to replace shortening, not only could the level of saturated and *trans* fats be greatly reduced in the gravy, but the product would also have the healthy benefits associated with DAG oil consumption.

Traditional Barbecue Sauce Made Using Vegetable Oil. Barbecue sauce was chosen as the third model system to be investigated to determine whether there were any differences in functionality between DAG and TAG oils in low pH systems compared with neutral pH systems. No major differences were noted in flavor or viscosity between barbecue sauces made with DAG and TAG oils; therefore, DAG oil could be used as either a partial or a complete replacement for TAG oil in the model system tested with no changes in processing or formulation required.

Marinade for Meats Used in Frozen Entrees. The final model system used to evaluate performance of DAG oil in sauce and gravy type products was as a replacement for TAG oil in marinades. Xanthan gum was added to the marinade to provide body, cling, and make it suitable for frozen applications. There were no major differences in flavor or viscosity between marinades made with DAG and TAG oils; therefore, DAG oil could be used as either a partial or a complete replacement for TAG oil in the model system tested with no changes in processing or formulation required.

Summary

Diacylglycerol oil can be used as either a partial or a complete replacement for triacylglycerol oil, depending on the individual product application and characteristics desired. In some cases, DAG oil can even be used as a partial replacement for shortening, reducing the level of saturated and *trans* fat in the end product. By adjusting the ratio of diacylglycerol oil to shortening in cookies, one can create textures ranging from a typical snap type to more of a soft batch–type cookie. Scratch cakes made with DAG and TAG oils had similar volume, but softer textures were noted in cakes made with DAG oil, especially at higher inclusion levels of oil. Box mix cakes made with DAG and TAG oils had similar volume and texture, with DAG oil cakes be statistically more resilient than TAG oil cakes. No significant differences were observed in muffins or brownies made with DAG oil.

Nutritional supplements and meal replacement beverages formulated with DAG oil showed similar or improved flavor profile and mouthfeel compared to those made with TAG oil, depending on the formulation and processing conditions tested. Using diacylglycerol oil in nutrition bars helped retard water absorption by the protein fraction, delaying hardening and improving the shelf life of the bar. DAG oil may be used in high-protein nougats or caramels or as an extrusion aid in these bars. Formulating sauces and gravies with DAG oil reduces the amount of saturated and *trans* fats in the product through replacement of shortening or dairy fat with liquid oil. The flavor profile of products made with DAG oil is similar to the controls made with shortening or dairy fat; however, in some cases, it may be necessary to use additional stabilizers/ thickeners in products made with liquid oil (DAG or TAG) to match the viscosity and mouthfeel of the controls made with a more solid fat source.

References

Abbott Nutrition (Website: http://abbottnutrition.com/)

Anonymous. *Approved Methods of the American Association of Cereal Chemists.* **1998**; Vol. 1, 9th edn., Method 10-50 D, pp. 1-6.

Archer Daniels Midland Food Ingredients Catalog, ADM, Decatur, IL. **2002/2003** edn; pp.10-13.

Askanazi, J, Elwyn, D., Silverberg, P., Rosenbaum, S., and Kinney, J. Respiratory Distress Secondary to a High Carbohydrate Load: A Case Report. *Surgery* **1980**, *87,* 596-598.

Bach, A. and Babayan V.K. Medium-Chain Triglycerides: An Update. *Am. J. Clin. Nutr.* **1982**, *36,* 950-962.

Baldwin, R., Moody, D., Cloninger, M., and Korschgen, B. Stability of Gravies to Freezing. *J. Am. Diet. Assoc.* **1972**, *60,* 218-222.

Barber, M., Ross, J., Voss, A., Tisdale, M., and Fearon, K. The Effect of an Oral Nutritional Supplement Enriched with Fish Oil on Weight-Loss in Patients with Pancreatic Cancer. *Br. J. Cancer.* **1999**, *81,* 80-86.

Benson, R. *Muffin Technology.* American Institute of Baking (AIB) *Technical Bulletin* **1998**, *10,* 1-4.

Calle, E., Rodriguez, C., Walker-Thurmond, K., and Thun, M. Overweight, Obesity, and Mortality from Cancer in a Prospectively Studied Cohort of U.S. Adults. *N. Engl. J. Med.* **2003**, *348,* 1625-1638.

Calle, E., Thun, M., Petrelli, J., Rodriguez, C., and Heath, C. BMI and Mortality in Prospective Cohort of U.S. Adults. *N. Engl. J. Med.* **1999**, *341,* 1097-1105.

Covelli, H., Black, J., Olsen, M., and Beekman, J. Respiratory Failure Precipitated by High Carbohydrate Loads. *Ann. Intern. Med.* **1981**, *95,* 579-581.

Das, S.K., and Roberts, S. Energy Metabolism, in *Present Knowledge in Nutrition.* **2001**; 8th edn. (Bowman, B. and Russell, R., eds.) pp. 1-12, ILSI Press, Washington, DC.

Ditschuneit, H., Flechtner-Mors, M., Johnson, T., and Adler, G. Metabolic and Weight-Loss Effects of a Long-Term Dietary Intervention in Obese Patients. *Am. J. Clin. Nutr.* **1999**, *69,* 198-204.

Flechtner-Mors, M., Ditschuneit, H., Johnson, T., Suchard, M., and Adler, G. Metabolic and Weight Loss Effects of Long-Term Dietary Intervention in Obese Patients: Four-Year Results. *Obes. Res.* **2000**, *8,* 399-402.

Garg, A., Bonanome, A., Grundy, S., Zhang, Z., and Unger, R. Comparison of a High-Carbohydrate Diet with a High-Monounsaturated-Fat Diet in Patients with Non-Insulin-Dependent Diabetes Mellitus. *N. Engl. J. Med.* **1988**, *319,* 829-834.

Garrow, J., and Webster, J. Quetelet's Index (W/H2) as a Measure of Fatness. *Int. J. Obes.* **1985**, *9,* 147-153.

Heymsfield, S., van Mierlo, C., van der Knaap, H., Heo, M., and Frier, H. Weight Management Using a Meal Replacement Strategy: Meta- and Pooling Analysis from Six Studies. *Int. J. Obes.* **2003**, *27,* 537-549.

Hoseney, R.C. *Principles of Cereal Science and Technology,* **1994**; 2nd edn., American

Association of Cereal Chemists, St. Paul, MN.

Kawai, S and Konishi, Y. **2000**; PCT Patent WO 00/78162 A2.

Kulp, K. *Cookie Chemistry and Technology*, **1994**; American Institute of Baking, Manhattan, Kansas.

Matz, S. *Formulas and Processes for Bakers*, Pan-Tech International. **1987**. McAllen, Texas.

National Institutes of Health. *The Practical Guide: Identification, Evaluation, and Treatment of Overweight and Obesity in Adults.* **2000**; NIH Publication no. 00-4084, 1-75.

National Starch and Chemical (Website: www.nationalstarch.com.)

Prendergast, K. The new way with soups and sauces? *Food Flavouring Ingredients and Processing*, **1980**, 26-27.

Przybyl, H., Boruch, M., and Urbaniak, M. The Influence of Some Parameters of Processing of Frozen Sauces on the Colloidal-Emulsive Medium Stability After Defreezing, 16th International Congress of Refrigeration Proceedings, **1983**; Vol. III, 705-710.

Przybyla, A. Frozen Food Mandate: Get the Calories Out. *Prepared Foods,* **1983**, 84-86.

Pyler, E.J. *Baking Science and Technology*, **1973a**; Vol. 1. Siebel Publishing, Chicago.

Pyler, E.J. *Baking Science and Technology*, **1973b**; Vol. 2, Siebel Publishing, Chicago.

Smith, R. *Formulation and Production of Brownies*, American Institute of Baking (AIB) *Technical Bulletin* **2001**, *23,* 1-9.

Willyard, M. *Muffin Technology (Update)*, American Institute of Baking (AIB) *Technical Bulletin* **2000**, *22,* 1-6.

Zelch, R. *Batter Cakes-Part 1. Ingredients and Formulations*, American Institute of Baking (AIB) *Technical Bulletin* **2001**, *23,* 1-10.

Part VI

Processing Technologies

Chapter 23

Manufacturing Process of Diacylglycerol Oil

Takaaki Watanabe[a], Hiroaki Yamaguchi[a], Naoto Yamada[b], and Inmok Lee[c]

[a]Processing Development Research Laboratories, Kao Corporation, Japan; [b]Production Quality Management Division, Kao Corporation, Japan; and [c]James R. Randall Research Center, Archer Daniels Midland Company, Decatur, IL

Introduction

Partial glycerides, such as monoacylglycerol (MAG) and diacylglycerol (DAG), have long been used as emulsifiers and stabilizers in the food industry. Pure MAG and DAG are used for the synthesis of more complex lipids that are useful for various pharmaceutical and cosmetic applications (Haftendorn & Ulbrich-Hofmann, 1995; Berger & Schneider, 1993). In recent years, there have been studies that reported the nutritional benefits of the oils containing ≥80% DAG (see Parts II–IV of this book). Mixtures of mono- and diacylglycerols are generally synthesized from triacylglycerol (TAG) oils and glycerol via a chemical glycerolysis reaction at temperatures >200°C using alkaline catalysts (Sonntag, 1982). The distribution of mono-, di-, and triacylglycerols in the reaction mixture depends on the molar ratio of glycerol to triacylglycerol used in the reaction. The DAG component in the resulting reaction mixture has physicochemical characteristics that are intermediate between those of MAG and TAG; this makes it difficult to separate DAG from both MAG and TAG, for the purpose of making an oil product high in DAG, by conventional separation processes.

Enzymes are macromolecular compounds that have selective catalytic functions. Enzymatic reactions have advantages such as mild reaction conditions, high selectivity, and low energy consumption; the use of enzymes is increasing in the field of food processing because of these advantages. Lipases are used for various modifications of fats and oils. Regioselective lipases are an excellent tool for the production of oils rich in DAG as described below.

Enzymatic Production of DAG

An enzymatic reaction for the production of DAG has advantages over a chemical reaction. Some researchers have reported DAG production by lipases through various pathways, such as hydrolysis, glycerolysis, and esterification (Plou et al., 1996; Fureby et al., 1997; Yang et al., 1994; Yamane et al., 1994; Ota et al., 1997; Kristensen et al., 2005; Berger et al., 1992; Waldinger & Schneider, 1990; Ergan et al., 1990).

Plou et al. (Plou et al., 1996) reported a high-yield production of a MAG/DAG mixture by lipase hydrolysis of TAG. From their screening test results of various solid supports with different hydrophilicities, they concluded that porcine pancreatic lipase that was adsorbed on Celite was the most effective biocatalyst. The time courses of TAG hydrolysis with free and immobilized lipases were analyzed for calculation of the reaction rate constants. The differences in the rate constants were related to differences in the regioselectivity of the enzymes and/or the hydrophilicity of the supports. Fureby et al. (Fureby et al., 1997) reported the preparation of DAG by lipase-catalyzed alcoholysis of TAG. A screening of lipases showed that the lipase from *Penicillium roquefortii* that was immobilized on porous polypropylene particles produced a high content of 1,2-DAG from TAG in organic media. The yield of DAG was 75%; furthermore, 95% of the total DAG was 1,2-DAG. Alcohol concentration and chain length of the alcohol had a significant effect on yield and enzyme reactivity.

Yamane et al. (Yamane et al., 1994) reported the glycerolysis of TAG, in which DAG was prepared by the reaction of hydrogenated beef tallow and glycerin in the presence of *Pseudomonas* lipase. The yield of DAG depended on the reaction temperature. The optimal reaction temperature gave a yield of ~90 % DAG. Kristensen et al. (Kristensen et al.,2005) reported glycerolysis of rapeseed oil by several lipases; a DAG yield of approx 60wt% was achieved with immobilized *Candida antarctica* lipase and immobilized *Pseudomonas cepacia* lipase. Berger et al. (Berger et al., 1992) reported the production of high-purity DAG by esterification of fatty acid and glycerol in an organic solvent. Enzymatic esterification of glycerol and different acyl donors in the presence of various 1,3-selective lipases produced DAG in high yield. The reactions were carried out in n-hexane, diethyl ether, or tert-butylmethyl ether.

Ergan et al. reported the production of glycerides from glycerol and fatty acid by immobilized lipase in nonaqueous media. Immobilized *Mucor miehei* lipase catalyzed the synthesis reaction between glycerol and oleic acid without organic solvent. The system indicated more production of triolein than diolein (Ergan et al., 1990). Rosu et al. (Rosu et al., 1999) reported enzymatic synthesis of symmetrical 1,3-DAG by direct esterification in a solvent-free system. The synthesis of 1,3-DAG was carried out by esterification of glycerol with relatively low-melting fatty acids and 1,3-selective immobilized *Rhizomucor meihei* lipase. Some of the system's parameters were optimized for the production of 1,3-DAG, and the maximal yield reached 98%. Those studies indicate that an esterification reaction with lipases is an efficient way to produce oils high in DAG.

Lipases that have a 1,3-regioselectivity can be used for the production of oils high in 1,3-DAG (Rosu et al., 1999; Watanabe et al., 2003; Kim & Lee, 2006). Pancreatic lipase, which is a well-known digestive enzyme, is a 1,3-regiospecific lipase. Microbial lipases from various sources with 1,3-regiospecificity are commercially available. To determine the optimal condition of the esterification, it is important to know the kinetics of the reaction. A lipase esterification reaction scheme from fatty acids and glycerol as shown in Fig. 23.1 is a likely scenario. This scheme also includes the acyl migration reaction in addition to the esterification reactions. Lortie et al. (Lortie et al., 1993) reported a kinetic study using a similar model in a triolein synthesis reaction.

It is generally thought that the lipase reaction takes place at the oil-water interface and that a reaction of fatty acids with glycerol occurs in the oil phase. To achieve an effective reaction, several methods were tested. Berger et al. (Berger et al., 1992) and

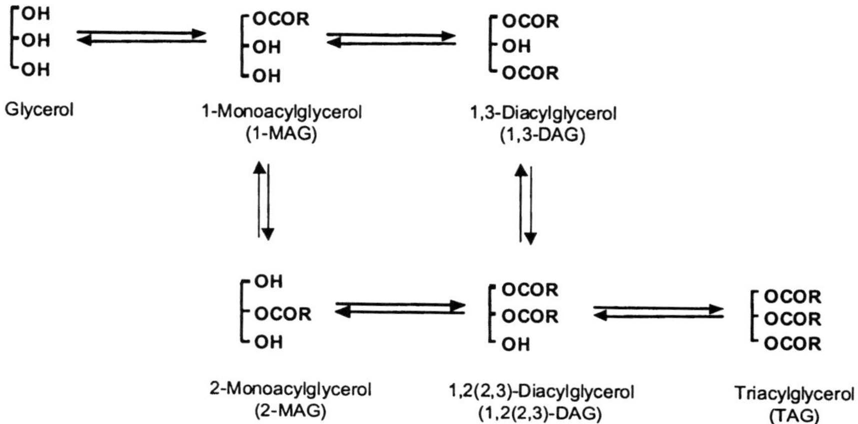

FIG. 23.1. Reaction scheme for the 1,3-DAG synthetic reactions.

Castillo et al. (Castillo et al., 1997) reported that the use of adsorbed glycerol on silica gels was an effective method for the esterification of fatty acid with glycerol in organic solvents. In a more recent study, the esterification of fatty acids and glycerol in a solvent-free system with 1,3-regioselective lipase was investigated (Watanabe et al., 2003). The experimental apparatus that was used in the study is shown in Fig. 23.2. The system is composed of a stirred batch reactor and a vacuum pump. Water in the reaction mixture was removed by maintaining a reduced pressure in the reactor. Typical esterification results using the stirred batch reactor are shown in Fig. 23.3. A high content of DAG was obtained with this system.

Generally, controlling water during the esterification reaction is an important issue because the reaction equilibrium can be shifted to a direction of esterification with a decrease in water content. Several methods for controlling water content are available; these include the use of a stream of nitrogen gas, dehydration agents such as silica gel, molecular sieves, and the use of a vacuum pump (Rosu et al., 1999). Water content also plays an important role in maintaining enzyme structure and catalytic activity (Zaks & Klibanov, 1988; Marty et al., 1997). Thus, extremely low water content in the reaction mixture can lead to deactivation of the enzyme; thus, it is important to conduct the esterification reaction under conditions that maintain the optimal water level in the reaction system.

Acyl migration is also an important issue in an esterification reaction, especially for producing DAG high in 1,3-DAG glycerides. Acyl migration, in which 1,3-DAG is converted to 1,2-DAG or 1-MAG is converted to 2-MAG, occurs during the esterification reaction. Acyl migrations often leads to an increase in TAG; as the result, the purity of DAG decreases. The effects of various factors, such as water content, temperature, enzyme load, and reaction time on acyl migration in transesterification reactions were investigated by Xu et al. (Xu et al., 1998a; Xu et al., 1999). They showed that all of these factors affect acyl migration. In the case of esterification of partial glycerides, optimization of these factors is important for efficient production of the desired products. The

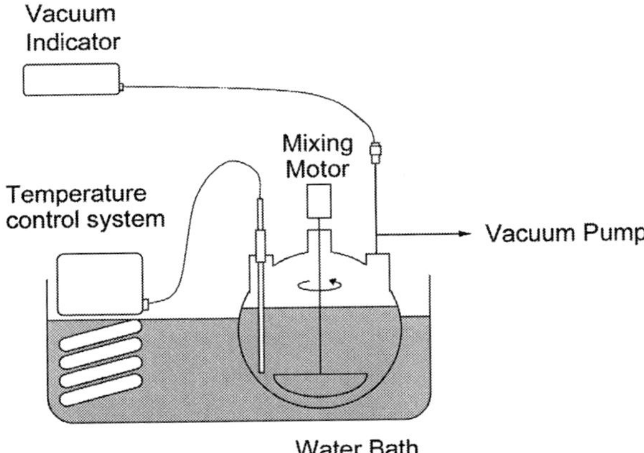

FIG. 23.2. Diagram of the stirred batch reactor used for the esterification reaction in diacylglycerol synthesis.

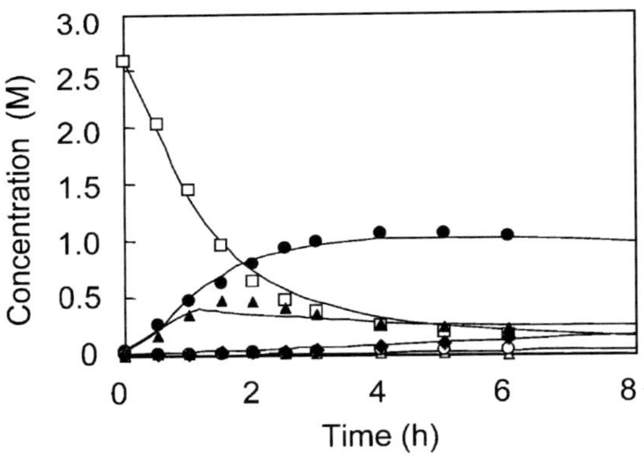

FIG. 23.3. Time courses for the esterification reaction using a lipase from fatty acid and glycerol. The reaction was performed at 50°C under a 3-mmHg vacuum: (□), FA; (▲), 1-MAG; (▲), 2-MAG;(●), 1,3-DAG; (○), 1,2(2,3)-DAG; (◆), TAG. Solid lines are calculated results. Source: Watanabe et al, 2003.

effects of various factors on the extent of acyl migration were also investigated for esterification reactions, using the stirred batch reactor shown in Fig. 23.2. The results indicated that a higher reaction temperature promoted acyl migration. Therefore, it was concluded that optimal reaction temperature as well as water removal in a reaction system are critical factors for the effective production of DAG with high 1,3-DAG content.

In a batch reactor, enzyme is suspended in reaction mixture. Then, the enzyme is separated from the reaction mixture at the end of the reaction. A packed-bed reactor, which is filled with enzymes immobilized on a carrier, is used for production of various food products. Packed-bed reactor has some advantages, especially on an industrial scale: (i) separation process of enzyme from the reaction mixture is not needed; (ii) mechanical loss of enzyme due to separation is prevented; and (iii) reactor equipment is relatively compact, and continuous operation is possible.

Enzymatic reactions using lipases in packed-bed reactors were investigated by several investigators (Xu et al., 1998b; Xu et al., 2000; Shimada et al., 1996; Colombie et al., 1998; Mensah & Catra, 1999; Arcos et al., 2000; Watanabe et al., 2005), in the areas such as transesterification (Xu et al., 1998b; Xu et al., 2000; Shimada et al., 1996) and alcoholysis of fatty acid with various alcohols (Colombie et al., 1998; Mensah & Catra, 1999). Water is generated during an esterification reaction. Thus a system for water removal is necessary for an efficient esterification reaction. Various methods were reported to prevent the accumulation of water in a packed-bed reactor for esterification reaction in organic solvents; these include using a polar solvent (Colombie et al., 1998), periodical drying by airflow (Colombie et al., 1998), or the use of a pure solvent (Mensah & Catra, 1999). A method using excess glycerol was also reported for an esterification reaction without solvent (Acos et al., 2000). In that method, an excess amount of glycerol was used to dissolve the water generated by esterification reaction, resulting in a conversion of >90%. Esterification by circulating the reaction mixture between a packed-bed reactor and a water removal vessel was reported. In that method, water was removed by drying under reduced pressure in the vessel, resulting in approximately 70% of 1,3-DAG (Watanabe et al., 2005). Again, it is very important to optimize the factors that affect the efficiency of esterification reaction, especially water removal and acyl migration. Understanding the kinetics of the enzyme is necessary for the design of an efficient reactor system as well as for determining optimal operation conditions.

Analytical Methods for Determining DAG Content

After esterification with glycerol and fatty acids, the reaction mixture is made up of complex mixtures of glycerides. Much effort and ingenuity have been expended, therefore, in determining the appropriate analyses for the detection and composition of glycerides. Analytical methods using thin-layer chromatography (TLC), gas-liquid chromatography (GLC), and a combination of gas chromatography and mass spectrometry (GC-MS) were developed for the assessment of the molecular species of DAG (Rnkonen, 1965; Kuksis & Marai, 1967; Myher & Kusis, 1982). GLC is the most widely accepted method and has been standardized by the IUPAC Commission (Bruschweiler & Dieffenbacher, 1991) and AOCS (AOCS, 2003). The method consists of direct injection of silylated oils into a capillary GC. According to AOCS official methods, partial glycerides are converted with bis(trimethylsilyl) trifluoroacetamide (BSTFA) and trimethylchlorosilane (TMCS) in pyridine into the more volatile trimethylsilyl ether derivatives and determined by capillary GLC. Furthermore, many analytical methods for quantitative determination of DAG isomers in edible oils have been investigated (Ruiz-Gutierrez & Perez-Camino, 2000; Perez-Camino et al., 1996; Perona & Ruiz-Gutierrez, 2003; Pacheo et al., 1998; D'Alonzo et al., 1982; Adlof, 1996; Février et al.,2001). Perez-Camino et. al (Perez-Camino et al., 1996) reported on an analytical system that combined the use of a diol-bonded phase column and GC. The procedure includes the isolation of polar fraction including DAG by solid-phase extraction using a bonded diol phase and further analysis of the silyl derivatives by capillary GC on 65% phenylmethylsilicone. It was conducted successfully for quantitative determination of DAG in edible oil. The procedure is easy, fast, and reproducible, allowing the quantitation and separation of DAG according to their carbon number, isomeric structure, and degree of unsaturation. Analytical methods for determination of the positional distribution of fatty acids on DAG by using lipase and phospholipase A2 (Fisher et al., 1973; William et al., 1995) were also reported.

Fine determinations of DAG in living cells were investigated. Determinations of molecular species of DAG in human adipose tissue and VLDL lipoprotein by solid-phase extraction, GC (Pacheco et al., 1998), and reversed-phase liquid chromatography (Perona & Ruiz-Gutierrez, 2003) were also reported. Measurement of cellular DAG content was reported in an investigation of signal transduction mechanisms (Bolleag & Griner, 1998).

Recently, trans-fatty acid content in fats and oil products has drawn attention worldwide because the fatty acid may lower HDL cholesterol (see Chapter 1). An AOCS official method shows the determination of cis- and trans-fatty acids in hydrogenated and refined oils and fats by capillary GLC (AOCS, 2002). This method consists of GLC conditions optimized to identify and quantify the trans-fatty acid isomers in vegetable oils and fats. The fatty acid methyl esters (FAME) of the sample are separated on a capillary GC column with a highly polar stationary phase according to their chain length, degree of (un) saturation, and geometry and position of the double bonds.

Characterization of DAG Oil Produced by Enzymatic Esterification Process

DAG oil is currently available commercially in Japan and the United States as a cooking oil. The oil is made from soybean and canola oils, and the typical fatty acid composition is described in Chapter 19. The DAG cooking oil contains <0.5 g trans-fatty acids/serving. Its physicochemical properties are described in detail in Chapters 18 and 19. The oil that is available in the United States is kosher, having been made from kosher materials at a kosher production facility.

TABLE 23.1
Characteristics of the Finished DAG Oil

General appearance	Pale yellow transparent liquid
Lovibond color (10R + Y)	<20
Moisture	≤0.1%
Acid value	≤0.5 mg KOH/g
Peroxide value	≤0.5 mEq/kg
DAG content	≥80%
Heavy metals (as Pb)	≤10 ppm

High 1,3-DAG yield and purity are attained by using a 1, 3-selective enzyme immobilized on a food-grade, phenolic-type resin carrier for esterification. Non-DAG components in the reaction mixture are separated after esterification to achieve >80% DAG content in the final DAG oil product. This product, measured by a slightly modified AOCS official method (AOCS, 2003), contains <20% TAG and only a small amount of MAG, as shown in Chapter 19. The ratio of 1,3-DAG to 1,2-DAG is ~7:3. Conventional oil-refining procedures are used for further refining of the DAG oil, and general quality of the final DAG oil is shown in Table 23.1. Antioxidants can be added to improve the shelf-life stability of the oil; commercial DAG cooking oil typically contains 30% of the U.S. recommended daily allowance for vitamin E. DAG cooking oil is packaged in a laminated film container, which was especially designed to maintain the freshness of the DAG oil during shelf storage. Various long-term storage tests with DAG oil, under both ambient and accelerated conditions, were done to determine the shelf-life stability of DAG oil in comparison with regular TAG salad oil (unpublished data). DAG cooking oil had a shelf-life stability equal to or better than that of the regular salad oil of similar fatty acid composition.

Summary

Among various methods reviewed in this chapter, the enzymatic esterification reaction with 1,3-selective lipase was an efficient way to produce an oil with a high 1,3-DAG content. Removing water from the reaction mixture and suppression of acyl migration were important factors for high DAG yield and purity. The esterification reaction mixture is further processed to improve the DAG purity and sensory quality of the DAG oil. Conventional oil-refining procedures are used for DAG oil refining, and the commercial DAG cooking oil had a shelf-life stability equal to or greater than that of regular salad oil.

References

Adlof, R.O. Analysis of Fatty Acid Mono- and Diacylglycerol Positional Isomers by Silver Ion High-Performance Liquid Chromatography. *J. Chromatogr.* **1996,** A 741, 135-138.

American Oil Chemists' Society Determination of *cis*-and *trans*-Fatty Acids in Hydrogenated and Refined Oils and Fats by Capillary GLC, Method Ce 1f-96, in Official Method and Recommended Practices of the American Oil Chemist's Society, **2002**; 5th edn., AOCS Press, Champaign, IL.

American Oil Chemists' Society Determination of Mono- and Diglycerides by Capillary Gas Chromatography, Method Cd 11b-91, in Official Methods and Recommended Practices of the American Oil Chemist's Society, **2003**; 5th edn., AOCS, Champaign, IL.

Arcos, J.A., Garcia, H.S., and Hill, C.G. Continuous Enzymatic Esterification of Glycerol with (Poly)Unsaturated Fatty Acid in a Packed Reactor. *Biotechnol. Bioeng.* **2000,** *68*, 563-570.

Berger, M., and Schneider, M.P. Regioisomerically Pure Mono and Diacylglycerols as Synthetic Building Blocks. *Fat Sci. Technol.* **1993,** *95*, 169-175.

Berger, M., Laumen, K., and Schneider, M. Enzymatic Esterification of Glycerol. I. Lipase-Catalyzed Synthesis of Regioisomerically Pure 1,3-sn-DAGs. *J. Am. Oil Chem. Soc.* **1992,** *69*, 955-960.

Bolleag, W.B., and Griner, R.D. Measurement of Cellular Diacylglycerol Content, in *Methods in Molecular Biology,* **1998**; Vol. 105: Phospholipid Signaling Protocols (Bird I.M., ed.) pp. 89-98, Humana Press, Totowa, NJ.

Bruschweiler, H., and Dieffenbacher, A. Determination of Mono- and Diglycerides by Capillary GC. *Pure Appl. Chem.* **1991,** *63*, 1153

Castillo, E., Dossat, V., Marty, A., Condoret, J.S., and Combes, D. The Role of Silica Gel in Lipase-Catalyzed Esterification Reactions of High-Polar Substrates. *J. Am. Oil Chem. Soc.* **1997,** *74*, 77-85.

Colombie, S., Tweddell, R.J., Condoret, J., and Marty, A. Water Activity Control: A Way To Improve the Efficiency of Continuous Lipase Esterification. *Biotechnol. Bioeng.* **1998,** *60*, 362-368.

D'Alonzo, R.P., Kozarek, W.J., and Wade, R.L. Glyceride Composition of Processed Fats and Oils As Determined by Glass Capillary Gas Chromatography. *J. Am. Oil Chem. Soc.* **1982,** *59*, 292-295

Ergan, F., Trani, M., and Andre, G. Production of Glycerides from Glycerol and Fatty Acid by Immobilized Lipase in Non-Aqueous Media. *Biotechnol. Bioeng.* **1990,** *35*, 195-200.

Février, P., Binet, A., Dufosse, L., Gree, R., and Yvergnaux, F. Separation of Glyceride Positional Isomers by Silver Ion Chromatography. *J. Chromatogr.* **2001,** *A 923*, 53-57.

Fisher, W., Heinz, E., and Zeus, M. The Suitability of Lipase from Rhizopus arrhizus (delemar) for the Analysis of Fatty Acid Distribution in Dihexosyl Diglycerides, Phospholipids and Plant Sulfolipids. *Hoppe-Seyler's Z. Physiol. Chem.* **1973,** *354*, 1115-1123

Fureby, A.M., Tian L., Adlercreutz, P., and Mattiasson, B. Preparation of Diglycerides by Lipase-Catalyzed Alcoholysis of Triglycerides. *Enzyme Microb. Technol.* **1997,** *20*, 198-206.

Haftendorn, R., and Ulbrich-Hofmann, R. Synthesis of 2-Modified 1,3-Diacylgrycerols, *Tetrahedron* **1995,** *51*, 1177-1186.

Kim, I.H., and Lee, S.M., Synthesis of Diacylglycerols Containing CLA by Lipase-Catalyzed Esterification, *J. Food Science,* **2006,** *71*, 378-382

Kristensen, J.B., Xu X., and Mu H. Diacylglycerol Synthesis by Enzymatic Glycerolysis: Screening of Commercially Available Lipases. *J. Am. Oil Chem. Soc.* **2005,** *82*, 329-334

Kuksis, A., and Marai, L. Determination of the Complete Structure of Natural Lecithins. *Lipids* **1967,** *2*, 217-224.

Lortie, R., Trani, M., and Ergan, F. Kinetic Study of the Lipase-Catalyzed Synthesis of Triolein. *Biotechnol. Bioeng.* **1993,** *41*, 1021-1026.

Marty, A., Dossat, V., and Condoret, J.S. Continuous Operation of Lipase-Catalyzed Reactions in Nonaqueous Solvents: Influence of the Production of Hydrophilic Compounds. *Biotechnol. Bioeng.* **1997,** *56*, 232-237.

Mensah, P., and Catra, G. Adsorptive Control of Water in Esterification with Immobilized Enzymes. Continuous Operation in a Periodic Counter-Current Reactor. *Biotechnol. Bioeng.* **1999,** *66*, 137-146.

Myher, J.J., and Kusis, A. Resolution of Diacylglycerol Moieties of Natural Glycerophospholipids by Gas-Liquid Chromatography on Polar Capillary Columns. *Can J. Biochem.* **1982,** *60*, 638-650.

Ota, Y., Takasugi, T., and Suzuki, M. Synthetic of Either Mono- or Diacylglycerol from High-Oleic Sunflower Oil by Lipase-Catalyzed Glycerolysis. *Food Sci. Technol.* **1997,** *3*, 384-387.

Pacheco, Y.M., Perez-Camino, M.C, Cert, A., Montero, E., and Ruiz-Gutierrez, V. Determination of the Molecular Species Composition of Diacylglycerols in Human Adipose Tissue by Solid-Phase Extraction and Gas Chromatography on a Polar Phase. *J. Chromatogr.* **1998,** *B 714*, 127-132

Perez-Camino, M.C, Moreda, W., and Cert, A. Determination of Diacylglycerol Isomers in Vegetable Oils by Solid-Phase Extraction Followed by Gas Chromatography on a Polar Phase. *J. Chromatogr* **1996,** *A 721*, 305-314

Perona, J.S., and Ruiz-Gutierrez, V. Simultaneous Determination of Molecular Species of Monoacylglycerols, Diacylglycerols and Triacylglycerols in Human VeryLow-Density Lipoproteins by Reversed-Phase Liquid Chromatography. *J. Chromatogr.* **2003,** *B 785*, 89-99.

Plou, J.F., Barandiarn, M., Calvo, V.M., Ballesteros, A., and Paster, E. High-Yield Production of Mono- and Di-oleylglycerol by Lipase-Catalyzed Hydrolysis of Triolein. *Enzyme Microb. Technol.* **1996,** *18*, 66-71.

Rnkonen, O. Individual Molecular Species of Different Phospholipid Classes. Part II.A Method of Analysis. *J. Am. Oil Chem. Soc.* **1965,** *42*, 298-304.

Rosu, R., Yasui, M., Iwasaki, Y., and Yamane, T. Enzymatic Synthesis of Symmetrical 1,3-DAGs by Direct Esterification of Glycerol in Solvent-Free System. *J. Am. Oil Chem. Soc.* **1999,** *76*, 839-843.

Ruiz-Gutierrez, V., and Perez-Camino, M.C. Update on Solid-Phase Extraction for the Analysis of Lipid Classes and Related Compounds. *J. Chromatogr.* **2000,** *A 885*, 321-341

Shimada, Y., Sugihara, A., Nakano, H., Yokota, T., Nagao, T., Komemushi, S., and Tominaga, Y. Production of Structured Lipids Containing Essential Fatty Acids by Immobilized *Rhizopus delemar* Lipase. *J. Am. Oil Chem.Soc.* **1996,** *73*, 1415-1420.

Sonntag, N.O.V. Glycerolysis of Fats and Methyl Esters-Status, Review, and Critique. *J. Am. Oil Chem. Soc.* **1982,** *59*, 795A-802A.

Waldinger, C., and Schneider, M. Enzymatic Esterification of Glycerol III. Lipase-Catalyzed Synthesis of Regioisomerically Pure 1,3-*sn*-DAGs and 1(3)-*rac*-Monoacylglycerols Derived from Unsaturated Fatty Acids. *J. Am. Oil Chem. Soc.* **1990,** *73*, 1513-1519.

Watanabe, T., Shimizu, M., Sugiura, M., Sato, M., Kohori, J., Yamada, N., and Nakanishi, K. Optimization of Reaction Conditions for the Production of DAG Using Immobilized 1,3-Regiospesific Lipase Lipozyme RM IM. *J. Am. Oil Chem. Soc.* **2003,** *80*, 1201-1207.

Watanabe, T., Sugiura, M., Sato, M., Yamada, N., Nakanishi, K., Diacylglycerol production in a packed bed bioreactor. *Process Biochem.* **2005,** *40*, 637-643

Williams, J.P, Khan, M.U., and Wong, D. A Simple Technique for the Analysis of Positional Distribution of Fatty Acids on Di- and Triacylglycerols Using Lipase and Phospholipase A2. *J. Lipid Res.* **1995,** *36*, 1407-1412.

Xu, X., Skands, A.R.H., Hoy, C.E., Mu, H., Balchen, S., and Nissen, J.A. Production of Specific-Structured Lipids by Enzymatic Interesterification: Elucidation of Acyl Migration by Response Surface Design. *J. Am. Oil Chem. Soc.* **1998a,** *75*, 1179-1186.

Xu, X., Balchen, S., and Hoy, C.E. Production of Specific-Structured Lipids by Enzymatic Intereseterification in a Pilot Continuous Enzyme Bed Reactor. *J. Am. Oil Chem. Soc.* **1998b,** *75*, 1573-1579.

Xu, X., Fomuso, L.B., and Akoh, C.C. Synthesis of Structured Triacylglycerols by Lipase-Catalyzed Acidolysis in a Packed Bed Bioreactor. *J. Agric. Food Chem.* **2000,** *48*, 3-10.

Xu, X., Skands, A.R.H., Hoy, C.E., and Nissen, J.A. Parameters Affecting DAG Formation During the Production of Specific-Structured Lipids by Lipase Catalyzed Interesterification. *J. Am. Oil Chem. Soc.* **1999,** *76*, 175-181.

Yamane, T., Kang, S.T., Kawahara, K., and Koizumi, Y. High-Yield DAG Formation by Solid-Phase Enzymatic Glycerolysis of Hydrogenated Beef Tallow. *J. Am. Oil Chem. Soc.* **1994,** *71*, 339-342

Yang, B., Harper, W.J., Parkin K.L., and Chen J. Screening of Commercial Lipase for Production of Mono- and DAGs from Butter Oil by Enzymatic Glycerolysis. *Int. Dairy J.* **1994,** *4*, 1-13.

Zaks, A., and Klibanov, A.M. The Effect of Water on Enzyme Action in Organic Media. *J. Biol. Chem.* **1988,** *263*, 8017-8021.

Index

C

D

I

J

K

L

M

N

O

P

R

T

U

V

W